Nanotechnology in the Life Sciences

Series Editor

Ram Prasad, Department of Botany
Mahatma Gandhi Central University
Motihari, Bihar, India

Nano and biotechnology are two of the 21st century's most promising technologies. Nanotechnology is demarcated as the design, development, and application of materials and devices whose least functional make up is on a nanometer scale (1 to 100 nm). Meanwhile, biotechnology deals with metabolic and other physiological developments of biological subjects including microorganisms. These microbial processes have opened up new opportunities to explore novel applications, for example, the biosynthesis of metal nanomaterials, with the implication that these two technologies (i.e., thus nanobiotechnology) can play a vital role in developing and executing many valuable tools in the study of life. Nanotechnology is very diverse, ranging from extensions of conventional device physics to completely new approaches based upon molecular self-assembly, from developing new materials with dimensions on the nanoscale, to investigating whether we can directly control matters on/in the atomic scale level. This idea entails its application to diverse fields of science such as plant biology, organic chemistry, agriculture, the food industry, and more.

Nanobiotechnology offers a wide range of uses in medicine, agriculture, and the environment. Many diseases that do not have cures today may be cured by nanotechnology in the future. Use of nanotechnology in medical therapeutics needs adequate evaluation of its risk and safety factors. Scientists who are against the use of nanotechnology also agree that advancement in nanotechnology should continue because this field promises great benefits, but testing should be carried out to ensure its safety in people. It is possible that nanomedicine in the future will play a crucial role in the treatment of human and plant diseases, and also in the enhancement of normal human physiology and plant systems, respectively. If everything proceeds as expected, nanobiotechnology will, one day, become an inevitable part of our everyday life and will help save many lives.

Vishnu D. Rajput • Arti Goel
Ram Prasad • Tatiana Minkina
Editors

Contemporary Suitability of Nanobionics in Agriculture

Nanotechnology in Horticultural Crops

Springer

Editors
Vishnu D. Rajput
Academy of Biology and Biotechnology
Southern Federal University
Rostov-on-Don, Russia

Arti Goel
Amity Institute of Microbial Technology
Amity University
Noida, Uttar Pradesh, India

Ram Prasad
Department of Botany
Mahatma Gandhi Central University
Motihari, Bihar, India

Tatiana Minkina
Academy of Biology and Biotechnology
Southern Federal University
Rostov-on-Don, Russia

ISSN 2523-8027 ISSN 2523-8035 (electronic)
Nanotechnology in the Life Sciences
ISBN 978-3-031-82778-5 ISBN 978-3-031-82779-2 (eBook)
https://doi.org/10.1007/978-3-031-82779-2

This Springer imprint is published by the registered company Springer Nature Switzerland AG
The registered company address is: Gewerbestrasse 11, 6330 Cham, Switzerland

Preface

Nanotechnology in agriculture has gained significant thrust in the last decade and is attributed to a distinctive nature of farm produce that ultimately functions as an accessible system whereby matter and energy can be freely exchanged. The input materials demand scale is always being large as compared to industrial nanoproducts due to absence of control over the input of the nanomaterials in contrast with industrial nanoproducts. Nanotechnology has been responsible to give new agrochemical agents as well as delivery mechanisms for crop productivity improvement by reducing the applications of pesticides. Nanotechnology has the significant potential to increase agricultural production. Eminent experts of nanotechnology hope that the benefits of nanotechnology for food and agriculture need to be weighed against apprehensions for the environment. Nanomaterials are deemed to have specific and unique properties compared to bulk materials. Agricultural waste products have gained attention for being renewable raw materials source. Agricultural attempts could include a varied range of selective procedures. Agricultural applications of nanomaterials aim to decrease the spraying concentration of plant defense products as well as enhance crop yields. Potential of nanotechnology in agriculture is large, but several issues related to the risk assessment need to be addressed yet. Nanotechnology has a significant role in all stages of production, processing, storing, packaging, and transport of agricultural produces. Nanotechnology will transform agriculture and food industry by enhancing the potential of plants to absorb nutrients, by detecting diseases, and by controlling pests. The current book will cover the current challenges of sustainability, food security, and climate change that are being explored by researchers in the area of nanotechnology in the headway of agriculture and horticulture. The application of nanotechnology in agriculture started with the increasing realization that traditional farming practices and technologies would neither be able to increase productivity nor restore ecosystems injured by current technologies. In the current scenario, not much information has been there about the safe use of nanomaterials in agricultural system. The current book will provide the insight of the same which will be helpful for the agricultural personnel. Despite the potential advantages, the application of nanotechnology in agriculture sector is still comparatively insignificant and has not yet made it to the market in terms of a range of products that are in health and electronics. Like other technologies, it also has some issues that need to be addressed yet. Hence, there is a need for persistent research in the agrochemical industry to utilize potential future benefits. Concern with respect to the regulation and assessment of nanomaterials safety also requires attention. The aim of this book is addressing the issue of knowledge gap. We, being the editors, hope that this book will interest students and researchers in the latest outcomes of nanotechnology research in the field of agriculture as well as allied sciences and encourage the young researchers to explore this emerging field. This book contains 16 chapters written by eminent subject specialists that covered the diverse aspects of nanotechnology to achieve sustainable agriculture. We are very grateful to many people who directly or indirectly helped to compile this book. We thank Ms. Natalie Munoz, Ms. Kenneth Teng, and Mr. Arun Siva Shanmugam from Springer for their help and patience in finalizing this book. We also like to thank our families, friends, colleagues, and research scholars for finalizing this book. We especially like to thank Dr. Ashok K. Chauhan, Founder President, Ritnand Baldev Foundation, Amity University, Noida, for the kind support and encouragement for academic endeavors.

Rostov-on-Don, Russia — Vishnu D. Rajput
Noida, Uttar Pradesh, India — Arti Goel
Motihari, Bihar, India — Ram Prasad
Rostov-on-Don, Russia — Tatiana Minkina

Contents

About the Editors

Vishnu D. Rajput working as Head of the International Laboratory on Nanobiotechnologies at Southern Federal University, Rostov-on-Don, Russia. His ongoing research is based on the investigation of effective remediation approaches using biochar/nano-biochar-based sorbents and nanomaterials. With long experience and experimental work, Dr. Rajput comprehensively detailed the state of research in environmental science in regard to "how nanoparticles/heavy metals interact with plants, soil, microbial communities, and the larger environment as well as possible remediation technology using nanoparticles/nano-biochar." Recently, he is working on nano-enhanced remediation of soil and water systems. He has published 350 scientific publications, 88 chapters, 28 books, and h-Index 59 (i10-index 249) until December 2023. He is an internationally recognized reviewer and received an Outstanding Reviewing Certificate from Elsevier and Springer. He is an editorial board member of various high-impact journals such as biochar. He is holding national and international projects, including Mega-Grant, BRICS. Dr. Rajput has received Certificate for Appreciation (2019, 2021), Certificate of Honor (2020), Diploma Award (2021, 2022), and Letter of Gratitude (2022, 2023) from Southern Federal University, Russia, for outstanding contributions in academic, creative research, and publication activities. He has also received "Highly Qualified Specialist" status from the Russian government. In 2023, he was included in the list of the 2% world's potential scientists by Elsevier BV publishing house.

Arti Goel is currently working as an assistant professor at the Amity Institute of Microbial Technology, Amity University, India. After receiving her master's degree in botany and doctorate in microbiology from Mohan Lal Sukhadia University, Udaipur (Rajasthan), India. She entered the Central Arid Zone Research Institute, Jodhpur, for a postdoctoral fellowship in microbial and agricultural nanotechnology as well as nanobiotechnology. She has 16 years of research and 13 years of teaching experience. Her research involved the study of antimicrobial properties of medicinal plants as well as synthesis, characterization, and applications of biologically synthesized metal and metal oxide nanoparticles in agriculture with special reference to plant disease management, soil health, and fertility improvement. Her publications in this field include research papers (national and international), articles (national and international), and patents (filed and granted). She is an assistant editor, regional editor, senior editor, and reviewer of various scientific journals of repute. She has published research/review papers (20) and book chapters (16) as well as she has obtained patents: 10 (filed) and 1 (granted). Currently, she is guiding three PhD students, and two students has been awarded PhD. Her h-index is 11, and i10-index is 11. She has been honored with FSASc Award by Society for Advancement of Sciences and Lifetime Achievement FMERC Award by Eudoxia Research Centre, Guwahati, India. She is a lifetime member of the Association of Microbiologists of India. She has got International Teaching and Research Excellence Award in Nanobiotechnology 2021, given by CPACE, a unit of IMRF registered with Government of India, NITI Aayog, NGO Darpan, Vijayawada, India. She got Best Researcher Award 2020, given by VD Good International Scientists Awards in the International Scientists Awards on Engineering, Science, and Medicine. Her profile has been recognized for listing in Research Excellence by RSquareL Indexing Agency. Her research includes nanoagriculture-related topics such as biogenic metal nanoparticles as a nanofungicide for *Alternaria* blight management, and mechanistic insight of the antifungal potential of biogenic metal nanoparticles against *Alternaria brassicae* and *Alternaria brassicicola*. Her current research is on analyzing the impact of biogenic nanoparticles on exopolysaccharides production by arid zone microorganisms and its agricultural application and combinatorial effect of bio-slurry and biogenic metal nanoparticles to ameliorate soil health and fertility.

Ram Prasad is Associate Professor, Department of Botany, Mahatma Gandhi Central University, Motihari, Bihar, India. His research interests include applied and environmental microbiology, plant-microbe interactions, nanobiotechnology, sustainable agriculture, and chemistry of biology. Dr. Prasad has more than 250 publications (total citations 14,316 with an h-index 62, i10-index 199) to his credit, including research papers, review articles and book chapters and seven patents issued or pending, and edited or authored several books. He has fourteen years of teaching experience. He has been awarded the Young Scientist Award; Prof. J.S. Datta Munshi Gold Medal by the International Society for Ecological Communications; Sir CV Raman Scientist Award; Fellow of Biotechnology Research Society of India; Fellow of Agricultural Technology Development Society, India; Fellow of the Society for Applied Biotechnology; the American Cancer Society UICC International Fellowship for Beginning Investigators, USA; Outstanding Scientist Award; BRICPL Science Investigator Award; and Research Excellence Award. He has been serving as an editorial board member of *Nanotechnology Reviews*, *Green Processing and Synthesis*, *BMC Microbiology*, *BMC Biotechnology*, *Current Microbiology*, *Archives of Microbiology*, *Annals of Microbiology*, *Journal of Nanomaterials*, *Nanotechnology for Environmental Engineering*, *Open Life Sciences*, *Journal of Agriculture and Food Research*, *SN Applied Sciences*, and *Agriculture*. He is also a series editor of *Nanotechnology in the Life Sciences*, Springer Nature, USA. Previously, he served as an assistant professor at Amity University Uttar Pradesh, India; visiting assistant professor, Whiting School of Engineering at Johns Hopkins University, Baltimore, United States; and research associate professor at School of Environmental Science and Engineering, Sun Yat-sen University, Guangzhou, China.

Tatiana Minkina is the head of Soil Science and Land Evaluation Department of Southern Federal University. The head of International Master's Degree Educational Program "Management and Estimation of Land Resources" (2015–2022, accreditation by ACQUIN). The area of scientific interest is soil science, biogeochemistry of trace elements, environmental soil chemistry, soil monitoring, assessment, modeling, and remediation using physicochemical treatment methods. She was awarded Diploma of the Ministry of Education and Science of the Russian Federation in 2015 for her long-term work on the development and improvement of educational process, significant contribution to the training of highly qualified specialists. Currently, she is handling projects funded by Russian Scientific Foundation, Ministry of Education and Science of the Russian Federation, and Russian Foundation of Basic Researches. She is a member of Expert Group of Russian Academy of Science, International Committee on Contamination Land, Eurasian Soil Science Societies, International Committee on Protection of the Environment, and International Scientific Committee of the International Conferences on Biogeochemistry of Trace Elements. Total scientific publications: 757 (389 in English); Web of Science: 141; Scopus: 195; Citing index: Web of Science: 803; h-index: 16; Scopus: 1281, h-index: 19. The invited editor of an Open Access Journal by MDPI "Water", ISSN 2073-4441 (Impact factor: 2.524). She is an editorial board member of *Geochemistry: Environment, Exploration, Analysis*; ISSN: 1467-7873 (Impact factor: 1.109; 5yr IF: 1.769; SJR: 0.334; SNIP: 0.598) and *Eurasian Journal of Soil Science*; e-ISSN: 2147-4249.

Chapter 1
Nanobionics: An Innovative Drive Towards Sustainable Agriculture

Surbhi Dabral, Mahima Das, Ananya Jayent, Divyanshi Raturi, Arti Goel, and Swati Tripathi

Abstract Continuous rise in the world population insists on expanded crop and food production, hence adapting new and sustainable approaches is the need of the time. In addition to the limited space and growing food demands, the management of biotic and abiotic stress factors also needs to be addressed, as these stresses significantly reduce crop yield and quality. To overcome this, novel techniques and strategies related to nanobionics, an emerging field that combines nanotechnology and plant sciences are required to be explored and adapted.

Nanoparticles when introduced into plant systems are known to augment various beneficial agronomic traits of plants which include the increased tolerance against various stresses, a high rate of photosynthesis, elevated nutrition acquisition, increased crop productivity, targeted fertilizer, and pesticide delivery. Despite these positive attributes, the approaches of nanobionics are still not adapted, as scarce information is available about the environmental and health dangers of nanoparticles. Their toxicity to nontarget organisms and the unknown long-term environmental repercussions can cause considerable concerns. Thus, nanoparticles have the potential to significantly enhance sustainable agricultural practices; however, additional research is essential to guarantee their safe and sustainable application.

Keywords Nanobionics · Abiotic stress · Biotic stress · Nanoparticles

1.1 Introduction

A substantial rise in population has intensified the demand for crops and food production. Traditional agricultural practices are incapable of matching the increasing demand, therefore sustainable agricultural practices are vital. An effective and

S. Dabral · M. Das · A. Jayent · D. Raturi · A. Goel · S. Tripathi (✉)
Amity Institute of Microbial Technology, Amity University, Noida, Uttar Pradesh, India
e-mail: stripathi2@amity.edu

V. D. Rajput et al. (eds.), *Contemporary Suitability of Nanobionics in Agriculture*, Nanotechnology in the Life Sciences, https://doi.org/10.1007/978-3-031-82779-2_1

strategic approach is essential to reduce the impact of various stress conditions on plants. Stress conditions can be broadly classified into abiotic and biotic stress, which trigger several responses in plants thus affecting the production and quality of the crops. Different approaches have been used to find a firm and effective solution that improve the efficiency and productivity of crops. One such approach is the introduction of nanobionics to improve plant yield. Nanobionics is an area in plant sciences that deals with the use of nanoparticles (NPs) and their interactions with the plant system. The goal of this emerging field of nanobionics is to bridge the gap between biology and nanotechnology (Bhaskar et al., 2024).

Nanoparticles used in agricultural systems have both positive and negative effects on the environment. Enhanced crop productivity and targeted delivery of fertilizers and pesticides to crop systems are some of its advantages. However, concerns arise regarding environmental and health risks, as NPs may be considered toxic to the environment as their long-term effects on ecosystem and human health remain largely unknown. Regulatory challenges also exist due to the rapid development of nanotechnology outpacing safety assessments, while the high costs may limit accessibility for smallholder farmers. Overall, NPs hold great promise for advancing agricultural practices therefore careful consideration of their potential risks and thorough research are essential for sustainable implementation. Based on their chemical properties NPs can be classified as organic or inorganic. Inorganic NPs include copper (Cu), aluminum (Al), silver (Ag), zinc oxide (ZnO), silica (Si), cerium oxide (Ce_2O_3), and titanium oxide (TiO_2). For example, silica NPs are efficient in improving soil aeration and water holding capacity of soil. On the other hand, organic nanomaterials for agriculture include polymers, lipids, and carbon nanotubes which offer unique benefits and promote plant nutrition (Kumari & Sarkar, 2021).

1.2 Abiotic Stress

Abiotic stress refers to the negative effects caused by physical and non living factors in a specific environment. In plants, abiotic stress can be induced by various environmental factors, which are discussed briefly in Fig. 1.1.

1.2.1 Heavy Metals (HMs) Stress

One of the primary abiotic stressors that hinder plant growth and productivity worldwide is HM contamination. In plants, HMs inhibit germination, growth reduction, reduce productivity, reduce water uptake, alter the transpiration rate, accumulate growth inhibitors, decrease protein synthesis, cause the breakdown of macromolecules, alter gene expression, and give rise to the formation of reactive oxygen species (ROS) species. Heavy metals acts as toxic agents for plants i.e. they

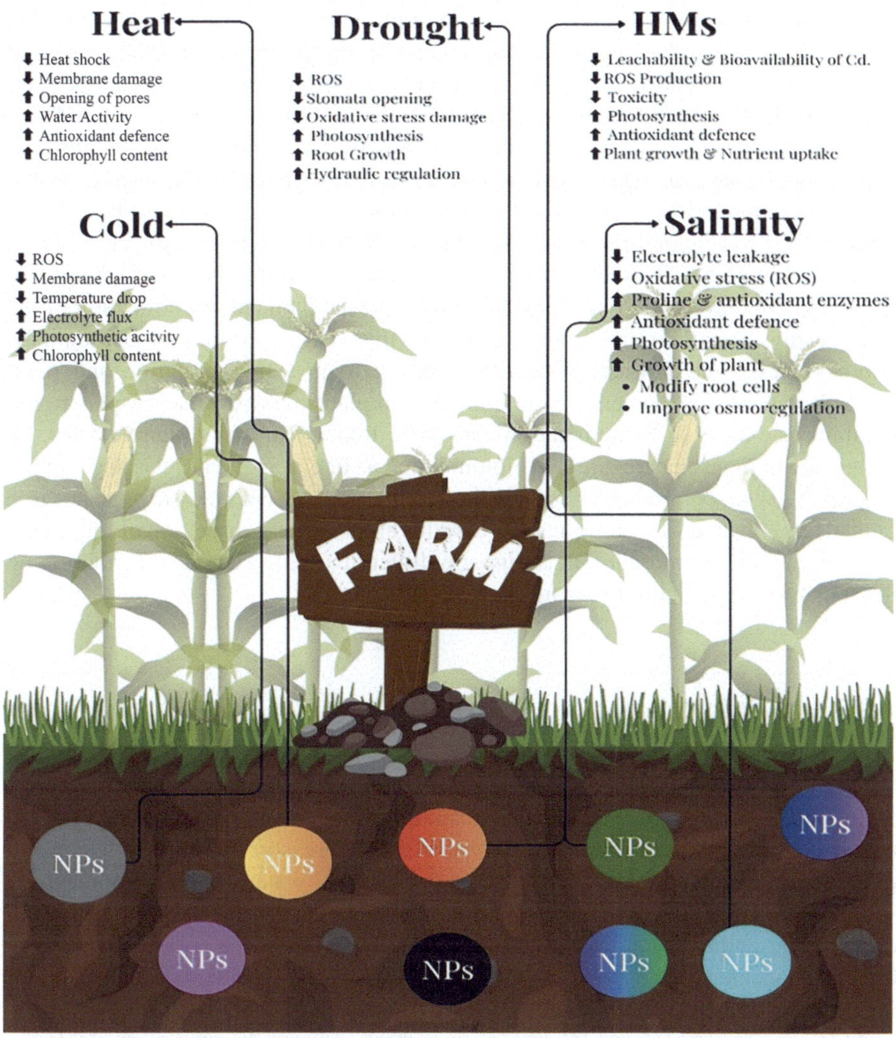

Fig. 1.1 Effect of nanoparticles on different abiotic stress conditions

alter the imperative cell molecules (Hossain et al., 2012) and enzymes (Ali et al., 2013) involved in normal functioning of plant cells hence undesirably affecting the membrane of plant cells. They can also hamper the metabolic process of plants in many ways like degradation and displacement of plant proteins (Hall, 2002) and affecting the vital processes like respiration and photosynthesis in plants (Hossain et al., 2012). Heavy metals are supplemented in the atmosphere due to various anthropogenic deeds, which lead to their deposition in the environment by air and water. Due to various human activities, the accumulation of HMs in soil occurs through municipal and industrial wastes, smelting, mining, contaminated irrigation water, use of fertilizers and pesticides and the discharge of wastes from leather industries (Arunakumara et al., 2013). Cadmium (Cd) is one of the prominent HMs found in soil, which has many negative effects on the environment. Plants exposed to Cd toxicity are reported to have decreased biomass, photosynthesis, and productivity, which is ascribed to slower morphological, physiological and molecular function. Another heavy metal pollutant is lead (Pb) which can infiltrate the soil through various pathways causing substantial toxicity to living organisms. It remains stable in soil for extended periods, is hard to break down and accumulates in living cells through the food chain. In addition Pb hinders cell division, growth, and chlorophyll production in plants. It affects seed germination by disrupting protease and amylase enzymes. Low levels of Pb also lead to mitochondrial damage, vacuolization of the endoplasmic reticulum and stunted root growth. Lead exposure can also damage photosynthesis and nutrient allocation, leading to growth retardation and plant death. Iron (Fe) aids in energy production within plants. However, a low pH growth medium or excessive Fe application can cause toxicity in many plant species. Fe deficiency in plants can result in symptoms such as interveinal chlorosis and stunted growth. Leaf discoloration occurs as the plant produces enzymes to manage free radicals in elevated Fe conditions. Fe toxicity can be either direct, involving metal absorption and accumulation in plants or indirect involving Fe plaque formation on the root surface, which impairs nutrient absorption.

1.2.2 Salinity Stress

Soil salinity is a condition where excessive salt is accumulated in the soil, poses a major challenge to modern agriculture, leading to stunted and impaired plant growth and generating both hypertonic and hyperosmotic stressors, ultimately leading to plant death (Mahajan et al. 2005). It causes osmotic, water, oxidative, nutritional stress, reduces cell division and induces oxidative stress by generating ROS which can damage cellular structures, proteins, lipids, and DNA (Tariq et al., 2021. The high salt concentration in the soil leads to a lower water potential, making it difficult for plants to absorb water, thereby causing dehydration. Osmotic stress forces the plant to use more energy to take up water, which reduces the energy available for growth. Salinity also causes ionic imbalance by increasing sodium (Na^+) and chloride (Cl^-) ions, which can be toxic at high concentrations and disrupt essential nutrient uptake like potassium (K^+), calcium (Ca^{2+}), and magnesium (Mg^{2+}), leading to nutrient deficiencies (Hussein & Abou-Baker, 2018).

1.2.3 Drought Stress

Drought stress is one of the most important abiotic stresses that plants face, affecting various physiological, biochemical and molecular processes. It is caused by lack of water in soil, low rainfall, high temperature, high evaporation and dry air. Soil salinity can also inhibit water absorption because high levels of salinity create osmotic stress that impairs the plant's ability to absorb water. In response to drought, plants close stomata to reduce water loss through transpiration, posing a major challenge to plant growth and agricultural productivity. Understanding these complex interactions is essential for developing effective management strategies to improve the drought resistance of crops and other plants to ultimately ensure food security in an unpredictable climate.

1.2.4 Heat Stress

Heat stress in plants is caused by high temperatures which can lead to physiological and biochemical damage. This stress disrupts cellular function, causing protein denaturation, cell membrane instability and increased production of ROS, resulting in oxidative stress. The severity of heat stress varies according to the plant species, developmental stage, and environmental conditions, with significant impacts occurring during reproductive periods such as flowering and grain setting. Physiological responses include decreased photosynthesis, reduced nutrient supply and changes in plant growth, which reduce biomass and yield (Hasanuzzaman et al., 2021). In addition, heat stress can compromise plant health and productivity by accelerating senescence and inhibiting important processes such as germination and root development (Mishra et al., 2023).

1.2.5 Cold Stress

Contrary to heat stress, cold stress in plants occurs when exposed to low temperatures which can be classified as cold stress (temperature between 0 °C and 20 °C) and freezing stress (below 0 °C). These conditions lead to physiological changes including membrane fluidity, cell desiccation and ice formation in plant tissues. Ice formation causes damage by perforating cell membranes and destroying cell integrity, causing energy leakage and loss of nutrients. In addition, cold stress can inhibit photosynthesis by reducing chlorophyll synthesis and destroying the photosynthetic apparatus, resulting in reduced production and growth. Many plants, especially those in tropical and subtropical regions are not able to adapt to cold therefore, they are vulnerable to low temperature related damage. In contrast, simple plants enhance cold tolerance by accumulating protective molecules such as sugars and proteins.

However, prolonged exposure to low temperatures can lead to stunted growth, reduced yield and even plant death.

1.3 Effect of Nanoparticles on Different Abiotic Stress Conditions

1.3.1 Heavy Metals

Certain NPs, like iron oxides, titanium dioxide and silica-based NPs can absorb heavy metals thus reducing their availability to plants. This helps in immobilizing toxic metals like arsenic (As), Pb and Cd thus preventing their uptake by plants (Prakash 2023). Studies revealed that nano-silica (SiO_2) decreased the leachability and bioavailability of Cd in agricultural soil, thus limiting the translocation of Cd content in wheat grains (Chen et al., 2018). These NPs also stimulate antioxidant enzymes, reduce ROS production, and enhance plant stress tolerance by stimulating root growth and improving nutrient uptake, particularly phosphorus (P) and potassium (K), which helps plants to resist heavy metal stress (Wang et al., 2020), as shown in Fig. 1.1.

Hydroxyapatite (Hap), a calcium phosphate mineral with the chemical formula $Ca_5(PO_4)_3(OH)$, is significant in soil because it influences nutrient availability and heavy metal stabilization. It binds to HMs and prevents its absorption by the roots. Hap also benefits in soil maintenance by immobilizing metals and preventing them from leaking into groundwater. Furthermore, it provides vital nutrients such as calcium (Ca) and P, which promote plant growth. Its usage in soil management promotes soil health and reduces contamination concerns (Table 1.1).

The various Hap have varying effects on Cu and Cd bioavailability and microbial community composition. Studies revealed how micro-hydroxyapatite (MAP), nano-hydroxyapatite (NAP), and normal hydroxyapatite (AP) enhanced soil pH by decreasing exchangeable acid. In comparison to the control, MAP was the most efficient in lowering the $CaCl_2$-extractable Cu and Cd, while most effective in increasing soil pH and decreasing the bioavailability of Cu and Cd (Cui et al., 2018).

TiO_2 NPs can adsorb Pb ions from the soil, hence reducing the availability of Pb to plants and soil organisms. TiO_2 NPs can impact chemical changes in soils and help in the conversion of Pb to less hazardous forms, such as lead sulfide (PbS) or lead carbonate ($PbCO_3$), which leads to enhanced plant growth, nutrient uptake, improved physiological functions, and higher yields (Rizwan et al., 2021). They promote plant photosynthesis (Emamverdian et al., 2022) and antioxidant system activity in plants (Li et al., 2015).

Zinc NPs aid plants under heavy metal stress by increasing antioxidant defenses, minimizing oxidative damage, and preserving cell membrane stability. They regulate metal intake, reducing toxicity and promoting increased growth and biomass (Adrees et al., 2021).

Table 1.1 Effects of abiotic stresses on plants and respective mechanisms associated with nanoparticles

Stress on plants	Effects on plants	NPs	Mechanisms	References
HM (Cd)	Decreased biomass, photosynthesis, and productivity	SiO_2	Stimulated antioxidant enzymes, reduced ROS production, enhanced plant stress tolerance and improved nutrient uptake	Wang et al. (2015)
HM (Pb)	Hinder cell division, growth, and chlorophyll production; disrupt protease and amylase enzymes	TiO_2	Pb converted to PbS and $PbCO_3$	Rizwan et al. (2021)
HM (Cu)	Causes oxidative stress, enzyme inhibition, photosynthesis impairment, and membrane damage	MAP	Increased soil pH and decreased bioavailability of Cu	Cui et al. (2018)
Salinity	Generate both hypertonic and hyperosmotic stress, reduce cell division, and induce oxidative stress, and disruption of essential nutrients (K^+, Ca^{2+}, Mg^{2+})	K NPs	Decreased electrolyte leakage and increased antioxidant enzymes	El-Sharkawy et al. (2017)
		CeO_2	Modified root cells for enhanced nutrient uptake	Rossi et al. (2017)
		Ag NPs	Increased antioxidant activity	Khan et al. (2020)
Drought	Protein denaturation, cell membrane instability, increased production of reactive oxygen species, and accelerated senescence	Si NPs	Enhanced photosynthetic activity, nutritional status, physical and chemical parameters, phenolic content, osmolality, antioxidant enzymes and abscisic acid	El-Sharkawy et al. (2017)
		TiO_2 Fe_3O_4 ZnO	Improved physical and metabolic functions in plants	Alabdallah et al. (2021)

(continued)

Table 1.1 (continued)

Stress on plants	Effects on plants	NPs	Mechanisms	References
Heat	Disrupts cellular function, causing protein denaturation, cell membrane instability, and increased production of reactive oxygen species (ROS), resulting in oxidative stress	SeNPs	Effects of heat stress reduced, increased water potential, chlorophyll content, and increased antioxidant activity in plants	Omar et al. (2023)
		TiO_2	Reduced the effect of thermal stress by controlling the opening of pores	Khalid et al. (2022)
		CeO_2	Modulated gene expression of heat shock proteins, HSP90	
Cold	Lead to physiological changes, including membrane fluidity, cell desiccation, and ice formation in plant tissues, and also inhibit photosynthesis by reducing chlorophyll synthesis and destroying the photosynthetic apparatus, resulting in reduced production and growth	Ti NPs	Improved electrolyte flux, photosynthetic activity, and membrane damage under cold conditions	Mohammadi et al. (2013)
		Chitosan NPS	Reduction in ROS due to the accumulation of osmo protectants	Wang et al. (2021)
		ZnO NPs	Reduced chilling stress through antioxidant systems and transcription factors	Song et al. (2021)
		Si NPs	Improved photosynthetic capacity of sugarcane plants	Elsheery et al. (2020)

1.3.2 Drought Stress

Nanoparticles are known for their specific shape, pore size and high reactivity with enhanced surface area. They are considered an effective and promising tool for regulating crop yield and overcoming current and future limitations in agricultural production by increasing tolerance mechanisms in plants under abiotic stress conditions. The mitigating effect of NPs on drought stress is caused by inducing physiological and biochemical regulation and balancing the expression of genes relating to drought tolerance. They enhance the photosynthetic activity, root growth, upregulation of aquaporins, altered intracellular water metabolism, accumulation of compatible solutes and ionic homeostasis of drought-affected plants. The NPs also reduce leaf water loss by accumulating abscisic acid (ABA) through stomatal closure and ameliorating oxidative stress damage by reduction of ROS and activation of antioxidant defense systems. Studies have revealed that NPs increase plant root hydraulic regulation and water uptake therefore increasing drought resistance and expression of relevant proteins which are responsible for influencing various morphological, physiological and biological processes (Kandhol et al., 2022). The use of metal oxide NPs such as TiO_2, ZnO, and Fe_3O_4 is reported to be effective in improving the physical and metabolic functions of plants under drought (Alabdallah et al., 2021). Application of Si NPs to drought-induced pomegranate plants enhanced the photosynthetic activity, nutritional status, physical and chemical parameters (especially those related to fruit damage), phenolic content, osmolality, antioxidant enzymes and ABA (El-Sharkawy et al., 2017). Green ZnO-NPs applied in small doses were able to increase the drought tolerance of tomatoes (El-Zohri et al., 2021). In addition to nano fertilizers, green synthetic Fe_3O_4 NPs were effective in reducing drought effects on fenugreek plants. Potter et al. (2021) emphasized on the potential benefits of using NPs to improve drought resistance in plants under specific environmental conditions.

1.3.3 Heat Stress

The negative effects of heat stress on plants are significantly detrimental, often resulting in reduced yield and productivity. Usage of selenium (Se) NPs at certain concentrations reduced the effects of heat stress by increasing water potential, chlorophyll content, and increased antioxidant activity in plants (Omar et al., 2023).

Plants are known to synthesize heat shock proteins and molecular chaperones during heat stress. It has already been reported that multiwalled carbon tubes can modulate the gene expression of heat shock proteins, HSP90. In addition, maize plants exposed to CeO_2 NPs showed increased H_2O_2 production and upregulation of HSP70. In addition, treatment with TiO_2 NPs reduced the effect of thermal stress by controlling the opening of pores (Khalid et al., 2022).

1.3.4 Cold Stress

Cold stress slows down plant growth because absorption and other protein functions are reduced at cold temperatures. Many processes in these plants are affected by low temperatures, including processes related to secondary metabolism, respiration, defense, proteins, and nucleic acid production. Chitosan particles and TiO_2 NPs have been widely used in various studies for their effectiveness in cold stress. Ti NPs were found to improve electrolyte flux, photosynthetic activity, and membrane damage under cold conditions in chickpea plants through transcriptional regulation (Mohammadi et al., 2013). The tolerance of bean plants to cold stress can be developed by controlling negative temperature drop stress by altering metabolism for plant growth by introducing TiO_2 particles in the plant system. The use of TiO_2 NPs in licorice reduces the negative effects of cold stress and improves glycyrrhizin content (Amini et al., 2017). The use of chitosan NPs was also effective in reducing ROS due to the accumulation of osmoprotectants in banana under cold conditions (Wang et al., 2021). The foliar application of ZnO NPs in rice plants can reduce chilling stress through antioxidant systems and transcription factors involved in chilling responses (Song et al., 2021). Similarly, the use of SiNPs can also improve the photosynthetic capacity of sugarcane plants under cold stress (Elsheery et al., 2020).

1.3.5 Salinity Stress

NPs improve salt tolerance in various plants. Ag NPs, SiO_2 NPs, C NPs, and K NPs were effective in minimizing the detrimental impacts of salt stress on a variety of plants. In salt-sensitive *Medicago sativa*, foliar application of K NPs increased salt tolerance through decreased electrolyte leakage and increased proline and antioxidant-enzyme content, especially catalase (El-Sharkawy et al., 2017). By modifying the root cells and enhancing the uptake of minerals, cerium-oxide NPs have been found to be beneficial in enhancing photosynthetic activity in *Brassica napus* (Rossi et al., 2017). Ag NPs decreased ROS and increased antioxidant activity in pearl millet by seed priming (Khan et al., 2020). Overall, the use of NPs helps plants buffer the negative impacts of salt conditions which aids in plant adaptation to stress. Also, nano-Zn fertilizer improves osmoregulation, enhances antioxidant defense, regulation of ion homeostasis. NanoSiO_2 (Avestan et al., 2019) and nano-TiO_2 (Khan et al., 2016) boost plant resilience to salt stress by increasing antioxidant defense, water retention and ion homeostasis. It preserves cell membranes, boosts photosynthesis and lowers salt-induced toxicity by boosting selective ion uptake and stabilizing cellular structures. Nano SiO_2 increases stress-related gene expression allowing plants to cope with salt-induced oxidative and physiological stress. This leads to enhanced plant growth and yield in saline circumstances (Yassen et al., 2017).

1.4 Biotic Stress

Living organisms that cause biotic stress in plants include bacteria, fungi, viruses, nematodes, insects, arachnids, and weeds. These organisms can cause various infections in plants, which lower crop yield and quality (A. Gull et al., 2019). Fungal infections are a major cause of illnesses in plants. For example, *Fusarium andiyazi* is known to induce wilt disease, which causes yellowing and wilting of plants. Fusarium wilt is also caused by *Fusarium oxysporum* (Namiki et al. 1994), another fungal infection that affects a variety of crops, such as cotton, tomatoes and bananas. Similarly, *Puccinia striiformis*, commonly known as stripe rust, infects cereals like wheat and barley leading to rust disease characterized by yellow or orange pustules on leaves and stems which disrupt photosynthesis and reduce grain yield.

The health of plants is threatened by various bacterial infections like bacterial leaf blight (BLB) is caused by *Xanthomonas oryzae*, a serious disease that causes rice leaves to wilt and dry up thus significantly reducing rice yield (Sanya et al., 2022). This disease spreads quickly in warm and humid environments. Other devastating diseases like black shank and black root rot are also caused by soil-borne pathogens like *Phytophthora nicotianae* and *Thielaviopsis basicola*, respectively. *T. basicola* attacks a range of plants, causing blackened and rotted roots that impede the plant's capacity to absorb water and nutrients. *P. nicotianae* infects tobacco and other crops (Nel et al., 2019), causing root and stem rot.

Another significant cause of biotic stress, particularly in stored crops, is related to insects and pests. Red flour beetles (*Tribolium castaneum*) and rice weevils (*Sitophilus oryzae*) are responsible for infesting grains that have been stored, resulting in large post-harvest losses. Other harmful insects that infest stored grains include the granary weevil, *Sitophilus granarius* and *Rhyzopertha dominica* (Trematerra et al., 2012). These insects lower grain quality and quantity, which can result in significant economic losses.

Weeds are known to compete with plants for vital resources like light, water, and nutrients. In addition, they are also responsible for biotic stress. Species of *Elsholtzia ciliata* and *Bidens pilosa* (Gao et al., 2024) are highly problematic, as they compete with crops for nutrients and space, thus lowering crop yields. Another invasive weed that can overtake the plant community is *E. ciliata*, which can stunt plant growth and lower down the biodiversity of a particular region. These weeds may result in higher weed management and control expenses and possible decreases in crop yield.

Overall, these diverse organisms seriously harm crops, as shown in Fig. 1.2, resulting in biotic stress which can have major effects on agricultural output and food security. For the management of the above-described problem, it is essential to comprehend their influence and put effective management techniques into place (Tortella et al., 2023).

1.5 Mechanisms of NPs in amelioration of Biotic Stress

Recently, NPs have drawn interest in agriculture as they can reduce the harmful effects caused by biotic stress through various mechanisms by which they work in plant systems, as shown in Fig. 1.3.

Chitosan NPs are known to work against wilt disease caused by *F. andiyazi* by inducing the upregulation of PR-proteins and antioxidant genes, which play a role in plant defense against pathogen attack (Jain et al., 2023). Plants produce a class of proteins known as PR-proteins in reaction to pathogen invasion. These comprise enzymes such as β-1,3-glucanases (PR-2 family), which play an essential role in plant defense management as they target fungal pathogen cell walls (Saboki Ebrahim et al., 2011). Apart from this, antioxidant genes are also activated, which reduce the

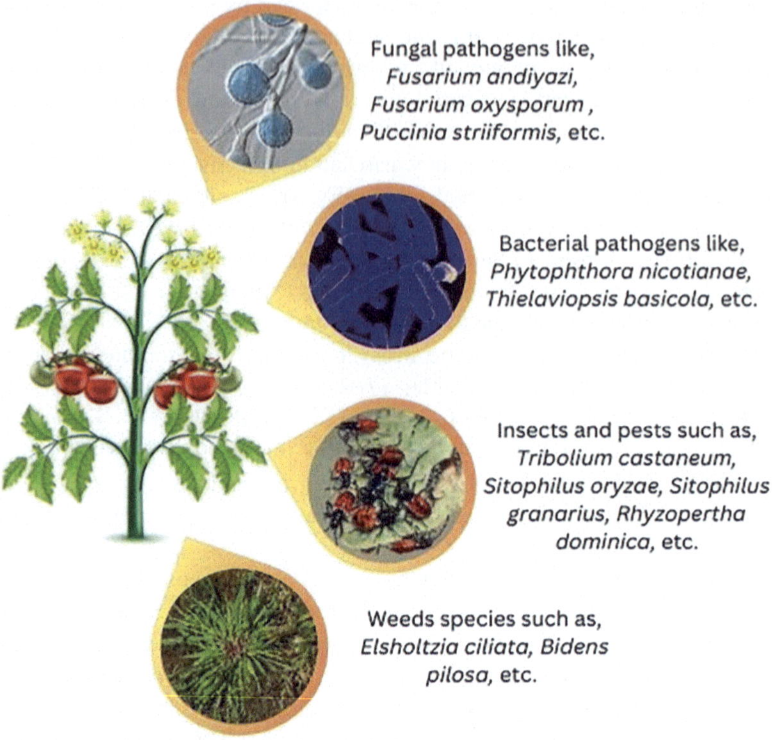

Fig. 1.2 Stress agents in plants responsible for biotic stress

oxidative stress induced by pathogen attack (Birben et al., 2012). Additionally, chitosan alters the amounts and types of hydrolytic enzymes, proteinase inhibitors, phytoalexin biosynthetic enzymes and cell wall proteins which strengthen the host defenses against wilt disease (Elagamey et al., 2022).

Chitosan, along with salicylic acid, is reported to fight infections caused by *P. striiformis.* PR1-PR5 and PR10 genes are involved in plant defense systems, and transcription levels are known to be activated by chitosan which assists in plant disease resistance (Elsheery et al., 2022).

Studies reveal that chitosan is also used against insects like *S. oryzae* and *T. castaneum.* Acetylcholinesterase (AChE) is a critical enzyme in insects for activating neurotransmission signals. AChE breaks down acetylcholine, a neurotransmitter thus terminating nerve impulses. Inhibiting AChE leads to the accumulation of acetylcholine and eventual paralysis and death of the pest. In addition, chitosan loaded with the essential peppermint oil can prevent insects by inhibiting AChE, which plays an important role in plant defense against pathogen invasion (Pangestuti et al., 2010).

Paraquat is an herbicide known to assist in weed control, therefore chitosan encapsulated is also reported to stimulate its weed control activity by increasing the efficacy of loaded herbicide (Grillo et al., 2014). Chitosan-encapsulated paraquat is reported to be less cytotoxic and genotoxic as compared to its free form. Therefore,

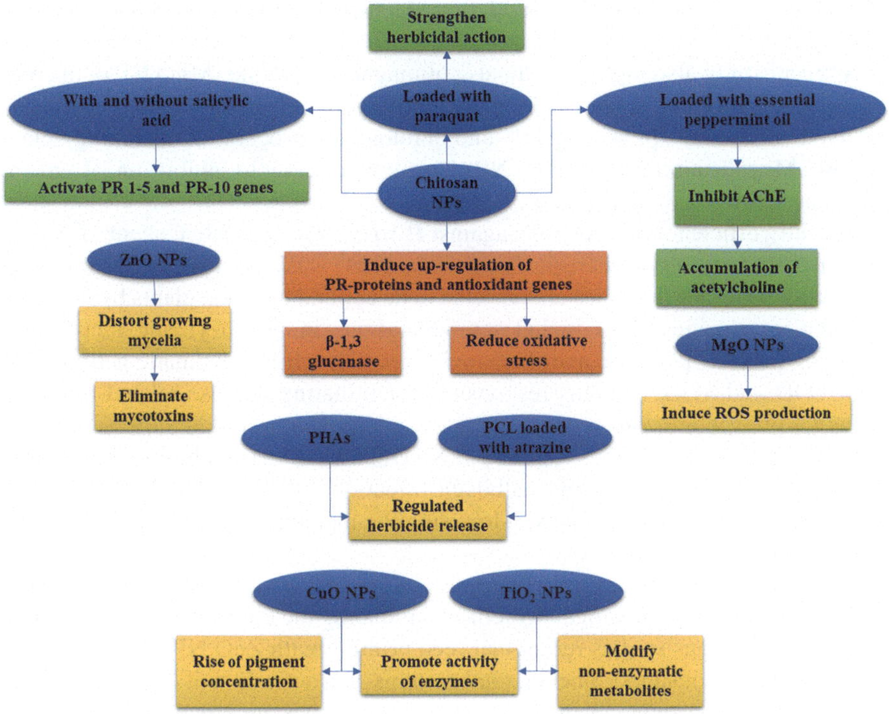

Fig. 1.3 Mechanism of different NPs against various biotic stress

it is less dangerous for nontarget organisms and the environment (Maruyama et al., 2016).

In addition, NPs working against biotic stress includes nanoencapsulation of atrazine within poly(ε-caprolactone) (PCL). These nanocapsules enhance herbicidal activity against weeds like *Bidens pilosa* (Sousa et al., 2018). This method allows for slow release of herbicide over a certain period leading to a more sustained effect. This method has been reported in soybean plants. Apart from the above methods, herbicides can be encapsulated using polyhydroxyalkanoates (PHAs), such as P(3HB) and P(3HB/3 HV), which are known to improve the herbicides stability. This approach is helpful for controlling species of weed such as *E. ciliata* as it permits a regulated herbicide release, hence decreasing the frequency of applications (Kiselev et al., 2020). *E. ciliata* extracts have been studied using various encapsulation techniques, including microencapsulation.

One major factor causing harm to stored grains are insects and pests. These insects were controlled using various chemicals that have long-term detrimental effects on grain and consumers (Upadhyay and Ahmad, 2011). Various NPs are now being used for this purpose and CuO NPs are proving to be beneficial against this problem. CuO NPs are also reported to ameliorate the effect of biotic stress on crops, as they act against insects by promoting the activity of enzymes like superoxide dismutase (SOD), peroxidase (POD), and ascorbate peroxidase (APX), which play a crucial role in scavenging ROS, thus shielding the plant from stress (Badawy et al., 2021).

A recent study also highlights the use of magnesium oxide (MgO NPs), in combating black root rot diseases. Magnesium (Mg) is an essential mineral that is involved in many physiological and biological processes, including plant defense. Magnesium derived MgO NPs have the potential to inhibit fungal growth and spore germination (Chen et al., 2020).

Recently efficiency of TiO_2 NPs against *P. striiformis*, causative agent of yellow stripe rust disease in wheat crops (Raza et al., 2024) was also reported. To diminish it's effects different strategies are reported to be applied in wheat plants by inducing biochemical changes. This includes modifications in nonenzymatic metabolites including proline, phenolic and flavonoid levels as well as enzymatic activity like CAT, POD and SOD. Proteins responsible for initiating defense-related reactions include 6-phosphogluconate dehydrogenase, which is involved in various pentose-phosphate pathway reactions to produce NADPH. NADPH-oxidase is the primary ROS-producing enzyme during pathogen infection. TiO_2 NPs application shows dose- and time-dependent effects in lowering the incidence and severity of the disease (Satti et al., 2022).

Use of ZnO NPs is also reported in inhibition of *F. oxysporum*. These NPs act directly by distorting the fungal mycelia and reducing the effects of mycotoxins like fusaric acid. Various studies reveal that seed priming with ZnO NPs has resulted in the accumulation of sugars, phenol, total proteins, and activation of defense enzymes such as SOD, POD and CAT in plants, thus creating resistance against the pathogen (Farhana et al., 2022).

1.6 Conclusion

Nanoparticles when introduced in plant systems are known to insert various benefits on plant health, including a high rate of photosynthesis, elevated nutrition acquisition, increased crop productivity and targeted delivery of fertilizers and pesticides. Despite these positive attributes, the approaches of nanobionics are still not adapted as scarce information is available about the environmental and health dangers of NPs. Their toxicity to nontarget organisms and the unknown long-term environmental repercussions provide substantial concerns. Therefore, studies related to NPs' impact on biotic and abiotic stress are very much required. The above-mentioned studies have given us an idea of how NPs can help in the alleviation of stress conditions. However, there is a massive scope in the field of nanobionics, and implementation of these ways can innovate the agriculture prospects.

References

Adrees, M., Khan, Z. S., Hafeez, M., Rizwan, M., Hussain, K., Asrar, M., et al. (2021). Foliar exposure of zinc oxide nanoparticles improved the growth of wheat (*Triticum aestivum* L.) and decreased cadmium concentration in grains under simultaneous Cd and water deficient stress. *Ecotoxicology and Environmental Safety, 208*, 111627.

Alabdallah, N. M., Hasan, M. M., Hammami, I., Alghamdi, A. I., Alshehri, D., & Alatawi, H. A. (2021). Green synthesized metal oxide nanoparticles mediate growth regulation and physiology of crop plants under drought stress. *Plants, 10*(8), 1730.

Ali, H., Khan, E., & Sajad, M. A. (2013). Phytoremediation of heavy metals—Concepts and applications. *Chemosphere, 91*(7), 869–881.

Amini, S., Maali-Amiri, R., Mohammadi, R., & Kazemi-Shahandashti, S. S. (2017). cDNA-AFLP analysis of transcripts induced in chickpea plants by TiO_2 nanoparticles during cold stress. *Plant Physiology and Biochemistry, 111*, 39–49.

Arunakumara, K. K. I. U., Walpola, B. C., & Yoon, M. H. (2013). Current status of heavy metal contamination in Asia's rice lands. *Reviews in Environmental Science and Bio/Technology, 12*, 355–377.

Avestan, S., Ghasemnezhad, M., Esfahani, M., & Byrt, C. S. (2019). Application of nano-silicon dioxide improves salt stress tolerance in strawberry plants. *Agronomy, 9*(5), 246.

Badawy, A. A., Abdelfattah, N. A., Salem, S. S., Awad, M. F., & Fouda, A. (2021). Efficacy assessment of biosynthesized copper oxide nanoparticles (CuO-NPs) on stored grain insects and their impacts on morphological and physiological traits of wheat (*Triticum aestivum* L.) plant. *Biology, 10*(3), 233.

Bhaskar, R., Pandey, S. P., Kumar, U., Kim, H., Jayakodi, S. K., Gupta, M. K., & Han, S. S. (2024). Nanobionics for sustainable crop production: Recent development to regulate plant growth and protection strategies from pests. *Open Nano, 15*, 100198.

Birben, E., Sahiner, U. M., Sackesen, C., Erzurum, S., & Kalayci, O. (2012). Oxidative stress and antioxidant defense. *World Allergy Organization Journal, 5*, 9–19.

Chen, R., Zhang, C., Zhao, Y., Huang, Y., & Liu, Z. (2018). Foliar application with nano-silicon reduced cadmium accumulation in grains by inhibiting cadmium translocation in rice plants. *Environmental Science and Pollution Research, 25*, 2361–2368.

Chen, J., Wu, L., Lu, M., Lu, S., Li, Z., & Ding, W. (2020). Comparative study on the fungicidal activity of metallic MgO nanoparticles and macroscale MgO against soilborne fungal phytopathogens. *Frontiers in Microbiology, 11*, 365.

Cui, H., Shi, Y., Zhou, J., Chu, H., Cang, L., & Zhou, D. (2018). Effect of different grain sizes of hydroxyapatite on soil heavy metal bioavailability and microbial community composition. *Agriculture, Ecosystems & Environment, 267*, 165–173.

Elagamey, E., Abdellatef, M. A., & Arafat, M. Y. (2022). Proteomic insights of chitosan mediated inhibition of *Fusarium oxysporum* f. sp. *cucumerinum*. *Journal of Proteomics, 260*, 104560.

El-Sharkawy, M. S., El-Beshsbeshy, T. R., Mahmoud, E. K., Abdelkader, N. I., Al-Shal, R. M., & Missaoui, A. M. (2017). Response of alfalfa under salt stress to the application of potassium sulfate nanoparticles. *American Journal of Plant Sciences, 8*(8), 1751–1773.

Elsheery, N. I., Sunoj, V. S. J., Wen, Y., Zhu, J. J., Muralidharan, G., & Cao, K. F. (2020). Foliar application of nanoparticles mitigates the chilling effect on photosynthesis and photoprotection in sugarcane. *Plant Physiology and Biochemistry, 149*, 50–60.

El-Zohri, M., Al-Wadaani, N. A., & Bafeel, S. O. (2021). Foliar sprayed green zinc oxide nanoparticles mitigate drought-induced oxidative stress in tomato. *Plants, 10*(11), 2400.

Emamverdian, A., Ding, Y., Barker, J., Liu, G., Hasanuzzaman, M., Li, Y., et al. (2022). Co-application of 24-epibrassinolide and titanium oxide nanoparticles promotes *Pleioblastus pygmaeus* plant tolerance to Cu and Cd toxicity by increasing antioxidant activity and photosynthetic capacity and reducing heavy metal accumulation and translocation. *Antioxidants, 11*(3), 451.

Farhana, Munis, M. F. H., Alamer, K. H., Althobaiti, A. T., Kamal, A., Liaquat, F., et al. (2022). ZnO nanoparticle-mediated seed priming induces biochemical and antioxidant changes in chickpea to alleviate Fusarium wilt. *Journal of Fungi, 8*(7), 753.

Gao, F. L., Roiloa, S., Xia, J., Ren, J., Zou, M., Zhao, X., & Yu, F. H. (2024). Effects of native species richness on reproduction of invasive *Bidens pilosa* vary with nutrient supply. *Ecological Processes, 13*(1), 46.

Grillo, R., Pereira, A. E., Nishisaka, C. S., De Lima, R., Oehlke, K., Greiner, R., & Fraceto, L. F. (2014). Chitosan/tripolyphosphate nanoparticles loaded with paraquat herbicide: An environmentally safer alternative for weed control. *Journal of Hazardous Materials, 278*, 163–171.

Gull, A., Lone, A. A., & Wani, N. U. I. (2019). Biotic and abiotic stresses in plants. *Abiotic and Biotic Stress in Plants, 7*, 1–9.

Hall, J. Á. (2002). Cellular mechanisms for heavy metal detoxification and tolerance. *Journal of Experimental Botany, 53*(366), 1–11.

Hasanuzzaman, M., Parvin, K., Bardhan, K., Nahar, K., Anee, T. I., Masud, A. A. C., & Fotopoulos, V. (2021). Biostimulants for the regulation of reactive oxygen species metabolism in plants under abiotic stress. *Cells, 10*(10), 2537.

Hossain, M. A., Piyatida, P., da Silva, J. A. T., & Fujita, M. (2012). Molecular mechanism of heavy metal toxicity and tolerance in plants: Central role of glutathione in detoxification of reactive oxygen species and methylglyoxal and in heavy metal chelation. *Journal of Botany, 2012*(1), 872875.

Hussein, M. M., & Abou-Baker, N. H. (2018). The contribution of nano-zinc to alleviate salinity stress on cotton plants. *Royal Society Open Science, 5*(8), 171809.

Jain, C., Gupta, S., Sharma, S. P., Sangha, M. K., Kalia, A., & Sarao, N. K. (2023). Effect of Fusarium wilt on proteinaceous content and pathogenesis-related proteins in *Cucumis melo. Biologia, 78*(11), 3329–3338.

Kandhol, N., Jain, M., & Tripathi, D. K. (2022). Nanoparticles as potential hallmarks of drought stress tolerance in plants. *Physiologia Plantarum, 174*(2), e13665.

Khalid, M. F., Iqbal Khan, R., Jawaid, M. Z., Shafqat, W., Hussain, S., Ahmed, T., et al. (2022). Nanoparticles: The plant saviour under abiotic stresses. *Nanomaterials, 12*(21), 3915.

Khan, M. N. (2016). Nano-titanium dioxide (nano-TiO_2) mitigates NaCl stress by enhancing antioxidative enzymes and accumulation of compatible solutes in tomato (*Lycopersicon esculentum Mill.*).

Khan, I., Raza, M. A., Awan, S. A., Shah, G. A., Rizwan, M., Ali, B., et al. (2020). Amelioration of salt induced toxicity in pearl millet by seed priming with silver nanoparticles (AgNPs): The oxidative damage, antioxidant enzymes and ions uptake are major determinants of salt tolerant capacity. *Plant Physiology and Biochemistry, 156*, 221–232.

Kiselev, E. G., Boyandin, A. N., Zhila, N. O., Prudnikova, S. V., Shumilova, A. A., Baranovskiy, S. V., et al. (2020). Constructing sustained-release herbicide formulations based on poly-3-hydroxybutyrate and natural materials as a degradable matrix. *Pest Management Science, 76*(5), 1772–1785.

Kumari, S., & Sarkar, L. (2021). A review on nanoparticles: Structure, classification, synthesis & applications. *Journal of Scientific Research, 65*(8), 42–46.

Li, J., Naeem, M. S., Wang, X., Liu, L., Chen, C., Ma, N., & Zhang, C. (2015). Nano-TiO_2 is not phytotoxic as revealed by the oilseed rape growth and photosynthetic apparatus ultra-structural response. *PLoS One, 10*(12), e0143885.

Mahajan, S., & Tuteja, N. (2005). Cold, salinity and drought stresses: An overview. *Archives of Biochemistry and Biophysics, 444*(2), 139–158.

Maruyama, C. R., Guilger, M., Pascoli, M., Bileshy-José, N., Abhilash, P. C., Fraceto, L. F., & De Lima, R. (2016). Nanoparticles based on chitosan as carriers for the combined herbicides imazapic and imazapyr. *Scientific Reports, 6*(1), 19768.

Mishra, S., Spaccarotella, K., Gido, J., Samanta, I., & Chowdhary, G. (2023). Effects of heat stress on plant-nutrient relations: An update on nutrient uptake, transport, and assimilation. *International Journal of Molecular Sciences, 24*(21), 15670.

Mohammadi, R., Maali-Amiri, R., & Abbasi, A. (2013). Effect of TiO_2 nanoparticles on chickpea response to cold stress. *Biological Trace Element Research, 152*, 403–410.

Namiki, F., Shiomi, T., Kayamura, T., & Tsuge, T. (1994). Characterization of the formae speciales of Fusarium oxysporum causing wilts of cucurbits by DNA fingerprinting with nuclear repetitive DNA sequences. *Applied and Environmental Microbiology, 60*(8), 2684–2691.

Nel, W. J., Duong, T. A., De Beer, Z. W., & Wingfield, M. J. (2019). Black root rot: A long known but little understood disease. *Plant Pathology, 68*(5), 834–842.

Omar, A. A., Heikal, Y. M., Zayed, E. M., Shamseldin, S. A., Salama, Y. E., Amer, K. E., et al. (2023). Conferring drought and heat stress tolerance in wheat *(Triticum aestivum L.)* genotypes and their response to selenium nanoparticles application. *Nanomaterials, 13*(6), 998.

Pangestuti, R., & Kim, S. K. (2010). Neuroprotective properties of chitosan and its derivatives. *Marine Drugs, 8*(7), 2117–2128.

Potter, M., Deakin, J., Cartwright, A., Hortin, J., Sparks, D., Anderson, A. J., et al. (2021). Absence of nanoparticle-induced drought tolerance in nutrient sufficient wheat seedlings. *Environmental Science & Technology, 55*(20), 13541–13550.

Prakash, P. (2023). Nano-phytoremediation of heavy metals from soil: A critical review. *Pollutants, 3*(3), 360–380.

Raza, A., Khandelwal, K., Pandit, S., Singh, M., Kumar, S., Rustagi, S., et al. (2024). Exploring the potential of metallic and metal oxide nanoparticles for reinforced disease management in agricultural systems: A comprehensive review (Vol. 22, p. 100998). Environmental Nanotechnology.

Rizwan, M., Ali, S., ur Rehman, M. Z., Riaz, M., Adrees, M., Hussain, A., et al. (2021). Effects of nanoparticles on trace element uptake and toxicity in plants: A review. *Ecotoxicology and Environmental Safety, 221*, 112437.

Rossi, L., Zhang, W., & Ma, X. (2017). Cerium oxide nanoparticles alter the salt stress tolerance of *Brassica napus L.* by modifying the formation of root apoplastic barriers. *Environmental Pollution, 229*, 132–138.

Saboki Ebrahim, K. U., & Singh, B. (2011). Pathogenesis related (PR) proteins in plant defense mechanism. *Science Against Microbial Pathogens, 2*, 1043–1054.

Sanya, D. R. A., Syed-Ab-Rahman, S. F., Jia, A., Onésime, D., Kim, K. M., Ahohuendo, B. C., & Rohr, J. R. (2022). A review of approaches to control bacterial leaf blight in rice. *World Journal of Microbiology and Biotechnology, 38*(7), 113.

Satti, S. H., Raja, N. I., Ikram, M., Oraby, H. F., Mashwani, Z. U. R., Mohamed, A. H., et al. (2022). Plant-based titanium dioxide nanoparticles trigger biochemical and proteome modifications in *Triticum aestivum L.* under biotic stress of *Puccinia striiformis. Molecules, 27*(13), 4274.

Song, Y., Jiang, M., Zhang, H., & Li, R. (2021). Zinc oxide nanoparticles alleviate chilling stress in rice *(Oryza Sativa L.)* by regulating antioxidative system and chilling response transcription factors. *Molecules, 26*(8), 2196.

Sousa, G. F., Gomes, D. G., Campos, E. V., Oliveira, J. L., Fraceto, L. F., Stolf-Moreira, R., & Oliveira, H. C. (2018). Post-emergence herbicidal activity of nanoatrazine against susceptible weeds. *Frontiers in Environmental Science, 6*, 12.

Tariq, M., Choudhary, S., Singh, H., Siddiqui, M. A., Kumar, H., Amir, A., & Kapoor, N. (2021). Role of nanoparticles in abiotic stress. *Technology in Agriculture, 323.*

Tortella, G., Rubilar, O., Pieretti, J. C., Fincheira, P., de Melo Santana, B., Fernández-Baldo, M. A., et al. (2023). Nanoparticles as a promising strategy to mitigate biotic stress in agriculture. *Antibiotics, 12*(2), 338.

Trematerra, P., Throne, J., Fernandez, M., & Knox, R. (2012). Insect and mite pests of durum wheat. Durum wheat, chemistry and technology (2nd ed., pp. 73–83). AACC International, St. Paul.

Upadhyay, R. K., & Ahmad, S. (2011). Management strategies for control of stored grain insect pests in farmer stores and public ware houses. *World Journal of Agricultural Sciences, 7*(5), 527–549.

Wang, S., Wang, F., & Gao, S. (2015). Foliar application with nano-silicon alleviates Cd toxicity in rice seedlings. *Environmental Science and Pollution Research, 22*, 2837–2845.
Wang, Y., Liu, Y., Zhan, W., Zheng, K., Lian, M., Zhang, C., et al. (2020). Long-term stabilization of Cd in agricultural soil using mercapto-functionalized nano-silica (MPTS/nano-silica): A three-year field study. *Ecotoxicology and Environmental Safety, 197*, 110600.
Wang, A., Li, J., Al-Huqail, A. A., Al-Harbi, M. S., Ali, E. F., Wang, J., et al. (2021). Mechanisms of chitosan nanoparticles in the regulation of cold stress resistance in banana plants. *Nanomaterials, 11*(10), 2670.
Yassen, A., Abdallah, E., Gaballah, M., & Zaghloul, S. (2017). Role of silicon dioxide nano fertilizer in mitigating salt stress on growth, yield and chemical composition of cucumber *(Cucumis sativus L.)*. *International Journal of Agricultural Research, 22*, 130–135.

Chapter 2
Application and Future Prospective of Nanomaterials Towards Nanobionics

Sagar Mondal, Swati Priya, Md Saad, Ajay Kumar Mahalka, Kavya Singh, Seema Ramniwas, Abeer Hashem, Elsayed Fathi Abd Allah, and Sumira Malik

Abstract Nanotechnology is an advanced field where invisible molecules are achieving remarkable results. It spans various disciplines, including electronics, agriculture, food science, pharmaceuticals, and medical diagnostics, with many researchers focussed on advancing technology through these innovations. Nanomaterials hold significant potential in agriculture due to their distinctive physicochemical properties. When plants interact with these nanoparticles, they can induce various morphological, physiological, and genotoxic changes. Understanding these interactions is crucial for fully realising nanotechnology's benefits in agriculture. Often referred to as "magic bullets", nanoparticles can significantly impact plant growth and photosynthesis. They affect critical growth processes, such as seed germination, root development, and photosynthesis. The emerging field of plant nanobionics involves integrating engineered nanomaterials with plant organelles to enhance plant functions. This chapter explores the diverse applications and future prospects of plant nanobionics in detail.

S. Mondal · S. Priya · M. Saad · A. K. Mahalka · K. Singh · S. Malik (✉)
Amity Institute of Biotechnology, Amity University Jharkhand, Ranchi, India
e-mail: smalik@rnc.amity.edu

S. Ramniwas
University Center for Research & Development (UCRD) Chandigarh University, Mohali, Punjab, India

A. Hashem
Department of Botany and Microbiology, College of Science, King Saud University, Riyadh, Saudi Arabia

E. F. A. Allah
Department of Botany and Microbiology, College of Science, King Saud University, Riyadh, Saudi Arabia

Department of Plant Production, College of Food and Agricultural Sciences, King Saud University, Riyadh, Saudi Arabia

V. D. Rajput et al. (eds.), *Contemporary Suitability of Nanobionics in Agriculture*, Nanotechnology in the Life Sciences, https://doi.org/10.1007/978-3-031-82779-2_2

Keywords Nanotechnology · Nanomaterials · Nanobionics · Agriculture · Crop development

2.1 Introduction

Nanoscience refers to the study of the characteristics of matter that fall between the dimensional range of 1–100 nm. Nanotechnology is the application of manipulation techniques at the nanoscale to construct entities of nanoscale dimensions. An essential characteristic of a nanomaterial is its presence at a key nanoscale in the range of 1–100 nm, which results in qualities that are not present within its molecular state alongside the bulk solid state. The growing significance of nanomaterials in diverse sophisticated technical applications has generated a distinct requirement to develop novel methods for acquiring nanoparticles with precise nanostructures and characteristics.

Nanomaterials represent the paramount technological advancement in the contemporary era. The material in question is characterised by their diameters being smaller than around 100 nm. These materials are of great interest to researchers due to their distinctive magnetic, optical, electrical, and wide range of other characteristics, which will have significant implications in several scientific disciplines. While some nanoparticles are naturally occurring, others are intentionally created for specific applications in commercial products including cosmetics, electronic devices, stain-resistant clothes, and medicine delivery systems. The enhanced surface area-to-volume ratios associated with these nanomaterials have a direct impact on their mechanical strength and chemical flexibility. Moreover, quantum effects exert influence on the material qualities and traits, therefore giving rise to innovative goods across several domains (Alagarasi, 2013). Engineered nanomaterials (ENMs) possess exceptional physiochemical properties that make them well suited for a wide range of applications in several research domains, including biotechnology, energy, cosmetics, and pharmaceuticals, as well as numerous other related areas. Engineered nanomaterials (ENMs) can be introduced into the environment either through recycling procedures or byproduct conversion during their production operations. Environmental nanomaterials (ENMs) have raised significant apprehension due to their growing range of applications, both in terms of their potential impacts on live organisms (biotic ecosystem components) and nonliving (abiotic) community components. Plants fulfil a crucial function in preserving ecological systems and play vital roles in the accumulation and movement of environmental nanomaterials (ENMs) in food chains at various trophic levels. Environmental nanomaterials (ENMs) can readily penetrate plant cells, where they interact and disrupt intracellular metabolic pathways along with structures.

Researchers exhibited both favourable and unfavourable reactions to the method of applying nanoparticles to the growth and development of plant species, as well as the underlying synthetic processes and characteristics of these nanomaterials. By their various forms and quantities, these substances exert either beneficial or detrimental effects on processes such as photosynthesis. Hence, plants are regarded as a fundamental component prevalent on Earth that will experience direct impacts from these environmental nanomaterials (ENMs), as they interact with the biosphere process (Hatami et al., 2016).

2.2 Roles of Nanomaterials in Plant Growth

Nanomaterials refer to a wide range of nanostructured materials that have at least a single dimension within the nanometric measurement range. Nanomaterials possess a special combination of characteristics that lie among the characteristics of bulk materials along with atoms. The characteristics exhibited by nanomaterials differ greatly compared to those of bulk substances including subatomic particles. Characteristic nanomaterials include nanocrystals and clusters measuring 1–10 nm in diameter; nanowires measuring 1–100 nm in diameter; nanotubes measuring 1–100 nm in diameter; dimensional nanoparticle arrays ranging from several square nanometres to square micrometres in area surfaces; thin films measuring 1–100 nm in thickness; and 3D structures such as or superlattices measuring several nanometres in all three dimensions.

Due to their reduced size, these nanoparticles possess significantly high surface area-to-volume ratios, resulting in a substantial portion of these materials exhibiting higher surface-dependent material characteristics. Assuming that the size of the nanomaterial is similar to its Debye length, the surface characteristics of these nanomaterials will impact the entire material. Research in nanotechnology and nanoscience is a highly multidisciplinary field, and development efforts have experienced rapid expansion globally in recent years.

The nanomaterials can exhibit nanoscale dimensions in a single dimension (1D, such as surface films), in two dimensions (2D, such as strands or fibres), or in three dimensions (3D, such as particles). They can also exist as individual, aggregated, even agglomerated fused structures with cylindrical, spherical, tubular, and irregular forms. Figure 2.1 illustrates Siegel's categorisation of nanostructured materials into zero-dimensional (0D—atomic clusters, filaments, and cluster assemblies), one-dimensional (multilayers), two-dimensional (ultrafine-grained over-layers along with buried layers), and three-dimensional (nano-phase materials composed of equiaxed nanometre-sized grains), 1D, 2D, and 3D nanostructures (Seigel et al., 1997). The predominant categories of nanomaterials include nanotubes, quantum dots (QDs), dendrimers, and fullerenes (C60 compound). Various nanomaterials, such as Ag nano, Au nano, carbon nanotubes (CNTs), fullerene, photocatalysts, carbon nano, and silica, exhibit distinct physical–chemical properties that make them highly suitable for use in nanotechnology.

Nanomaterials are fundamental elements of nanotechnology and nanoscience, exhibiting profound potential to transform the methods of nanodevice and nanoproduct fabrication and the range of functions that can be explored. Their commercial influence has already been significant, and this is expected to substantially grow

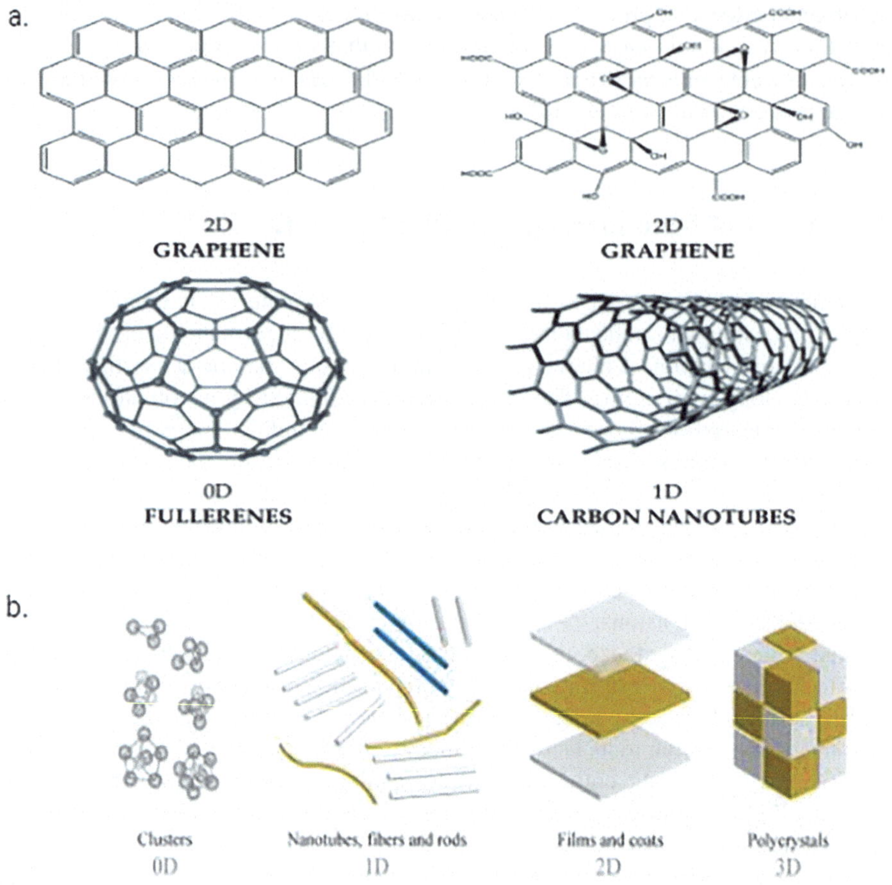

Fig. 2.1 Classification of nanomaterials

in the future. Unconventional physicochemical characteristics of nanomaterials provide limitless possibilities in the agricultural sector. Understanding the morphological, physiological, including genotoxic changes that occur when nanoparticles interact with plants is crucial for the practical application of nanotechnology for agriculture. Given the remarkable progress in nanotechnology, electron nanomaterials (ENMs) have found extensive use in several aspects of plant systems, resulting in remarkable improvements in quality, growth, nutritional value increase, gene preservation, and more. Numerous published reports have examined the effects of ENMs on growth enhancements and development inhibition, as well as the specific impacts of ENMs on plants. These studies have investigated the uptake, growth responses, translocation, biotransformation, and stress modulation mechanisms of ENMs in plant systems. Investigating the biochemical, physiological, and molecular mechanisms underlying nanoparticle interactions with plants has been a recent area of research aimed at understanding the dangers associated with applying nanomaterials to agriculturally significant crops.

Table 2.1 Effects of engineered metal oxide nanoparticles on the growth of plants

Nanoparticles	Plants	Effects on plants
Aluminum oxide (Al_2O_3)	*Daucus carota* and *Brassica oleraceae*	Reduced root growth
Cerium oxide (CeO_2)	*Solanum lycopersicum, Zea mays*, and *Cucumis sativus*	Reduced shoot growth
Iron oxide (Fe_2O_3)	*Syzygium aromaticum*	Reduced aboveground and belowground biomass
Cerium oxide (CeO_2)	*Oryza sativa L.*	Reduced hydrogen peroxide generation in roots and shoots
Combination of SiO2 and TiO2	*Glycine max*	Increased germination and shoot growth, enhanced antioxidant system, increased nitrate reductase activity
Magnetite (iron oxide)	*Glycine max*	Increased chlorophyll
Nanoanatase	*Spinacia oleracea L.*	Enhanced messenger RNA expression, protein activity, and photosynthetic activity
Inorganic bentonite clay (TiO2)	*Zea mays*	Inhibited hydraulic conductivity, leaf growth, and transpiration

Plants undergo a confluence of both abiotic and biotic pressures, commencing with seed germination, the initial phase of plant development. The identification of specific nanomaterials that contribute to the effective growth of living things with regard to nitrogen metabolism, photosynthesis, as well as other benefits is documented in Table 2.1. The effective infiltration of nanomaterials leads to the activation of genes that encode water channel proteins. As a result, efficient control of the cell cycle facilitates improved cell development. Efficacious seed germination relies on the utilisation of stored food resources through enzymes like amylase and protease to ensure the survival of young plants till they can produce their own energy resources via photosynthesis (Khan et al., 2017).

2.3 Plant Nanobionics

The overarching goal is to develop a diverse range of indigenous plants that possess the ability to perceive inanimate objects in their surroundings, serve as infrared communication devices, operate as autonomous light sources, and also work as autonomous groundwater sensors. Plants possess unparalleled suitability for carrying out these functions due to their capacity to produce energy from sunlight through the process of photosynthesis. Within the realm of nanobiotechnology, scientists strive to create bionic plants that demonstrate optimal photosynthetic efficiency and biochemical sensing capabilities.

Plant nanobionics is a nascent discipline within the realm of bioengineering, characterised by the introduction of nanoparticles into plants, therefore modifying or enhancing the functioning of plant tissues including organelles (Ghorbanpour &

Fahimirad, 2017). Plants facilitate ongoing gas exchange with their environment. Phytotranspiration is a physiological mechanism in which plants transport water through the roots and release water vapour through their leaves. This process exhibits several specific features, including (a) the stabilisation and movement of water under high pressures, (b) the continuous heat transfer resulting from water evaporation, and (c) the continual extraction of water coming from sources that are already saturated (Wong et al., 2017). Biomaterials developed using traditional methods depend on the replication of biological structures to improve the photosynthetic activity process by incorporating synthetic nanoparticles into engineered materials together with plant organelles.

2.3.1 Analysis of the Plant Cell Wall Architecture Using Nanobionics

Plant cells and carbon nanotubes (CNTs) combined in a bionic material demonstrate superior temperature sensitivity, surpassing the performance of the most advanced synthetic materials. The fundamental process under consideration is the ionic conductivity that links microfibrils of cellulose in the plant cell wall together with the pectin backbone. Carbon nanotubes (CNTs) are effective in inhibiting cell death and enhancing the structural integrity of desiccated cells under elevated temperatures. The plant cell wall is characterised by a sophisticated macromolecular architecture that exhibits highly dynamic structural properties. Extending beyond the plasma membrane, the cell wall consists of microfibrils. Ion transport inside the cell wall is affected by several parameters including charge, porosity, pH equilibrium, and surface charge. Pectins are composed of negatively charged saccharides that engage cations to create cross-links, therefore enhancing the structural integrity of the cell wall (Di Giacomo et al., 2015). The primary mechanism responsible for mediating photosynthetic activity in chloroplasts is the thylakoid membrane. Approximately half of the incoming solar radiation energy is absorbed by chloroplasts in the visible range within the light spectrum (Ghorbanpour & Fahimirad, 2017). In 2014, Giraldo and colleagues developed a system consisting of highly charged single-walled carbon nanotubes (SWCNTs) coated in DNA and chitosan. These nanotubes were capable of penetrating chloroplasts (Giraldo et al., 2014).

2.3.2 Analysis of Supercharged Photosynthesis: Mechanisms Involved in the Photosynthetic Machinery

Plant chloroplasts are the primary organelles responsible for the biological process called photosynthesis. The first step involves the absorption of light by chlorophyll, which stimulates electrons and induces their movement across the thylakoid membrane that covers chloroplasts. The produced electrical energy is utilised by the plant to synthesise sugars during the second phase of photosynthesis. Upon removal from plants, chloroplasts retain their ability to carry out these activities. However, after a certain period, they deteriorate due to the detrimental effects of light and oxygen on the photosynthetic proteins. Therefore, researchers have employed a method called lipid interchange envelope penetration (LEEP) to transport cerium oxide nanoparticles, sometimes referred to as nanoceria and acting as potent antioxidants, into the interior of chloroplasts in order to safeguard them from harm. Encapsulating these particles in polyacrylic acid, a highly charged polymer, enables the nanoparticles to permeate the lipid hydrophobic membranes that envelop the chloroplasts. Immediately, the degree of injury diminished. Through vascular infusion, scientists have administered nanoparticles to the bottom of leaves and found that carbon nanotubes (CNTs) penetrated the chloroplasts, resulting in a 30% increase in electron flow during photosynthesis (Enamala et al., 2019).

2.3.3 Engineered Nanomaterials and Photosynthesis

While certain nanomaterials occur naturally, ENMs are particularly noteworthy as they find application in a wide range of commercially accessible products and processes, including sunscreens, cosmetics, stain-resistant clothes, electronics, sporting goods, tires, and several other everyday items. Furthermore, they find application in the field of medicine for the objectives of imaging, diagnosis, and therapy administration. Energy nanomaterials (ENMs) are materials specifically engineered at the molecular (nanometric) scale to exploit their small dimensions and unique characteristics that are often absent in their classical, bulk-form equivalents. These characteristics mostly arise from the larger relative surface area along with quantum effects.

Plants utilise light energy (photons) from sunshine to synthesise their own food substrates and distribute it to other cellular components. Photosynthesis refers to the biological process by which living organisms convert light energy into chemical energy. Furthermore, photosynthesis is accountable for the formation of fossil fuel energy reserves, which have accumulated over millions of years (Szabó et al., 2015). Plants efficiently utilise only 2–4% of the overall energy allocated for the development of the succeeding generation (Siddiqui et al., 2015). Therefore, scientists are currently investigating RuBisCO, an enzyme that is recognised as a crucial catalyst for accelerating these chemical reactions, in order to enhance their rate. Chloroplasts house light-sensitive pigments that harness solar energy and serve as sites for intricate photosynthetic processes, such as light absorption and charge dissociation facilitated by light and electron transport between photosystems. This energy undergoes conversion into several sugars including maltose, triose phosphate, and glucose. Additionally, this photosynthetic process is employed in plants to produce adenosine triphosphate (ATP). Extracellular metabolites (ENMs) have been discovered to exert both beneficial and detrimental impacts on the process of photosynthesis. The adverse impacts encompass decreases in chlorophyll and carotenoid levels, impaired enzyme functions resulting from membrane lipid peroxidation, including oxidative stress induced by the ENMs. Chlorophyll is the primary pigment required by photoautotrophic organisms. It is the most photosensitive form of chlorophyll and readily loses its structural integrity when exposed to light. Electron microarrays (ENMs) have the potential to greatly contribute to the development of robust biomarkers (Hatami et al., 2016).

Plants cultivated under circumstances of intense light and abundant nutrients exhibited reduced rates of photosynthesis and CO2 assimilation when exposed to cerium oxide, titanium oxide, and copper hydroxide at varying concentrations. Environmental nanomaterials (ENMs) interfered with photosynthesis in plants cultivated under extreme stress circumstances, such as intense light and limited nutrients, by inhibiting the photo-oxidation process in the photosystem II reaction centre. Numerous scientists are currently engaged in efforts to enhance the efficiency of photosynthesis by employing gene editing methods. This involves integrating cyano-bacteria carbon-fixing genes in tobacco plants, therefore substituting the RuBisCO genes in the plants. Researchers are currently undertaking the development of nanobionic plants, which exhibit enhanced photosynthetic efficiency and biochemical sensing capabilities. The integration of single-walled carbon nanotubes (SWCNTs) into plants has been documented to increase photosynthetic activity threefold, facilitate plants in detecting nitric oxide, and improve electron transport rates (Siddiqui et al., 2015).

2.4 Applications of Nanomaterials

2.4.1 Food and Agriculture

Nanotechnology and nanobiotechnology are burgeoning disciplines with immense capacity to revolutionise agriculture and other sectors. Contemporary nanotechnology research in agriculture mostly centres on farming and encompasses the application of nanoparticles to enhance agricultural and livestock productivity. Industrial applications of nanotechnology include agricultural production, animal nutrition, food additives, food processing, along with food contact materials (Prasad et al., 2017a, b). An extensive variety of nanomaterials has been documented in various agricultural and food-related uses, with ongoing research of new nanomaterials and applications. The utilisation of nanomaterials in agricultural and food goods is projected to grow in the next years, leading to a higher level of nanomaterial exposure in both humans and the environment.

Advanced nanotechnology applications, including processing, packaging, and food supplements, are now under investigation and experimentation in all areas of food technology, including agriculture to the food business. Indeed, in the field of agriculture, scientific advancements play a crucial role in tackling worldwide issues

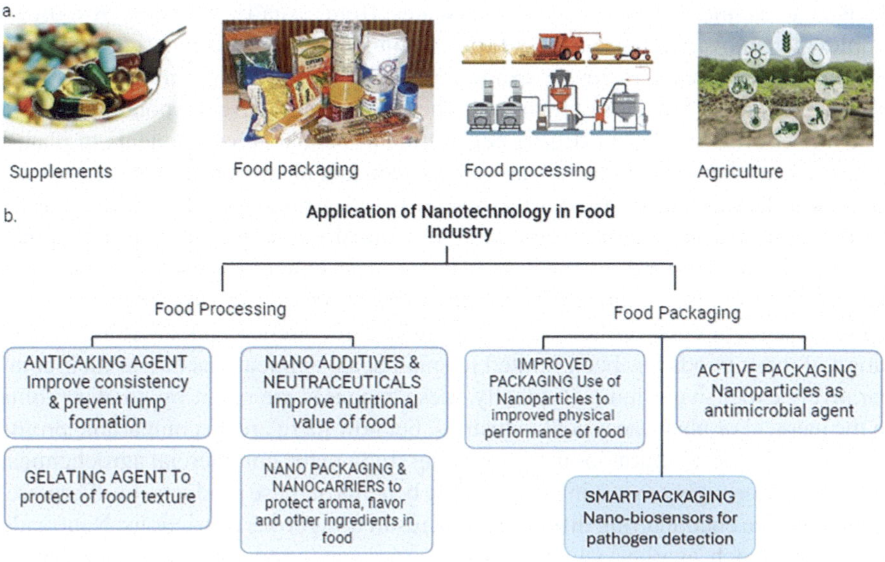

Fig. 2.2 Nanotechnology applications in (**a**) the food industry sector and (**b**) the agriculture sector and other food sectors

including population increase, climate change, and the restricted supply of vital plant nutrients like phosphorus and potassium. Research on agricultural uses of nanotechnology has been active for almost ten years, and it has the potential to play a crucial role in agricultural productivity. Furthermore, this also addresses the matter of nanotechnology in rapidly developing nations. Examples used in the agricultural and food industries are illustrated in Fig. 2.2.

Notwithstanding the potential benefits of nanotechnology in the farming industry, its implementation remains somewhat limited and it has not yet established a significant presence in the market, when compared to other areas of nanotechnology. The great potential of nanotechnology in the farming industry is accompanied by some unresolved concerns. These include the necessity for enhancements in production, procedures, and cost reduction, together with the evaluation of risks. Given these concerns, researchers find nanoparticles produced from biopolymers, such as proteins and carbohydrates, to be especially appealing due to their minimal effects on human health and the environment. For example, research organisations are thoroughly assessing the potential of starch-based nanoparticles as sustainable delivery systems/carriers for agrochemicals and biostimulants, as they are inherently nontoxic. Agricultural waste has gained relevance as a sustainable supply of raw materials for processing in several applications and as raw materials for the synthesis of nanomaterials, which can serve as alternatives to fossil fuel resources. The qualities of biomaterial-based nanocomposites are far superior to those of conventional micro- and macro-composite materials. Furthermore, their manufacturing is environmentally friendly. Currently, numerous manufacturing techniques are being devised to acquire valuable nanocomposites from usually collected resources, employing innovative approaches. From a commercial perspective, current agrochemical businesses are closely examining the possibilities of nanotechnologies, specifically nanosized active chemicals. These substances have the potential to provide higher effectiveness or deeper penetration of beneficial components in plants. Nevertheless, they have failed to showcase significant enhancements in their goods, such as reductions in costs and the ability to sustain large-scale manufacturing. Technology-oriented smaller firms have strategically marketed certain nanoproducts for the agricultural sector. For instance, soil enhancer products have been developed to facilitate uniform water distribution and storage, thereby leading to water conservation. Nevertheless, the current size of commercial market implementation of these goods has been limited to small-scale applications due to their comparatively expensive pricing. Typically, these expenses are offset by greater profits in the medical or pharmaceutical industries, but now there are no comparable profits possible in the agricultural industry. Ongoing study in the commercial agrochemical industry is thoroughly assessing the possible benefits of these products. Furthermore, companies are encountering difficulties as a result of rigorous restrictions. Nanoscale formulants, such as clay, silica, polymers, pigments, and macromolecules, have been widely employed for several decades and are also present in numerous common household products.

2.4.2 Gene Delivery Systems

Gene therapy is a contemporary technological approach employed by numerous researchers to treat various ailments by the substitution or correction of missing genes. Although the theoretical knowledge is commendable, there exist certain constraints in practical implementations that require rectification. The practical implementations of gene delivery devices have a somewhat low success rate. The development of secure and effective vectors for delivering genes into intracellular locations with degradation is one of the specific obstacles that gene therapy encounters. There are two main categories of gene delivery systems: viral and nonviral vectors. Initial distribution of therapeutic genes was initially accomplished utilising viral vectors due to their ability to efficiently transport genes and guarantee sustained expression at the specific location (Sun et al., 2017).

Drug delivery systems including nanomaterials are being employed in diverse therapeutic domains. The utilisation of these drug delivery systems can increase the pharmacokinetics of medications by enabling regulated distribution of essential components to the intended locations and optimise the drug release process (Jacob et al., 2018). Physiochemical features such as DNA-binding capacity, chemical alterations, and minimal immunogenicity make polymeric vectors very viable sources of nonviral vectors. The various generations of cationic polymers encompass both first-generation and second-generation cationic derivatives. Among the first-generation polymers is poly-L-lysine, which has the ability to attach to DNA and undergo spontaneous condensation, resulting in the formation of compact structures. Polyethyleneimine (PEI) and polyamide amine dendrimers (PAMAMs) are examples of second-generation polymers (Jacob et al., 2018). Numerous well-established drug delivery technologies exist, and among them, inorganic nanoparticles have garnered significant attention as promising delivery techniques.

Particles derived from plants provide several advantages compared to nanoparticles like liposomes, dendrimers, and quantum dots (QDs). The advantages encompass versatile techniques for preparation, human safety, biodegradability, and the simplicity of conducting genetic and chemical alterations.

There exist several categories of nanoparticles, including polymers, lipids, carbon nanotubes (CNTs), and viruses. Within this range of sorts, viral nanoparticles are the most commonly employed due to their numerous advantages, as previously stated. Viruses are minute particles composed of nucleic acids, including DNA and RNA, which contain genetic information and diverse protein molecules capable of assembling into a hollow packing structure responsible for safeguarding crucial information. An inherent benefit of these viruses is their ability to endure elevated temperatures and maintain their integrity at different pH levels when exposed to organic solvent–water composites. The widespread use and significant influence of viral nanoparticles can be attributed to their many characteristics, including size homogeneity, unique morphologies, and resistance to a wide range of temperatures and pH levels. An additional benefit of viral capsids is their ability to accommodate the insertion of foreign components, such as peptides or ligands, through the usage of numerous locations. Development of polyvalent scaffolds can improve the targeting capabilities of viral particles (Alemzadeh et al., 2018).

Protein pharmaceuticals (PDs) can be introduced into lyophilised plant cells due to their long-term stability at room temperatures, which is achieved by maintaining their folding and therapeutic effectiveness. Upon ingestion, these medications are sequestered by a protective covering enclosed within the sheet. The sheet serves as a barrier against the acidic intestinal fluids, allowing the medicine to be released when it comes into contact with microorganisms in the gut lumen that break down the plant cell wall. When an undamaged plant cell carrying a plant disease (PD) enters the gastrointestinal tract, the intestinal microorganisms break down the cell wall of the plant and liberate the PD. The human digestive enzymes lack the ability to fully hydrolyse all glycosidic linkages present in the cell walls of plants. Fusion of receptor-binding proteins with PDs allows their passage across the intestinal epithelium and subsequent delivery to the immune system. Recently, newly created sole protein receptor-binding sites have been identified. Mucopolysaccharides are the chemical compounds produced by gut bacteria that break down the mucin glycoprotein, enabling the mucous barrier to be penetrated (Kwon & Daniell, 2016).

VLPs, in general, exhibit exceptional stability; those variants capable of withstanding significant fluctuations in temperature and additional parameters are highly advantageous for both the production and distribution processes. Vaccine logistics packages (VLPs) are specifically designed to endure storage and transportation at various temperatures. They can greatly simplify the logistics involved in distributing vaccines, especially in regions throughout the developing world where maintaining a cold chain is challenging. Simultaneously, enhancement in the stability of vaccine liquid polymers (VLPs) leads to a decrease in the cost of vaccines by enabling an extended duration of storage and also decreases the energy expenses linked to cold refrigeration. Additional modifications have been made to improve the structural stability of VLPs. The aforementioned alterations not only enable the engineering of recombinant VLPs but have also been carried out with consideration for various downstream applications. Hepatitis B core antigen very large particles (VLPs) exhibit increased stability. Lyophilisation or spray drying of vitreous lipid polymers (VLPs) after manufacture can improve their thermal stability and enable extended storage.

2.5 Microbial Nanobionics

In general, VLPs have high stability. Specifically, those varieties that can endure significant fluctuations in temperature and other environmental factors are highly advantageous for both the production and distribution processes. Temperature-resistant VLPs are designed to endure storage and transportation at various temperatures and help streamline logistics in vaccine distribution, especially in regions of the developing world where maintaining a cold chain is challenging. Simultaneously, enhancement in the stability of vaccine liquid polymers (VLPs) serves to decrease the expenses related to vaccines by enabling an extended duration of storage and also diminishes the energy expenditures linked to cold storage. Modifications have been made to VLPs in order to improve their structural stability. In addition to facilitating the engineering of recombinant VLPs, these alterations have also been implemented with additional downstream uses in consideration. VLPs containing the core antigen of Hepatitis B exhibit increased stability. Advanced postproduction techniques, which include lyophilisation or spray drying of VLPs, can further improve their resistance to heat and enable extended storage.

2.5.1 Microbial Cell Factories

The performance of microbial cell factories in several processes enables the implementation of cost-effective biorefineries. The conversion of biomass obtained from pentose and hexose into specific compounds is achieved by metabolic pathway optimisation. Important factors to consider in the emergence of microbial cell factories include direct saccharification of bio-mass and stress tolerance under extreme circumstances during fermentation (Kondo et al., 2013). Microbial secondary metabolites were extensively utilised for diverse biological interventions to enhance human health. These chemicals are used as antibiotics, therapeutic agents, poisons, insecticides, and growth factors for plants and animals (Nielsen & Nielsen, 2017). Although there are many species, the study on secondary metabolites has primarily focussed on actinomycetes species due to the fact that most of the metabolites are derived from the genus *Streptomyces*. Microbial cell factories exhibit high reliability in terms of metabolic processes. To achieve optimal performance, engineering approaches are necessary to facilitate the integration and modification of metabolic pathways within these cell factories. *Streptomyces cerevisiae* is a well-suited host for biorefinery operations in the industrial production of bioethanol due to its considerable cell recycling capacity during fermentation and its ability to withstand many stressors like low pH and high temperature. This yeast organism serves as a highly studied model in the realm of metabolic engineering for the purpose of comprehensive genome sequencing. The extraordinarily intricate metabolic system of this species enables researchers to forecast the outcomes of metabolic alteration (Kondo et al., 2013). A significant proportion of the widely recognised microbial antibiotics, namely 45%, are derived from actinomycetes, 38% from fungi, and 17% from unicellular bacteria. Within this category of antibiotics, only a limited number are currently being used for human treatment (Nielsen & Nielsen, 2017). Several *Aspergillus* strains, strongly connected in their phylogenetic tree structure, have been employed for the synthesis of organic acids due to their high capacity to release substantial quantities of required organic acids. Certain organic acid compounds, particularly low-molecular-weight carboxylic acids, play a significant role in biotechnological manufacturing. A number of organic acids are now synthesised using biotechnological methods, including the generation of citric acid and itaconic acid by *Aspergillus niger* along with *Aspergillus terreus* (Yang et al., 2017).

In addition to the generation of diverse secondary metabolites, microbial electrosynthesis is a novel method that has the potential to substitute several synthesis approaches with contemporary methods. This method involves the reduction of substrates using microbial catalysts and an external electron supply with the aim of producing various biocommodities. In electrochemical processes, these bioelectrochemical systems (BESs) employ a consortia of microorganisms as bioelectrocatalysts. They possess both biological benefits, such as self-replication, and electrochemical benefits, such as a readily available source of redox equivalents without bulk production. At present, microbial fuel cells (MFCs) are currently employed in both microbial desalination and microbial electrosynthesis processes. Microbial electrosynthesis is a highly promising technology in a microbial cell factory due to its ability to generate organic substrates utilising microbial electrocatalysts (Shin et al., 2017). In addition to the diverse microbial consortiums used in microbial cell factories, eukaryotic species called cyanobacteria are also extensively employed in several technologies including biofuels, enzyme synthesis, and protein synthesis. Photosynthetic cell factories have established themselves as a highly promising resource for the sustainable and cost-effective generation of fuel and chemical synthesis. Efficient optimisation and evaluation of these photosynthetic cell factories under ideal conditions can facilitate the achievement of a consistent pattern of cell development and accumulation of products (Luan & Lu, 2018).

The clustered regulatory interspaced short palindromic repeat (CRISPR)–Cas system has emerged as a game-changing important component in metabolic engineering over the last two decades, in addition to the conventional microbial cell factories. Evidently, metabolic engineering remains a formidable undertaking in constructing or reconstructing metabolic pathways for the synthesis of diverse bio-based products, even when using model organisms. Development of novel pathways necessitates substantial alterations in both gene integration and deletion, as well as the careful management of essential enzymes for gene expression and the optimisation of non-native gene codons. Therefore, it is not unexpected that the development of the CRISPR–Cas system has initiated a new age in the domain of precise genetic engineering (Jakočiūnas et al., 2016). There is extensive usage of biopharmaceutical proteins in the therapy of cancer, diabetes, chronic viral hepatitis, inflammatory disorders, and autoimmune illnesses. At now, a significant number of these proteins are being synthesised via the expression of proteins by *Escherichia coli*, as it features tiny proteins with non-posttranslational modifications. Contemporary research has enhanced metabolic engineering through the implementation of intricate post-transcriptional alterations, such as the disulphide production pathway, in heterologous investigations. Contrasting with the bacterially derived vaccines utilised in recent years, vaccinations derived from viruses will have a significant impact on world health by minimising adverse effects and enhancing immunogenicity. Vector-like particles (VLPs) are protein complexes consisting of several subunits that have the ability to self-assemble, hence creating structures that mimic the three-dimensional conformation of natural viruses. They are devoid of viral genetic material, rendering them noninfectious and incapable of replication. Immuno-vaccines derived from attenuated or inactivated viruses are regarded to be safer than conventional ones (Rodríguez-Limas et al., 2013).

2.5.2 Microbial Fuel Cells

The devices known as microbial fuel cells employ microorganisms as biocatalysts to convert organic waste materials into electrical energy (Shah et al., 2019). This environmentally sustainable technique has been increasingly recognised worldwide in recent years as a cutting-edge technology for both power production and garbage management. The apparatus of a typical microbial fuel cell (MFC) comprises a dual-chamber system, including an anode chamber and a cathode chamber, which are divided by a proton exchange membrane (PEM). The bacterial consortium operating in the anodic chamber undergoes oxidation of energy, resulting in the production of electrons and protons. Following this, electrons are transported via an external circuit to the cathode chamber, ultimately resulting in the generation of a current (Zhao et al., 2013). With this configuration, it is necessary to provide oxygen in a constant supply during the reaction at the cathode, therefore supplying additional energy (Liu et al., 2015). Microbial fuel cells (MFCs) have the capability to harness renewable energy from organic sources and transform this energy from chemical energy into electrical energy while conducting wastewater treatment (Shah et al., 2019). Additional applications of these microbial fuel cells (MFCs) involve the breakdown of bioremediation pollutants (Fang et al., 2013). Efficient performance of a microbial fuel cell (MFC) relies on the electrochemical reactions between the low-potential organic substrate (e.g. glucose) and the high-potential final electron acceptor (e.g. oxygen). The performance of a microbial fuel cell (MFC) is influenced (Mishra et al., 2017) by factors such as the microbial species, fuel feedstock composition, pH level, temperature, and reactor configuration (Siegert et al., 2019). Velasquez-Orta et al. (2009) conducted an experiment and determined that micro-algae and macroalgae, which are used for energy production, can be cultivated on a wide range of substrates (including carbon dioxide), occupy minimal space, exhibit rapid growth rates, and, when employed for electricity generation, yield generally high power densities (Velasquez-Orta et al., 2009). Given the anticipated advancements in this technology, such as the incorporation of different substrates in the MFC for present production and waste-water treatment, there is potential for the identification of further substrates in the next years. These substrates would offer enhanced power output in terms of power generation, as well as improved wastewater treatment and pollution monitoring, which constitutes a significant contributor to global warming (Behera, 2015). The study of photosynthetic microorganisms, including microalgae and macroalgae, is highly interesting in the subject of generating a microbial current. This approach is based on the same principle used to generate current utilising bacterial species. The integration could be accomplished using three potential methods (Xiao & He, 2014).

(i) Phototrophic microalgae have the potential to serve as a substrate for the provision of electrons.
(ii) The anode reaction could be catalysed by photoheterotrophic bacteria.
(iii) Oxygen can be supplied as an electron acceptor for the cathode process by photoautotrophic bacteria.

2.5.3 Nanobionic Engineering of Plant Organelles

Significant advancements have been made in nanotechnology, enabling the customisation of nanoscale materials for targeted drug delivery to certain cells or tissues. The incorporation of nanoparticles into treatments has the potential to enhance drug stability, pharmacokinetics, and formulation, hence potentially mitigating toxicity. The creation of nanoparticle drug delivery systems is a time-consuming process, and the large-scale manufacture of these medications also requires a significant amount of time, which poses significant challenges to achieving success. Plant viruses are a developing category of biologics employed in pharmaceutical delivery systems (Czapar & Steinmetz, 2017). An inherent benefit of using these plant sources for bio-logics is that the plant cell wall shields the produced PDs from stomach acids and enzymes by means of bioencapsulation. Nevertheless, gut microorganisms have developed the ability to degrade all elements of plant cell walls. Consequently, when whole plant cells containing plant-derived proteins (PDs) enter the gut, the commensal microorganisms break down the cell walls and liberate the proteins. Upon fusion with PDs, receptor-binding proteins (tags) effectively traverse the intestinal epithelium, which is the largest mucosal region in the body, measuring 1.8–2.7 m^2, and are transported to either the circulatory system or the immune system (Kwon & Daniell, 2015). Vegetative cell walls consist of a variety of polysaccharides. The function of the cell wall is to act as a protective barrier, rejecting the infiltration of extraneous particles and nanoparticles into the cell, therefore avoiding potential harm to the internal components of the cell. A novel category of nanoparticles known as layered double hydroxides (LDHs) has been documented to efficiently transport biomolecules into plant cells. Utilising LDH nanoparticles has significant benefits, including the protection of RNA activity and the minimum harm caused by their entry through cell walls into intact cells.

Typically, LDHs are denoted as [M2 + 1 − xM3 + x(OH)2][An−]x/n.zH2O, where M2+ and M3+ are metal ions with divalent and trivalent configurations, respectively, and An is an anion located between layers. Intercalated with charge-balancing anions, the positively charged ion sheets form the tunable LDH complex, which may be delaminated into ultrathin nanostructures. LDHs exhibit very desirable characteristics including excellent compatibility, facile biodegradability, minimal cytotoxicity, and regulated drug release (Bao et al., 2017).

Engineered nanomaterials (ENMs) possess remarkable physicochemical characteristics that make them well-suited for a wide range of applications in medicine, biotechnology, energy, pharmaceuticals, cosmetics, and electronics. This is mostly due to their extremely small size and strong surface reactivity. Engineered nanoparticles (ENPs) have had a significant influence in recent decades, exhibiting characteristics such as a high affinity for target molecules, enhanced efficient release of molecules within target cells or to specific organelles, as well as non-destructive penetration of cell walls.

Environmental nanomaterials (ENMs) exert their influence by infiltrating plant cells and engaging with metabolic pathways along with intracellular structures, therefore potentially inducing toxicity or facilitating plant development and growth through several processes.

2.6 Conclusions and Future Work

Application of nanoparticles in the development of nanobionic plants is increasingly becoming a tangible possibility. The utilisation of plant-enabled sensors would prove advantageous in delivering instantaneous data. In the future, humans will have the capability to establish communication with plants on their growth processes, temperature, humidity levels, and pollutants through the integration of intelligent technology. Additionally, the causes of the impacts on plants will also be understood. Extensive research is ongoing to augment the quantity of sensors specifically designed for monitoring plant growth, plant signalling pathways, drought conditions, pest infestations, and other relevant factors. This technology will bring about a fundamental transformation in the agricultural sector. Furthermore, the genetic reaction of plants to the existence of nanoparticles is a significant subject of investigation. Nanoparticles exhibit multidirectional mobility and successfully penetrate vascular tissues. Thorough investigation is necessary to study the targeted distribution of nanoparticles to specific plant tissues. The implementation of nanotechnology in agriculture is expected to yield significant advantages for farmers and the food sector, since new products are being created. At present, nanosensors and nanobiosensors are employed for the detection of pathogens and assessment of soil quality. Additionally, there has been new research on nanoenabled biofertilizers. Further investigation is required to examine the application of nanomaterials in crop disease control. Based on the preceding explanation, it is evident that plant nanobionics will decisively dominate the future of the agricultural sector, and farmers will unquestionably reap several advantages from this.

References

Alagarasi, A. (2013). *Chapter-introduction to nanomaterials* (pp. 1–24). Indian Institute of Technology Madras.

Alemzadeh, E., Dehshahri, A., Izadpanah, K., & Ahmadi, F. (2018). Plant virus nanoparticles: Novel and robust nanocarriers for drug delivery and imaging. *Colloids and Surfaces B: Biointerfaces, 167*, 20–27.

Bao, W., Wan, Y., & Baluška, F. (2017). Nanosheets for delivery of biomolecules into plant cells. *Trends in Plant Science, 22*(6), 445–447.

Behera, P. (2015). Microbial fuel cells—A boon or a bane. *Research Review Journal Microbiology and Biotechnology, 4*(1), 1–11.

Czapar, A. E., & Steinmetz, N. F. (2017). Plant viruses and bacteriophages for drug delivery in medicine and biotechnology. *Current Opinion in Chemical Biology, 38*, 108–116.

Di Giacomo, R., Daraio, C., & Maresca, B. (2015). Plant nanobionic materials with a giant temperature response mediated by pectin-Ca2+. *Proceedings of the National Academy of Sciences, 112*(15), 4541–4545.

Enamala, M. K., Kolapalli, B., Divya Sruthi, P., Sarkar, S., Kuppam, C., & Chavali, M. (2019). Applications of nanomaterials and future prospects for nanobionics. *Plant Nanobionics: Volume 2, Approaches in Nanoparticles, Biosynthesis, and Toxicity*, 177–197.

Fang, Z., Song, H.-L., Cang, N., & Li, X.-N. (2013). Performance of microbial fuel cell coupled constructed wetland system for decolorization of azo dye and bioelectricity generation. *Bioresource Technology, 144*, 165–171.

Ghorbanpour, M., & Fahimirad, S. (2017). Plant nanobionics a novel approach to overcome the environmental challenges. *Medicinal Plants and Environmental Challenges*, 247–257.

Giraldo, J. P., Landry, M. P., Faltermeier, S. M., McNicholas, T. P., Iverson, N. M., Boghossian, A. A., Reuel, N. F., Hilmer, A. J., Sen, F., & Brew, J. A. (2014). Plant nanobionics approach to augment photosynthesis and biochemical sensing. *Nature Materials, 13*(4), 400–408.

Hatami, M., Kariman, K., & Ghorbanpour, M. (2016). Engineered nanomaterial-mediated changes in the metabolism of terrestrial plants. *Science of the Total Environment, 571*, 275–291.

Jacob, J., Haponiuk, J. T., Thomas, S., & Gopi, S. (2018). Biopolymer based nanomaterials in drug delivery systems: A review. *Materials Today Chemistry, 9*, 43–55.

Jakočiūnas, T., Jensen, M. K., & Keasling, J. D. (2016). CRISPR/Cas9 advances engineering of microbial cell factories. *Metabolic Engineering, 34*, 44–59.

Khan, M. N., Mobin, M., Abbas, Z. K., AlMutairi, K. A., & Siddiqui, Z. H. (2017). Role of nanomaterials in plants under challenging environments. *Plant Physiology and Biochemistry, 110*, 194–209.

Kondo, A., Ishii, J., Hara, K. Y., Hasunuma, T., & Matsuda, F. (2013). Development of microbial cell factories for bio-refinery through synthetic bioengineering. *Journal of Biotechnology, 163*(2), 204–216.

Kwon, K.-C., & Daniell, H. (2015). Low-cost oral delivery of protein drugs bioencapsulated in plant cells. *Plant Biotechnology Journal, 13*(8), 1017.

Kwon, K.-C., & Daniell, H. (2016). Oral delivery of protein drugs bioencapsulated in plant cells. *Molecular Therapy, 24*(8), 1342–1350.

Liu, T., Rao, L., Yuan, Y., & Zhuang, L. (2015). Bioelectricity generation in a microbial fuel cell with a self-sustainable photocathode. *The Scientific World Journal, 2015*(1), 864568.

Luan, G., & Lu, X. (2018). Tailoring cyanobacterial cell factory for improved industrial properties. *Biotechnology Advances, 36*(2), 430–442.

Mishra, P., Mishra, S., Datta, S., Taraphder, S., Panda, S., Saikhom, R., Laishram, M., Swain, D., & Nanotkar, R. (2017). Microbial fuel cell (MFC): Recent advancement and its application. *International Journal of Pure and Applied Bioscience, 5*(5), 911–923.

Nielsen, J. C., & Nielsen, J. (2017). Development of fungal cell factories for the production of secondary metabolites: Linking genomics and metabolism. *Synthetic and Systems Biotechnology, 2*(1), 5–12.

Prasad, R., Kumar, M., & Nanotechnology, V. K. (2017a). *An agricultural paradigm*. Springer.

Prasad, R., Kumar, V., & Kumar, M. (2017b). *Nanotechnology: Food and environmental paradigm*. Springer.

Rodríguez-Limas, W. A., Sekar, K., & Tyo, K. E. (2013). Virus-like particles: The future of microbial factories and cell-free systems as platforms for vaccine development. *Current Opinion in Biotechnology, 24*(6), 1089–1093.

Seigel, R., Hu, E., & Roco, M. (1997). *WTEC panel report on R & D status and trend in nanoparticles*. Nanostructured Materials, and Nanodevices Workshop.

Shah, S., Venkatramanan, V., & Prasad, R. (2019). Microbial fuel cell: Sustainable green technology for bioelectricity generation and wastewater treatment. *Sustainable Green Technologies for Environmental Management*, 199–218.

Shin, H. J., Jung, K. A., Nam, C. W., & Park, J. M. (2017). A genetic approach for microbial electrosynthesis system as biocommodities production platform. *Bioresource Technology, 245*, 1421–1429.

Siddiqui, M. H., Al-Whaibi, M. H., & Mohammad, F. (2015). Nanotechnology and plant sciences. *Springer International Publishing Switzerland, 10*, 978–973.

Siegert, M., Sonawane, J. M., Ezugwu, C. I., & Prasad, R. (2019). Economic assessment of nanomaterials in bio-electrical water treatment. *Advanced Research in Nanosciences for Water Technology*, 1–23.

Sun, Y., Yang, Z., Wang, C., Yang, T., Cai, C., Zhao, X., Yang, L., & Ding, P. (2017). Exploring the role of peptides in polymer-based gene delivery. *Acta Biomaterialia, 60*, 23–37.

Szabó, T., Magyar, M., Hajdu, K., Dorogi, M., Nyerki, E., Tóth, T., Lingvay, M., Garab, G., Hernádi, K., & Nagy, L. (2015). Structural and functional hierarchy in photosynthetic energy conversion—From molecules to nanostructures. *Nanoscale Research Letters, 10*, 1–12.

Velasquez-Orta, S. B., Curtis, T. P., & Logan, B. E. (2009). Energy from algae using microbial fuel cells. *Biotechnology and Bioengineering, 103*(6), 1068–1076.

Wong, M. H., Giraldo, J. P., Kwak, S.-Y., Koman, V. B., Sinclair, R., Lew, T. T. S., Bisker, G., Liu, P., & Strano, M. S. (2017). Nitroaromatic detection and infrared communication from wild-type plants using plant nanobionics. *Nature Materials, 16*(2), 264–272.

Xiao, L., & He, Z. (2014). Applications and perspectives of phototrophic microorganisms for electricity generation from organic compounds in microbial fuel cells. *Renewable and Sustainable Energy Reviews, 37*, 550–559.

Yang, L., Lübeck, M., & Lübeck, P. S. (2017). Aspergillus as a versatile cell factory for organic acid production. *Fungal Biology Reviews, 31*(1), 33–49.

Zhao, Y., Collum, S., Phelan, M., Goodbody, T., Doherty, L., & Hu, Y. (2013). Preliminary investigation of constructed wetland incorporating microbial fuel cell: Batch and continuous flow trials. *Chemical Engineering Journal, 229*, 364–370.

Chapter 3
Nanobionics and Nanomaterials: Impact on Soil Microbial Community and Functions

Viabhav Kumar Upadhayay, Dhruv Mishra, Manoj Kumar Prajapati, Manoj Kumar Chitara, Deepanjali Gupta, Prateek Ranjan Behera, and Marcelo Carvalho Minhoto Teixeira Filho

Abstract The advancement of agriusable nanocompounds has revealed their significant potential in enhancing crop performance. Nanomaterials (NMs), particularly those with dimensions ranging from 1 to 100 nm, whether metal-based or organic, not only impact crop growth but also influence the soil ecosystem upon application. These materials can modify the functional properties of soil, which serves as a habitat for a diverse range of microorganisms responsible for critical processes such as soil formation, nutrient cycling, and plant growth promotion. Understanding the effects of NMs on soil microbial communities, enzymatic activities, and nutrient dynamics is essential for optimizing their application. While some

V. K. Upadhayay (✉)
Department of Microbiology, Graphic Era Deemed to be University, Dehradun, Uttarakhand, India

D. Mishra
Department of Biological Sciences, College of Basic Sciences and Humanities, Govind Ballabh Pant University of Agriculture and Technology, Pantnagar, India

M. K. Prajapati
Department of Plant Pathology, PGCA, RPCAU Pusa (Samastipur), Bihar, India

M. K. Chitara
Department of Plant Pathology, College of Agriculture, Chandra Shekhar Azad University of Agriculture and Technology, Kanpur, Uttar Pradesh, India

D. Gupta
Department of Soil Science, College of Agriculture, Govind Ballabh Pant University of Agriculture and Technology, Pantnagar, India

P. R. Behera
Department of Plant Pathology, School of Agriculture, GIET University, Gunupur, Odisha, India

Marcelo Carvalho Minhoto Teixeira Filho
Department of Rural Engineering, Plant Health and Soils, São Paulo State University (UNESP),
Ilha Solteira, Brazil

V. D. Rajput et al. (eds.), *Contemporary Suitability of Nanobionics in Agriculture*, Nanotechnology in the Life Sciences, https://doi.org/10.1007/978-3-031-82779-2_3

studies highlight the beneficial impacts of NMs on soil health, others indicate adverse effects on microbial populations. Consequently, the careful selection of NMs is vital to maximize their efficacy while safeguarding the integrity of the soil ecosystem. Moreover, the integration of machine learning tools presents a novel approach to predicting the impacts of NMs on soil functions, enhancing our ability to assess and manage these interactions effectively. This chapter comprehensively illuminates the diverse effects of NMs on soil microbiota and functions, contributing to an understanding of their environmental significance and implications for sustainable agricultural practices.

Keywords Nanomaterials · Soil microbial communities · Enzymatic activities · Nutrient cycling · Sustainable agriculture

3.1 Introduction

The current climate change scenario is profoundly affecting crop production and soil health due to the growing prevalence of both biotic and abiotic stressors. Soil is a lively ecosystem that plays a vital role in supporting both plant and animal life. It drives essential processes that help manage and adapt to climate change. Therefore, preserving healthy soil is critical for achieving long-term food security. However, food security remains a persistent challenge, particularly in developing nations, where climate change and poor farming practices further exacerbate the issue (Rajput et al., 2023). In response, the application of nanomaterials (NMs) has garnered attention for promoting efficient crop growth. Due to their unique chemical properties, NMs can deliver nutrients more effectively and in lower doses compared to traditional fertilizers (Maithani et al., 2023; Upadhayay et al., 2024a). Nanofertilizers have shown the potential to reduce nutrient losses through leaching and volatilization while maintaining chemical stability. This enhances nutrient use efficiency and addresses environmental concerns (Upadhayay et al., 2024a). Numerous studies have highlighted their positive impact on crop growth and yield, with several commercial nanofertilizer products now available globally, providing essential nutrients such as nitrogen, phosphorus, potassium, zinc, and calcium. Despite the potential benefits of nanomaterials in enhancing crop yields, their accumulation in the soil through various pathways makes the soil a significant sink for these particles (Rajput et al., 2020). The impact of NMs on soil microbial communities, enzymatic activities, and nutrient cycling varies, depending on the type and dosage of the nanomaterials applied (Huang et al., 2022; Hafeez et al., 2024; Rani et al., 2024; Upadhayay et al., 2024a). Microorganisms are crucial to soil function and health, as they facilitate the turnover of organic matter and play an integral role in nutrient cycling. Soil-dwelling microbes are key agents in maintaining soil health by participating in organic matter decomposition, nutrient cycling, and influencing soil's physical properties (Nogueira et al., 2012). Additionally, microbial communities include beneficial microorganisms that colonize the rhizosphere and promote plant growth (Upadhayay et al., 2022a; Dhaigude et al., 2024; Khan et al., 2024). These microbes also enhance nutrient uptake, alleviate biotic and abiotic stress, and improve overall plant resilience (Upadhayay et al., 2023a, b, c). Better soil enzymatic activities reflect the presence and effectiveness of beneficial soil microbes (Upadhayay et al., 2022b). Moreover, soil microbiota helps particularly in nutrient cycles like the carbon and nitrogen cycles. Nanomaterials (NMs) can significantly influence microbial activities, abundance, and diversity in natural soils. They are critical for essential soil functions, including microbial nitrogen cycling, symbiotic relationships with organisms like arbuscular mycorrhizal fungi, iron metabolism through siderophores, and the production of antifungal compounds, all of which are important for crop production dependent on microbial processes (Karimi & Mohseni Fard, 2017). The effects of NMs on soil microbial communities and enzymatic activities can vary, exhibiting both positive and negative impacts. These variations depend on factors such as "organic matter content," "soil mineral composition," "the type, "size," and "dosage of NMs," "exposure duration," and "environmental conditions" (Ge et al., 2011; Upadhayay et al., 2024a). Moreover, machine learning (ML) has recently emerged as a tool to predict the ecological risk of NMs on soil microbial communities (Xu et al., 2022a, b; Zhang et al., 2024c). NMs used in agriculture need to be applied responsibly, keeping in mind factors like the right dose, exposure time, and the characteristics of the NM. This chapter outlines the impact of various NMs on soil microbial communities and functions.

3.2 Fundamentals of Nanobionics and Nanomaterials

Nanotechnology includes a multidisciplinary approach that integrates various fields of science, including physics, chemistry, biology, engineering, and materials science, to design and manipulate materials at the nanoscale (<100 nm) (Bayda et al., 2019; Upadhayay et al., 2023a, b, c, d, 2024a). Nanotechnology has gained significant attention due to the unique properties exhibited by nanomaterials, such as enhanced mechanical, electrical, optical, and magnetic characteristics (Baig et al., 2021; Mekuye & Abera, 2023). Nanobionics and nanomaterials epitomize a quickly evolving area of science that integrates nanotechnology with biological applications (Shaban et al., 2022). Nanobionics is an emerging field that combines nanotechnology with biological systems to enhance their functionality and performance (Enamala et al., 2019). It involves the integration of nanomaterials with living organisms, such as plants, to improve their growth, resilience, and productivity (Bhaskar et al., 2024). Nanobionics is a subbranch of nanotechnology, which mainly focuses on the integration of nanomaterials with biological systems to enhance their functionality (Lew et al., 2020a, b, c). It can help in charge storage and transport in molecular structures and could become a significant key value in molecular diagnostic tests and highly specific electro-analytical methods or devices (Hassanzadeh et al., 2022). Nanobionics have significantly been effective in modulating the photosynthesis capacity of plants (Ranjan et al., 2022). Such an approach has significant agriculture implications, particularly in improving plant resilience against environmental stress such as heavy metals, drought, and salinity under changing climatic conditions (Lew et al., 2021; Faizan et al., 2023). For example, the application of zinc oxide and silicon oxide nanoparticles has been improving plant growth, yield, and water use efficiency (Al-Selwey et al., 2023). Similarly, priming of multi-walled carbon nanotubes enhances plant growth under drought stress (Yousefi et al., 2017). Nanomaterials can be categorized into several types, including nanoparticles, nanocomposites, and nanostructures (Saleh, 2020). The synthesis of these materials employs various methods, which can be classified into physical, chemical, and biological nature (Saha et al., 2022; Mekuye & Abera, 2023). Physical and chemical methods allow for precise control over size and shape, they often produce hazardous byproducts, leading to a growing interest in green synthesis methods that utilize biological entities. The unique properties of nanomaterials arise from their small size and high surface area-to-volume ratio (Rajput et al., 2018). These properties make them suitable for a wide range of applications, including drug delivery (Teleanu et al., 2019), environmental pollution remediation (Roy et al., 2021), and energy storage (Pomerantseva et al., 2019). For instance, nanomaterials are being explored for their potential to create more efficient batteries and supercapacitors due to their ability to enhance ionic conductivity and charge storage capabilities. Amendment of engineered nanomaterials such as SiO_2, TiO_2, and Fe_3O_4 in maize (*Zea mays*) significantly influenced the soil microbial communities (Zhang et al., 2020a, b).

3.3 Agricultural Applications of Nanobionics and Nanomaterials

Nanobionics, the fusion of nanotechnology and biological systems, represents a cutting-edge frontier in agricultural innovation (Singh et al., 2024). By integrating nanomaterials directly into plant structures, nanobionics enhances plant functions at the molecular level, offering unprecedented control over biological processes (Mony et al., 2022). For instance, the incorporation of carbon nanotubes into chloroplasts has been shown to increase photosynthetic efficiency by up to 49%, enabling higher crop yields with less resource input (Khatri and Rathore, 2018). Furthermore, nanobionic sensors can detect early signs of stress or disease, allowing for real-time intervention and minimizing crop losses (Kumari et al., 2023a, b). The use of nanoscale materials, such as metal oxides and quantum dots, enhances nutrient absorption and water use efficiency by modifying root architecture and cellular uptake pathways (Liu et al., 2019). However, with ongoing research, nanobionics holds the promise of revolutionizing sustainable agriculture by enhancing productivity, reducing chemical usage, and addressing global food security challenges. Recent nondestructive sensors have transformed plant health monitoring by providing real-time digital insights, enabling growers to optimize conditions for increased productivity and reduced resource use (Ye et al., 2023). A key breakthrough in this area is electrochemical sensing technology, which detects plant hormones and signaling molecules with high sensitivity, portability, and low cost. Nanomaterials, including carbon-based nanostructures and metallic nanoparticles, significantly enhance sensor performance by improving electron transfer efficiency. This enhancement enables rapid and precise in situ detection of analytes. For instance, a paper-based electroanalytical device has been developed for the detection of hydrogen peroxide (H_2O_2) and salicylic acid (SA) in *Botrytis cinerea*-infected tomato leaves, providing crucial insights into plant defense mechanisms (Lew et al., 2020a, b, c). However, traditional methods often require invasive sampling, which can damage plant tissues. Furthermore, single-walled carbon nanotubes (SWNTs) have emerged as powerful optical nanosensors, capable of nondestructive detection of stress-induced signaling molecules like H_2O_2 across various plant species (Lew et al., 2020a, b, c; Kordrostami et al., 2021). Modulating photosynthesis is a critical approach to enhancing crop productivity. Efforts to improve photosynthesis have historically focused on optimizing RuBisCO (Prywes et al., 2023), converting C3 plants to the C4 pathway (Talukder et al., 2024), adjusting chlorophyll antenna size (Mao et al., 2023), broadening chlorophyll light absorption, and mitigating photoinhibition (Yangen et al., 2019). Nanoparticles (NPs), particularly as nanobionics, offer promising strategies for enhancing photosynthesis. For instance, titanium dioxide (TiO_2) NPs improve light absorption and protect chloroplasts from photochemical stress, subsequently enhancing RuBisCO activity and chlorophyll content in plants like *Lactuca sativa* L. (Weng et al., 2024), *Solanum lycopersicum* (Li et al., 2023), and *Triticum aestivum* (Chen et al., 2019). Other metal and metal oxide NPs, such as gold (Au), zinc oxide (ZnO), and silver (Ag), have been shown to positively influence photosynthesis by increasing chlorophyll content, electron transport rates, and photosystem efficiency (Tighe-Neira et al., 2018). However, these benefits are dose-dependent. High concentrations of NPs can impair photosynthetic machinery, as seen with Ag NPs, where optimal doses (e.g., 40–60 mg/L) enhance chlorophyll content in *Oryza sativa* (Gupta et al., 2018) and *T. aestivum* (Kannaujia et al., 2022), but higher doses significantly reduce it. TiO_2 NPs, especially when synthesized chemically and combined with plant growth-promoting rhizobacteria (PGPR), have demonstrated substantial increases in chlorophyll content (Zand et al., 2020). Recent studies confirm the role of NPs in modulating photosynthesis, but the effects vary by NP type, concentration, and application method. Nanobionics enhances nutrient use efficiency by utilizing smart nanocarriers that precisely deliver fertilizers, minimizing nutrient loss through controlled, stimuli-responsive release systems (UrRahim et al., 2021). These systems can trigger nutrient delivery based on environmental factors such as pH, temperature, and light (Lastochkina et al., 2022). Nanoparticles (NPs) under 100 nm can penetrate plant cell walls, entering via pathways like endocytosis or diffusion (Jaafar et al., 2020; Ďúranová et al., 2024). Inside the plant, NPs travel through the xylem and phloem, affecting processes like photosynthesis (Karny et al., 2018). For example, TiO_2 and ZnO NPs enhance photosynthesis by increasing light absorption. However, NP size, shape, and composition influence their effects on plant growth, with smaller NPs penetrating deeper into tissues, potentially leading to both beneficial and harmful outcomes (Kolenčík et al., 2020; Al-Juthery et al., 2021). Nanoparticle (NP)-based pesticides offer a promising alternative, with enhanced pest control and lower toxicity. Studies show that metal NPs, such as silica, alumina, and silver, disrupt pests by binding to proteins and nucleic acids, leading to cell death (Bhaskar et al., 2017; Landa, 2021). Ag (Benelli, 2018) and carbon NPs reduce enzyme activity, cause oxidative stress, and disrupt insect development, while SiO_2 and Al_2O_3 NPs dehydrate and kill insects by binding to their cuticle (Karunakaran et al., 2014). Additionally, TiO_2 NPs, known for enhancing seed germination, also exhibit antimicrobial properties that help prevent fungal and bacterial diseases (Parveen & Siddiqui, 2022). Copper oxide nanoparticles (CuO-NPs) inhibit fungi like *Fusarium oxysporum* and *Alternaria alternate* (Ashraf et al., 2021). ZnO nanoparticles (ZnO-NPs) are effective against fungi such as *Erythricium salmonicolor* and *Colletotrichum gloeosporioides* (Pariona et al., 2020).

3.4 Soil as a Major Sink for Environmental Nanomaterials

Nanoparticles (NPs) are becoming an integral part of various industrial and agricultural processes that will lead to their increasing presence in the environment. Among all the environmental parameters soil has emerged as a major sink for these tiny particles, where they accumulate and could potentially migrate. NPs can enter the soil through multiple pathways. Atmospheric deposition, where particles fall to the ground with rain or dust, is one common route. Direct absorption of gaseous compounds by soil particles, shedding of particles from leaves, and other natural processes also contribute to the accumulation of NPs in the soil (Rajput et al., 2020). However, human activities are the predominant source of NPs in the soil. In agriculture, the widespread use of NPs in fertilizers, pesticides, seed treatments, and hydroponic systems has led to their significant accumulation in cultivated soils. For instance, more than 95% of copper used in agricultural applications eventually settles in the soil, while other NPs like zinc oxide (ZnO) and cerium oxide (CeO2) are also found in increasing concentrations (Rajput et al., 2021). This makes soil a more significant receptor of NPs compared to that of air or water. Accumulation of NPs in the soil raises the possibility of negative effects on plants and the beneficial organisms that support the growth of the plant (Xu et al., 2022a, 2022b). Organic nanomaterials, such as liposomes, carbon-based nanoparticles, dendrimers, micelles, and inorganic nanomaterials, including metal and metal oxide nanoparticles, interact with soil microflora through distinct mechanisms (Wijesooriya et al., 2023). Nanoparticles enter the soil unintentionally through several routes. Industrial spills, the use of landfill sites, and the application of sewage sludge as fertilizer all are common sources of NP contamination in soil. Additionally, wastewater treatment plants and landfills that manage the waste produced by human activities release NPs through sludge and effluent discharge (Ogunkunle et al., 2021). These particles can also leak into the environment during their production, transportation, and use. Furthermore, the overuse of pesticides and fertilizers, as well as runoff from residential and industrial areas, contributes to the growing presence of NPs in the soil. On the other hand, some NPs are intentionally introduced for the remediation purposes of the soil. For example, for remediation of the soil, the nanoscale zero-valent iron (nZVI) particles are used to clean up contaminated soils by removing the pollutants (Li et al., 2021). The introduction of NPs like titanium dioxide (TiO_2) and carbon nanotubes (CNTs) into the soil whether through biosolids (treated sewage) or irrigation is present in a complex state (Zahra et al., 2020). While these particles can have beneficial effects, such as protecting crops and promoting plant growth, they also pose risks to the soil ecosystems. NPs can potentially harm soil microorganisms, disrupt nutrient cycles, and negatively impact plant health (Pullagurala et al., 2019).

3.5 Soil Microbial Communities: Structure and Functions

Numerous microorganisms that thrive in soil illustrate their adaptability and importance. They are key agents in maintaining balance in ecosystems and ensuring the vitality of soil and plants (Sahu et al., 2019). Soil microbiomes are composed of a vast array of bacteria, archaea, fungi, and other microorganisms, each with its unique capabilities and functionalities (Sahu et al., 2019). These microbes are responsible for a plethora of essential biogeochemical processes, from nutrient cycling and decomposition to carbon sequestration and climate regulation. The fate of soil organic carbon is a complex interplay between the activities of plants and the diverse microorganisms that inhabit the soil, with microbes playing a central role in determining whether carbon is retained or released into the atmosphere (Naylor et al., 2020).

Microbial diversity, distribution, and influencing factors across different groups—Archaea, Bacteria, and Fungi—in soil ecosystems. For Archaea, soil communities are primarily composed of 12 phyla, with Euryarchaeota and Crenarchaeota being dominant. Diversity shows an unimodal pattern with altitude, peaking between 1200 m and 1600 m, while Shannon's diversity index increases with altitude (Zhang et al., 2022). Exclusive species account for 12.3% and common species for 58.6% of the community. Influencing factors include soil properties like organic matter and pH, vegetation type, land use, climate, and topography (Zhang et al., 2022; Celestina et al., 2019; Kačergius et al., 2023). Bacteria are the most diverse group with 82 phyla and 11,576 species, predominantly *Proteobacteria, Acidobacteria, Actinobacteria, Bacteroidetes*, and *Firmicutes*. Bacterial species richness follows an unimodal pattern with altitude, but Shannon's diversity decreases with altitude (Zhang et al., 2022). Exclusive and common species make up 6.6% and 75.1% of the community, respectively. Key factors affecting bacterial communities include soil moisture, aggregation, root exudates, vegetation cover, land use, and climate, especially temperature (Celestina et al., 2019; Labouyrie et al., 2023; Kačergius et al., 2023). For Fungi, the community is dominated by Ascomycota and exhibits an unimodal pattern with altitude, with Shannon's diversity increasing with altitude (Zhang et al., 2022). Exclusive species account for 15.6% while common species make up 50.4%. Soil moisture, aggregation, vegetation cover, land use, and precipitation are key influencing factors (Celestina et al., 2019; Labouýrie et al., 2023; Kačergius et al., 2023). Advances in soil microbiology research have shed light on the remarkable adaptability of these microbes, as they respond to various environmental stressors, including climate change disturbances. Shifts in the taxonomic and functional profiles of soil microbial communities can have far-reaching consequences, affecting nutrient availability, plant health, and overall ecosystem resilience. The importance of interkingdom interactions, where microbes collaborate with plants and other organisms, has also been increasingly recognized as a critical factor in understanding the complex dynamics of soil ecosystems (Sahu et al., 2019). Soil microbial communities are integral to the health and functionality of terrestrial ecosystems, performing a myriad of essential roles that underpin soil fertility and plant growth. Soil microorganisms play a vital role in supporting ecosystem health through various key functions. They facilitate nutrient cycling by breaking down organic matter and transforming essential nutrients like nitrogen and phosphorus into forms that plants can readily absorb (Khan et al., 2020). Additionally, these microbes improve soil structure by producing polysaccharides and glomalin, which enhance soil aggregation and improve water-holding capacity (Kaviya et al., 2019). Furthermore, they promote plant growth by secreting growth hormones and siderophores that optimize nutrient uptake, particularly in nutrient-poor environments. Soil microbes also suppress diseases by producing antibiotics and competing with pathogens, while also degrading pollutants to detoxify harmful substances in the soil (Prajapati et al., 2024) and also help in the alleviation of abiotic stress like drought and salinity (Singh et al., 2024). Their ability to form symbiotic relationships, such as mycorrhizal associations, further enhances nutrient acquisition and water uptake (Upadhayay et al., 2019). Figure 3.1 illustrates the diverse roles of soil microbial communities, including their effects on plant health, soil structure improvement, and pollutant degradation.

3.5.1 Nutrient Cycling

Soil microbes, including bacteria, fungi, and protozoa, are essential for nutrient cycling in terrestrial ecosystems, facilitating the decomposition of organic matter and the mineralization of vital nutrients such as nitrogen, phosphorus, and sulfur (Sahu et al., 2017). These microorganisms convert complex organic substrates into bioavailable forms that plants can readily assimilate, promoting plant growth and supporting sustainable agricultural practices. In the ectomycorrhizosphere, ectomycorrhizal fungi and bacteria collaborate with plant roots to weather minerals, solubilize nutrients, and enhance soil structure, thereby driving nutrient cycling and improving soil fertility (Koshila Ravi et al., 2019). A study by Ding et al. (2024)

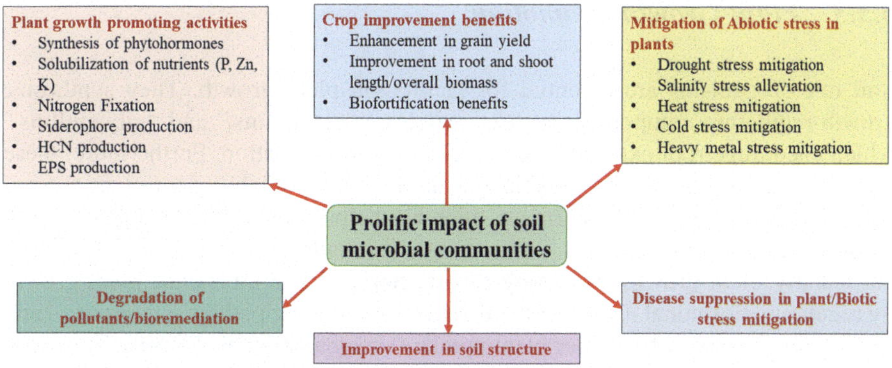

Fig. 3.1 Multifaceted roles of soil microbial communities

reveals that long-term continuous cropping alters soil microbial communities, leading to an increase in nutrient-cycling bacteria like *Sphingomonas*, driven by changes in soil pH, alkaline nitrogen, and organic carbon, which affect soil health and nutrient availability. Furthermore, Philippot et al. (2013) highlight the crucial role of microbial diversity in maintaining ecosystem functions, noting that reduced diversity, particularly among denitrifiers, significantly impairs nutrient cycling, especially nitrogen processes. Thus, a comprehensive understanding of microbial processes and the maintenance of diverse microbial communities are vital for optimizing agricultural practices and promoting sustainable land management.

3.5.2 Soil Structure Improvement

Soil microbes play a crucial role in enhancing soil structure through their metabolic processes, significantly contributing to various soil properties (Khan et al., 2020). Their secretion of organic acids facilitates the weathering of mineral particles, which increases soil porosity and enhances water infiltration capacity. Additionally, microbial activities promote the formation of stable soil aggregates, improving soil aeration, root penetration, and overall resilience (Kaviya et al., 2019). By fostering aggregation and enhancing pore connectivity, soil microbes improve water infiltration, nutrient flow, and aeration, which are vital for healthy plant growth. Sustainable agricultural practices that support microbial biodiversity can further strengthen soil structure, reducing erosion and building resilient agroecosystems (Hartmann & Six, 2023). Moreover, soil microbes influence the development of biopores and aid in nutrient cycling and the decomposition of crop residues, all of which help the development of "stable aggregates." This enhancement of soil structure ultimately leads to better crop growth, increased nutrient availability, and improved disease suppression.

3.5.3 Plant Growth Promotion

Soil microorganisms are essential for facilitating plant growth. They synthesize growth-regulating compounds such as "auxins," "cytokinins," and "gibberellins," which encourage root expansion, cell division, and elongation. Furthermore, these microbes can solubilize nutrients, increasing their bioavailability for plant absorption (Khan et al., 2019). Furthermore, microbes help suppress plant pathogens, protecting plants from diseases and promoting healthy growth. Soil microbes, particularly plant growth-promoting rhizobacteria (PGPR), enhance plant growth by regulating hormonal and nutritional balances, inducing pathogen resistance, and solubilizing nutrients for easier plant uptake (Upadhayay et al., 2024b). They also interact synergistically with other microorganisms in the rhizosphere, further boosting plant growth rates and supporting sustainable agriculture by reducing the need for synthetic fertilizers and pesticides (Khan et al., 2023; Trivedi et al., 2024). PGPR promotes plant development directly by producing plant hormones and solubilizing nutrients, and indirectly by suppressing soil-borne pathogens through mechanisms like siderophore production and antibiotic synthesis. PGPR also improves soil structure and organic matter content, reduces the need for synthetic fertilizers, and enhances nutrient retention and availability (Jaggi et al., 2023). Soil microbes, particularly P-solubilizing bacteria from genera such as "*Pseudomonas*," "*Bacillus,*" and "*Rhizobium,*" enhance plant growth by increasing phosphorus uptake and crop yield. They achieve this primarily through the production of organic acids and acid phosphatases, which solubilize mineral and organic phosphates in the soil. In addition, mycorrhizae also form a symbiotic relationship with plants, mainly supplying plants with P and alleviating the abiotic stress effect (Upadhayay et al., 2019).

3.5.4 Disease Suppression

Soil microbes play a crucial role in plant disease suppression through various mechanisms, including direct antagonism, production of inhibitory compounds, and indirect mechanisms like plant growth promotion (Prajapati et al., 2024). By fostering a healthy soil microbiome, it is possible to minimize the reliance on chemical pesticides and promote sustainable agricultural practices. These microbes engage in mutual interactions with plants and other microorganisms, combating pathogens through the secretion of antimicrobial compounds, production of lytic enzymes, and physical interactions such as coiling around pathogenic hyphae. Additionally, they augment plant resilience against stresses (abiotic and biotic stresses), contributing to improved plant health and increased agricultural productivity. Furthermore, soil microbes interact with soil and plant factors to create disease-suppressive soils (DSS), which are influenced by both biotic and abiotic factors (Jayaraman et al., 2021). Understanding these microbial interactions and their mechanisms can help develop sustainable alternatives to chemical control, ultimately improving soil health and agricultural productivity.

3.5.5 Degradation of Pollutants

The soil microbes are essential for the degradation of pollutants as they have enzymes that can decompose a range of organic and inorganic pollutants, including hydrocarbons, heavy metals, and pesticides. Through biodegradation, these microbes can convert harmful substances into less toxic or even benign byproducts, aiding in environmental cleanup initiatives. Soil microbes are essential for the bioremediation of hazardous organic pollutants. They utilize enzymes such as oxidoreductases and hydrolases to break down pollutants, offering a cost-effective and environmentally friendly approach to reducing toxicity and recovering useful substances (Karigar & Rao, 2011). Additionally, soil microbiota facilitates the complete mineralization of hydrocarbons into carbon dioxide, water, and inorganic compounds, offering a cost-effective and efficient alternative to mechanical and chemical methods for treating contaminated sites (Das & Chandran, 2011). Furthermore, soil microbes are essential for degrading environmental pollutants by breaking down complex compounds into simpler forms using various enzymes. They help in bioremediation by optimizing conditions like nutrient availability and environmental factors to enhance degradation efficiency (Fouad et al., 2022). Soil microbiota also play an imperative role in degrading pollutants by adapting to toxic compounds through mutation or gene acquisition. This adaptation allows them to break down persistent pollutants, although the process can be slow and difficult to predict (van der Meer, 2006).

3.5.6 Abiotic Stress Alleviation

Soil microbes play a significant role in helping plants alleviate stress by producing osmoprotectants that aid in tolerating drought and salinity. They also enhance soil structure and solubilize nutrients, which helps plants cope with various environmental challenges. Specifically, plant growth-promoting bacteria contribute to alleviating abiotic stress by improving soil quality and supporting crops in adverse conditions, providing eco-friendly alternatives to chemical fertilizers and pesticides (Singh et al., 2024). Heavy metal-tolerant plant growth-promoting microbes having metal-tolerant ability are particularly important in managing abiotic stress caused by heavy metal contamination. These microbes help plants survive and grow in contaminated soils by accumulating, transforming, or detoxifying heavy metals, thus offering a sustainable and cost-effective approach to soil remediation (Mishra et al., 2017). Various soil microorganisms, including bacteria such as soil microbes such as *Azospirillum sp., Pseudomonas syringae, Pseudomonas fluorescens,* and *Bacillus sp.*, help plants manage stresses like salinity and drought. Additionally, *Bacillus polymyxa* and *Pseudomonas alcaligenes* aid in overcoming nutrient deficiencies. Fungi, like *Trichoderma harzianum* and *arbuscular mycorrhizal fungi (AMF),* including *Glomus mosseae, G. etunicatum, G. intraradices, G.*

fasciculatum, G. macrocarpum, and G. coronatum, enhance nutrient uptake and stress tolerance through biochemical and physiological changes (Chakraborty et al., 2015). Endophytes, AMF, and PGPR also help crops manage abiotic stresses (such as drought, salinity) and nutrient deficiencies. Furthermore, soil microorganisms, including PGPR, mycorrhizal fungi, actinomycetes, and cyanobacteria, help crops manage abiotic stresses like climate extremes. They enhance plant stress tolerance through mechanisms such as phytohormone production, ion homeostasis, osmoprotectant accumulation, improved nutrient uptake, and antioxidant activity (Adeleke et al., 2024).

3.6 Impact of Nanobionics and Nanomaterials on Soil Microbial Functions

Nanotechnology has shown a significant impact in various fields, including medicine and agriculture. In agriculture, NMs are utilized for various purposes, including enhancing plant growth and mitigating abiotic and biotic stresses. Although some physiological and molecular mechanisms are understood, the effects of NMs on soil ecosystems remain largely unstudied. It is essential to investigate how NMs affect both beneficial and nontarget soil microbes and to evaluate the overall health of the soil ecosystem. Factors such as organic matter and mineral content significantly influence the interactions between NMs and soil microbial communities. Plenty of research has documented both positive and negative impacts of NMs on soil microbial communities, including their functions, enzymatic activities, and nutrient cycling (Upadhayay et al., 2024a). Figure 3.2 depicts the influence of nanomaterials on soil microorganisms, encompassing positive effects such as greater microbial abundance and enhanced enzymatic activities, negative effects including reduced microbial populations and hindered nutrient cycling, as well as mixed effects. The following sections explain how NMs influence soil microbial functions and their broader environmental significance.

3.6.1 Impact on Soil Microbial Diversity

Soil harbors a complex and diverse microbial community that performs essential functions such as nutrient recycling and organic matter decomposition, contributing significantly to plant health and ecosystem stability. Among these microorganisms are "gram-positive" and "gram-negative" bacteria and fungi that show important roles in plant growth through beneficial interactions in the rhizosphere. However, conserving this biodiversity and protecting these vital soil microbial communities remain a significant challenge in agriculture (Nafees et al., 2020). The introduction of nanomaterials (NMs) into the soil—through agricultural amendments,

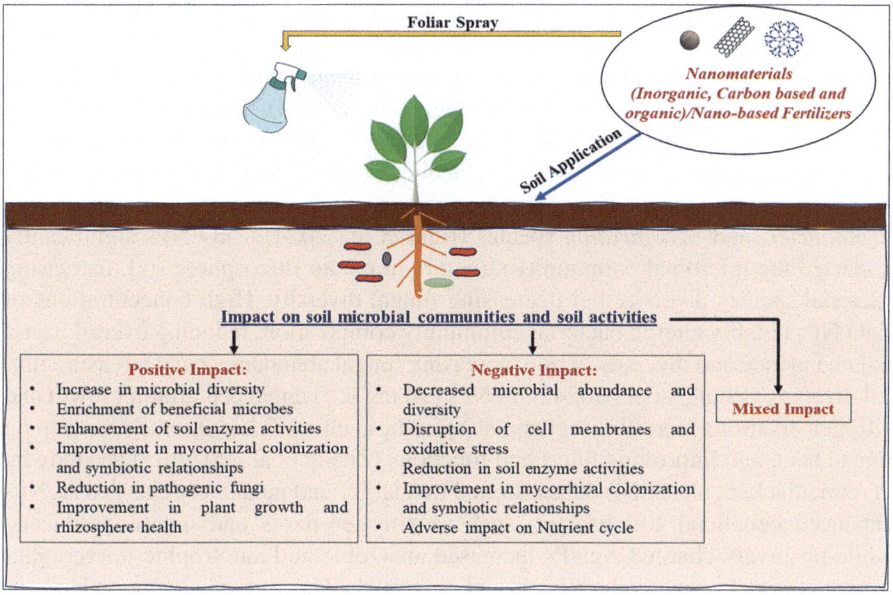

Fig. 3.2 Impact of nanomaterials on soil microbes: a comprehensive overview of positive, negative, and mixed effects

atmospheric deposition, landfills, or industrial spills—can disrupt microbial diversity, abundance, and activity. The effects of NMs on soil microbiota are influenced by their bioavailability and type, with organic NMs (e.g., micelles, dendrimers, liposomes, carbon-based) and inorganic NMs (e.g., metals, metal oxides) interacting with microorganisms in diverse ways that affect their functions differently (Wijesooriya et al., 2023). This section explores the positive and negative impacts of various NMs on soil microbial communities, underscoring the need for a thorough understanding of these effects to manage soil health effectively. NPs affect soil biota through mechanisms such as direct binding to cell membranes, oxidative stress, and ion release. For instance, metal-based NPs like silver nanoparticles can disrupt cell membranes, release toxic ions that generate reactive oxygen species (ROS), and impair microbial functions and community structures. Additionally, NPs may negatively impact mycorrhizal colonization and soil fertility by inducing oxidative stress and membrane damage (Mubeen et al., 2023). Engineered nanomaterials, including metal oxides (e.g., CuO and ZnO NPs) and C-based materials (e.g., reduced graphene oxide and multi-walled carbon nanotubes), have shown varied effects. These materials have been reported to reduce pathogenic *Fusarium* abundance by over 40%, decrease soil enzyme activities, enhance the Shannon index of microbial communities by 12.2–29.86%, and increase the abundance of beneficial microbes such as *Pseudomonas* by 29.7–96.2% and fungal communities by 10.18–29.86% (Li et al., 2024). Moringa chitosan NPs have been effective in reducing rice blast disease while enhancing soil microbial diversity. The application of this NM at 200 mg L^{-1} led to a significant increase in beneficial bacteria, including *Agarivorans*, *Azonexus*, *Bradyrhizobium*, *Nitrospira*, *Nocardioides*, *Sphingomicrobium*, and *Thiobacillus*, promoting plant health and contributing to a balanced rhizosphere microbiome (Hafeez et al., 2024). Similarly, AgNPs and chitosan-modified AgNPs at concentrations of 100 and 200 ppm have proven effective as plant growth promoters by enhancing nutrient uptake and increasing the abundance of beneficial soil microbes such as phosphate-solubilizing bacteria, *Azotobacter*, and *Azospirillum* species (Rani et al., 2024). ZnO NPs significantly impacted the microbial community structure in potato rhizosphere soil, increasing bacterial species diversity but decreasing fungal diversity. High concentrations of ZnO NPs notably altered bacterial community composition, reducing overall bacterial abundance and diversity, while increasing fungal abundance but decreasing fungal diversity (Zhang et al., 2024a). FeNPs (10 mg L^{-1}) enhanced alfalfa growth and nitrogen fixation, recruiting beneficial microbes, intensifying interactions among fungal taxa, and improving microbial functions (Zhang et al., 2024b). The study by Przemieniecki et al. (2024) demonstrated that larger and negatively charged AgNPs enhanced beneficial soil bacteria such as nitrogen-fixers and urease producers, while positively charged AgNPs increased anaerobic and autotrophic microorganisms, potentially negatively affecting plant growth. The surface charge and size of AgNPs are critical factors influencing their impact on soil microbiomes, potentially enhancing or disrupting microbial activity. In acidic soils, nanoscale zero-valent iron (nZVI) enhanced lettuce growth and positively impacted microbial communities by improving enzyme activity and increasing fungal genetic diversity, with a 0.5% nZVI treatment notably enhancing microbial functional diversity and positively influencing fungal populations (Gil-Díaz et al., 2024). Nano-hydroxyapatite application to e-waste contaminated sandy loam soil significantly altered bacterial community structure, increasing the abundance of *Stenotrophomonas* sp. and *Bacteroides*, and improving ryegrass growth, and reducing concentrations of Pb, Cu, Cd, and Zn (Wei et al., 2016). In soils contaminated with Pb, nano-ZnO ($\geq$10 mg/kg) reduced DTPA-extractable Pb by 10.6%–21.3% and decreased soil pH, improving microbial diversity and richness of metal-tolerant bacteria (Huang et al., 2022). Nanochitosan (40 mg L^{-1}) combined with bioinoculants *Pseudomonas taiwanensis* and *Pantoea agglomerans* significantly affected soil microbiome composition, enhancing maize growth and microbial health, resulting in a 54% increase in plant height and a 101% rise in total bacterial count. Metagenomic analysis identified shifts in the soil microbiome, with an increased abundance of beneficial bacterial genera such as *Proteobacteria*, *Actinobacteria*, *Chloroflexi*, and *Firmicutes* (Agri et al., 2022). The combined use of nanochitosan (50 mg L^{-1}), nanozeolite, and *Bacillus* sp. significantly increased soil microbial diversity and operational taxonomic units (OTUs), notably enhancing genes associated with *Actinobacteria*, *Bacteroidetes*, *Acidobacteria*, and *Chloroflexi* (Chaudhary et al., 2021). Silver nanoparticles (550 mg/kg) did not exhibit toxicity toward native soil microbial communities, including nitrogen-fixing genera, and led to a positive shift in community richness (Shah et al., 2014). Despite the growing use of NMs for their novel properties, their environmental introduction raises concerns regarding agriculture, food security, and health. Accumulation of NMs in soils can disrupt soil and plant systems, potentially threatening crop yields by altering soil composition, affecting microflora and macrofauna, and impacting plant growth through genotoxic and cytotoxic effects, as well as changes in oxidative stress responses (Shende et al., 2024). The introduction of silver silicate nanoparticles ($AgSiO_2$) into rice field soil significantly impacted the microbial community structure, decreasing the abundance of the WPS2 phylum from 8% to 0.09% while maintaining *Proteobacteria* and *Acidobacteria* levels. This selective impact suggests that $AgSiO_2$ affects specific bacterial groups crucial for nitrogen metabolism and carbon fixation (Singha et al., 2024). AgNPs also significantly altered soil microbial community structure and metabolite profiles, impacting key microbial functions related to carbon, nitrogen, and phosphorus cycling (Zhang et al., 2024c). Exposure to AgNPs at concentrations of 60 to 2000 mg kg^{-1} resulted in significant decreases in microbial biomass and activity, with metatranscriptomic analysis identifying the upregulation of heavy metal resistance genes (Meier et al., 2020). Table 3.1 also illustrates the effects of different NMs on soil microbial communities. Overall, the effects of nanobionics and nanomaterials on soil microbial diversity are multifaceted, with varying impacts based on particle type, concentration, and soil characteristics.

3.6.2 Impact on Soil Enzymatic Functions and Nutrient Cycling

The use of nano-hydroxyapatite in soil contaminated with e-waste resulted in a substantial increase in the activities of crucial soil enzymes, such as "urease," "phosphatase," and "dehydrogenase." This treatment also improved ryegrass biomass and altered the microbial community structure, highlighting its effectiveness in enhancing soil health (Wei et al., 2016).

Nanochitosan, when applied at 40 mg L^{-1} in combination with bioinoculants *Pseudomonas taiwanensis* and *Pantoea agglomerans*, notably impacted soil enzyme activities. The treatment resulted in a 94.88% increase in dehydrogenase activity, a 112% rise in fluorescein diacetate activity, and a 32.09% boost in alkaline phosphatase activity, underscoring its potential to improve soil health and microbial activity (Agri et al., 2022). Similarly, a combination of nanochitosan (50 mg L^{-1}), nanozeolite, and bioinoculant (*Bacillus* sp.) significantly improved soil enzyme activities. This treatment led to approximately a twofold increase in fluorescein diacetate hydrolysis, dehydrogenase, and alkaline phosphatase compared to untreated soil (Chaudhary et al., 2021). He et al. (2011) observed that the addition of iron oxide magnetic nanoparticles (IOMNP) at concentrations of 0.5, 1.0, and 2.0 mg g^{-1} enhanced soil urease and invertase activities. This suggests improved nitrogen and carbon cycling, as urease accelerates the conversion of urea to ammonia and invertase breaks down sucrose into simpler sugars, thereby boosting nutrient cycling.

Table 3.1 Impact of various nanomaterials on soil microbial communities and their functional attributes

S. No.	Nanomaterial(s)	Impact on soil microbial community	References
1.	Biologically synthesized silicon oxide nanoparticles	Enhanced microbial diversity; increased richness and abundance, especially in *Proteobacteria* and *Acidobacteria*	Ahmed et al. (2024)
2.	CeO_2, TiO_2, and SiO_2 NPs	Reshaped rhizosphere microbial community; promoted beneficial taxa (*Lysobacter*, *Bacillaceae*); suppressed pathogenic fungi (*Fusarium*, *Gibberella*)	Ren et al. (2024)
3.	Bio-synthesized Se NM	Significantly altered microbial community structure; enhanced abundance of functional microbes (e.g., ammonia oxidizers, nitrogen fixers, sulfate reducers)	Sonali et al. (2024)
4.	CuO NPs	Decreased indole acetic acid (IAA) production in *Bacillus megaterium* AGNTL MM7, indicating toxicity and negative impact on PGPRs	Tripathi et al. (2024)
6.	Silver nanoparticles	Induced significant shifts in the eukaryotic microbiome; enriched *Acidobacteriota* and beneficial bacteria (*Devosia*); minimal impact on mycorrhizal fungi	Gorczyca et al. (2024)
7.	Biochar-loaded nanoscale zero-valent iron	Improved soil pH, CEC, and catalase activity; altered microbial community richness and diversity, with greater differences compared to control	Song et al. (2024)
8.	Biochar (BC) and nano-biochar (NBC)	Increased microbial populations (TPC, PSB, NFB) with optimal concentrations; NBC more effective than BC in pesticide removal; SOC decline with higher concentrations	Chandi et al. (2024)
9.	Nanoscale sulfur	Reversed Hg-induced microbial diversity loss; improved abundance of *Myxococcota* and *Bacteroidota*	Zhuang et al. (2024)
10.	Fe7(PO4)6 NMs	Increased relative abundances of *Flavisolibacter, Ramlibacter, Pseudomonas, Mesorhizobium, Nitrospira, Rhodanobacter, Sphingomonas, Gemmatimonas, Massilia, Bryobacter, Chthoniobacter*, RB41, and *Rubellimicrobium*; improved nutrient uptake, soil fertility, and fruit quality	Jiao et al. (2024)
11.	TiO_2 NPs (P25, E171) and bulk TiO_2	Altered prokaryotic community structure significantly; no significant change in fungal diversity	Moll et al. (2017)
12.	TiO_2 NPs	Increased soil microbial biomass and enzyme activity (up to 20 mg kg^{-1}); significant decline beyond 20 mg kg^{-1}, with a marked reduction at 100 mg kg^{-1}. Soil respiration and metabolic quotient increased, indicating sub-lethal stress. PLFA biomass and microbial community structure showed a decline at 100 mg kg^{-1}, with temperature effects less severe at 40 °C compared to 25 °C	Bhattacharjya et al. (2021)

(continued)

Table 3.1 (continued)

S. No.	Nanomaterial(s)	Impact on soil microbial community	References
13.	AgNPs	Significant reduction in microbial growth, activity, and diversity; emergence of AgNP-tolerant bacterium *Rhodanobacter* sp. (at lower concentrations)	Samarajeewa et al. (2017)
14.	ZnO and FeO NPs	Increased microbial diversity and richness; altered abundance of *Proteobacteria*, *Actinobacteria*, and *Planctomycetes*; enriched with potential "streptomycin producers," "carbon and nitrogen fixers," and "lignin degraders"; decreased chitin degraders, ammonia oxidizers, and nitrite reducers	Afzal and Singh (2022)
15.	P-NF	No significant changes in microbial community composition or diversity; minimal effects on soil respiration and enzyme activities, with only protease activity significantly inhibited	Ciurli et al. (2022)
16.	AgNPs	Significant reduction in total soil dehydrogenase and urease activity; shifts in bacterial community composition with decreased *Acidobacteria* and *Verrucomicrobia*, increased *Proteobacteria*; reduced bacterial and archaeal amoA gene abundance; significant impact on fungal community structure with several ribotypes increasing	McGee et al. (2017)
17.	SWNTs	Transient effect on bacterial community with complete recovery after 14 days; persistent adverse impact on fungal community	Rodrigues et al. (2013)

Biogenic zinc nanoparticles, synthesized from *Parthenium hysterophorus* leaves and ranging from 16 to 108.5 nm in size, improved soil enzyme activity and plant physiological parameters over a 30 to 60-day period, indicating a beneficial effect on soil and plant health (Sri Sindhura et al., 2014). In contrast, prolonged exposure to silver nanoparticles at concentrations ranging from 0.01 to 1 mg/kg of soil notably decreased soil microbial biomass and functional activities. Quantitative analyses revealed a 25% decrease in microbial biomass, a 30% reduction in bacterial ammonia oxidizers, and a 40% reduction in nitrogen-fixing microorganisms. Additionally, leucine aminopeptidase activity decreased by 20%, highlighting AgNP's detrimental impact on soil enzyme functions and nitrogen cycling (Grün et al., 2018). Conversely, exposure to environmentally relevant concentrations of AgNPs did not significantly affect soil extracellular enzyme activities after 60 days. Despite alterations in microbial biomass and community composition, soil physicochemical properties and enzyme functions remained largely unaffected, suggesting minimal impact on soil biogeochemical cycles (Montes de Oca-Vásquez et al., 2020). In soil microcosm experiments, zinc oxide nanoparticles (ZnO NPs) affected soil enzyme activities based on the applied dosage. At a concentration of 5 mg per kg of soil, acid and alkaline phosphatase activities increased, while these activities decreased at 50 mg per kg of soil. Urease activity was enhanced across all ZnO NP concentrations, while dehydrogenase activity remained unchanged up to 25 mg kg^{-1}, indicating varying effects on soil enzyme functions and nutrient cycling processes (Srivastav et al., 2023). CuO-NPs significantly inhibited soil enzyme activities, including acid phosphatase and dehydrogenase. Smaller particles (10 nm) and higher concentrations (100 to 500 mg/kg) caused more substantial inhibition, though toxicity diminished after 30 days, suggesting that Cu ions are the primary factor affecting enzyme activities (Adhikari et al., 2024). Quantitative analysis of various nanoparticles showed that AgNPs and Ag^+ led to a dose-dependent decrease in dehydrogenase activity, with reductions up to 40%, whereas ZnO NPs and fullerene (C60) NPs exhibited less impact (Huang et al., 2023a, b). In sandy soils, carbon NMs applied (at 200–400 mg kg^{-1}) significantly augmented urease activity by 34–44% and microbial biomass carbon by 27–29% compared to fertilizer-only controls. In contrast, biochar application had no significant effect on soil enzyme activities (Nepal et al., 2023). In a 60-day microcosm study, nanoLi2O treatment at 474 μg Li/g soil increased CO_2 emissions by 3.45-fold and significantly altered soil enzyme activities, decreasing β-glucosidase while increasing urease activity. NanoMoO_3 and nanoNiO treatments temporarily suppressed β-glucosidase activity (Avila-Arias et al., 2019). Liu et al. (2014) assessed the impact of carboxymethyl-β-cyclodextrin-Fe_3O_4 magnetic nanoparticles-Diuron (CM-β-CD-MNPs-Diuron) on soil microbial activity and found a significant reduction in urease enzyme activity, confirmed by microcalorimetric analysis and qPCR, indicating toxicity to soil microorganisms. Maliszewska (2016) evaluated biogenic AuNPs of size 14 ± 3 nm and found no significant impact on β-glucosidase, urease, alkaline phosphatase, or dehydrogenase activities up to 33 mg/kg, suggesting minimal adverse effects on key soil enzymatic functions.

CuO NPs have shown a more substantial impact on soil enzyme activities compared to TiO_2 NPs, with CuO NPs causing significant reductions in urease, phosphatase, and dehydrogenase activities, while TiO_2 NPs led to comparatively smaller reductions (Xu et al., 2015). Similarly, green-synthesized ZnO NPs at medium concentrations (e.g., 50 mg/kg) led to increased soil enzyme activities (urease and catalase activities raised, while saccharase and acid phosphatase activities showed significant increases compared to controls) (Rehman et al., 2024). NM applications demonstrate varied impacts on soil enzyme activities: ZnO NPs increased urease activity by 1.42–1.57 times and phosphatase by 1.33–1.43 times, whereas copper hydroxide nanoparticles enhanced fluorescein diacetate hydrolase (FDAse) and N-acetylglucosaminidase (NAG) activities but decreased urease (URE) activity (Tang et al., 2021). Polystyrene nanoparticles (NPSt) at 400 mg/kg did not significantly affect soil nitrogen cycling, microbial communities, or carrot growth, indicating no harmful impact on these processes (Al-Gahtany et al., 2024).

The use of 100 mg/L γFe_2O_3 nanoparticles applied through soil drenching resulted in a notable increase in soil bacterial populations, while CuO NPs at 50 mg/L, delivered via dipping and soil drenching, and achieved the highest dehydrogenase activity (Fetyan et al., 2024). Exposure to TiO_2 nanoparticles significantly inhibited nitrification and denitrification enzyme activities and altered the structure of soil microbial communities, even at low concentrations (Simonin et al., 2016). Physiological assessments and DNA analyses revealed that single-walled carbon nanotubes (SWNTs) could adversely affect fungi and bacteria involved in the biogeochemical cycles of carbon and phosphorus (Rodrigues et al., 2013). The application of ZnO NPs severely disrupted nutrient cycling, leading to a reduction in CO_2 emissions by up to 43%, a decrease in organic nitrogen mineralization by as much as 32%, and diminished plant N recovery from manures (Shah et al., 2022). Graphene-based nanofertilizers, including "graphene oxide (GO)," "reduced graphene oxide (rGO)," "graphene nanoplatelet (GNP)," and "graphite nanoadditive (GNA)," were shown to decrease nitrification potential while increasing the abundance of denitrifying genes (*nirK*, *nirS*, *nosZ*), indicating their potential to suppress nitrification and reduce nitrate loss through enhanced denitrification (Das et al., 2022). Composite nanofertilizers (composed of 100% CF + 50% NC + 50% NCC) significantly boosted soil enzyme activities, resulting in a 22.4% increase in urease activity and a 17.8% rise in alkaline phosphatase activity compared to 100% composite fertilizers alone. In contrast, 70% composite nanofertilizers (70% CF + 35% NC + 35% NCC) showed smaller increases, with urease activity rising by 19.0% and alkaline phosphatase by 13.6% (Chen et al., 2023). CuO NPs were found to considerably reduce nitrogen cycle activities, including denitrification, nitrification, and soil respiration, particularly at concentrations of 100 mg/kg (Simonin et al., 2018).

The use of a "nano-chitosan-urea composite" led to a notable enhancement in soil dehydrogenase activity, organic carbon levels, and available potassium. Concurrently, it caused a reduction in ammonical nitrogen, nitrate nitrogen, urease activity, and the population of ammonia-oxidizing bacteria and archaea (Kondal et al., 2021). Carbon NPs used as a fertilizer synergist have been found to markedly improve corn growth and nutrient use efficiency. At an optimal application rate (200 mg CNPs kg^{-1}), significant increases in urease activity, dehydrogenase, and phosphatase activities were observed. However, higher application rates led to a decrease in enzyme activity, indicating a dose-dependent effect (Zhao et al., 2021).

The incorporation of nanomaterials, including carbon nanoparticles, nanobiochar (nB), and nanowater treatment residues (nWTR), has demonstrated significant enhancements in soil enzyme activities such as urease, dehydrogenase, and catalase. This boost in enzyme activity was associated with increased microbial biomass and biological activity, leading to considerable increases in crop yield. For instance, nWTR applied at 50 mg kg^{-1} resulted in a 150.64% increase in canola seed yield, while nB applied at 250 mg kg^{-1} improved growth parameters in degraded soils (Elsawy et al., 2022). Potassium-based nanomaterials, specifically K_2SiO_3-NPs and $K_{18}Mo_8O_{33}$-NPs, applied at concentrations of 20 and 50 mg kg^{-1}, significantly enhanced soil enzyme activities, including peroxidase, catalase, and urease. The highest urease activity was observed with 50 mg kg^{-1} K_2SiO_3-NPs, highlighting the effectiveness of these nanomaterials in improving soil health and enzyme functions (Wang et al., 2024).

3.7 Machine Learning Application in Depicting the Ecological Risk of Nanocompounds

Machine learning has become essential for analyzing microbial community data, enabling the prediction of disease states, environmental quality, and contamination through advanced pattern recognition techniques. However, the complexity of these models often limits their interpretability, as they operate as "black boxes" with unclear mechanisms behind their predictions (Ghannam & Techtmann, 2021). Machine learning encompasses a variety of methods grounded in probability theory and statistics, including both supervised and unsupervised learning approaches. Supervised learning utilizes labeled data for tasks such as prediction, employing techniques like regression and classification. In contrast, unsupervised learning methods, including clustering techniques like k-means, organize data without predefined labels. In microbial studies, these methodologies facilitate operational taxonomic unit (OTU) clustering and dimensionality reduction, with algorithms such as support vector machines (SVM), naïve Bayes, and random forests frequently used for classification and prediction (Qu et al., 2019). Machine learning significantly improves soil analysis by providing rapid, accurate, and robust measurements of soil texture and water content. Techniques such as "random forests," "support vector machines," and "neural networks" develop predictive models from soil and environmental data. Additionally, geostatistical methods like kriging enhance the representation of spatial data. AI and machine learning also address challenges such as false positives and data noise, offering substantial advantages over traditional statistical tools for effective and timely soil management (Awais et al., 2023).

A groundbreaking study by Xu et al. (2022a, b) utilized high-throughput sequencing of 365 soil samples combined with advanced machine-learning techniques to explore the impact of nanoparticles on soil microbial communities. Their research revealed that while nanoparticles do not significantly affect alpha diversity, they substantially alter beta diversity, leading to a reduction in key beneficial bacteria such as Dyella, Methylophilus, and Streptomyces. This decline in microbial health impairs essential soil functions, underscoring the urgent need for careful nanoparticle application to maintain soil ecosystem integrity. Machine learning analysis further quantified that silver nanoparticles (AgNPs) reduced soil enzyme activity by up to 25% in acidic soils ($pH < 5.5$) and conditions of low organic matter (<1.9% OM). Interestingly, clay soils mitigated AgNP toxicity by 15%, while sandy soils exhibited a 30% increase in toxicity, with sandy clay loam showing the least inhibition at 55.9% (Zhang et al., 2024c). Chen et al. (2024) found in their meta-analysis of 2,134 observations from 107 studies that nanomaterials (NMs) reduced soil microbial diversity by 0.96%, biomass by 14.01%, activity by 3.39%, and function by 14.44%, with metal NMs, especially silver (Ag) NMs, having the most pronounced effects, and machine learning models effectively predicted these impacts, enhancing the assessment of NM risks in soil. The integration of machine learning into microbial community analysis represents a significant advancement in our understanding of soil ecosystems. By providing insights into the effects of nanoparticles and improving predictive accuracy, machine learning not only enhances our comprehension of microbial dynamics but also informs sustainable agricultural practices. As research continues to evolve, the potential for machine learning to contribute to soil health and ecosystem integrity remains promising.

3.8 Nanotechnology: Regulations and Strategies

Numerous uses of nanotechnology are possible in agriculture, such as water filtration, fertilizers, irrigation, animal feed, food processing, packaging, and aquaculture. It has shown to be an efficient and reasonably priced sector for agricultural advancement in recent decades. This involves refining fertilizers, agrochemicals, biofertilizers, and bio-stimulants; utilizing nanoparticles as carriers for gene modification in crops; and boosting post-harvest storage and nutrient delivery. The increased need for food due to the growing worldwide population has prompted the use of nanotechnology in food packaging, antimicrobials, agrochemicals, and nutrition based on nanoparticles. These goods must, however, be confirmed to be safe for users and the environment before they may be sold. Countries all around the world are assessing their regulatory frameworks to make sure they can control nanotechnology, leading to the development of a variety of ways to control nano-based goods in agriculture, food, and feed. An overview of the international and national laws governing these products is given in this discussion (Kumari et al., 2023a, b).

International regulatory agencies have established laws and procedures to ensure the safe application of nanotechnology in agricultural goods.

Key regulations include the following.

- *Regulatory oversight*: Every nation has its own laws regulating the use of nanotechnology in agriculture. In the United States, pesticides are governed by the EPA, while agricultural biotechnology products are overseen by the USDA. The EFSA and ECHA regulate nanotechnology in food and pesticides in the EU.
- *Risk assessment*: Regulatory bodies require a risk assessment before approving any agricultural product based on nanotechnology. This examines the toxicity of nanomaterials, their environmental potential, and their impact on human health.
- *Labeling*: Labeling guidelines for agricultural products using nanoparticles are mandated to help consumers make informed choices.
- *International standards*: The ISO has developed standards related to nanotechnology, such as ISO/TS 80004-1 on terminology and definitions for nanoparticles.
- *Research and development*: Regulatory agencies fund R&D to ensure the safety and quality of nano-based agricultural products. Cooperation between researchers, authorities, consumers, and producers is necessary for the ethical application of nanotechnology in agriculture. Regular evaluation and monitoring of risks are essential to ensure the safety and effectiveness of nanotechnology in agricultural products (Arya et al., 2021; Kumari et al., 2023a, b).

In India, key government agencies managing nano-based agricultural products include the Food Safety and Standards Authority of India (FSSAI), the Department of Biotechnology (DBT), the Ministry of Environment, Forests, and Climate Change (MoEFCC), and the Indian Council of Agricultural Research (ICAR). Relevant Indian regulations include the Seeds Act (1966), Insecticides Act (1968), Environment (Protection) Act (1986), Food Safety and Standards Act (2006), and various guidelines from the DBT, ICAR, and FSSAI. More thorough and coordinated legal frameworks are needed to ensure the efficacy and safety of nano-based agricultural products in India (Kumari et al., 2023a, b).

3.9 Conclusion

The integration of nanotechnology in agriculture has shown considerable advantages in promoting crop growth; however, it is vital to assess the environmental consequences of nanomaterials (NMs). When NMs are introduced into the soil, they can produce both beneficial and harmful effects. Understanding how these materials affect soil microbial communities, enzymatic activities, and nutrient cycling is essential, given the critical role of soil microorganisms in nutrient dynamics. Several factors influence the impact of NMs on soil functions, including the characteristics and concentrations of the NMs, the type of soil, and the organic matter content. Recent advancements in machine learning have demonstrated potential in forecasting the effects of NMs on soil microbiomes. Comprehensive research into the effects of nanomaterials is necessary to refine their properties, aiming to minimize toxicity while fostering beneficial soil microbiomes. Future advancements in nanotechnology could lead to the development of NMs that not only enhance soil functions but also positively affect plant growth, all while reducing environmental toxicity.

References

Adeleke, B. S., Chaudhary, P., Ayilara, M. S., Ojo, F. M., Erinoso, S. M., Upadhayay, V. K., Adeyemo, A. I., & Akinola, S. A. (2024). Rhizosphere microbiomes mediating abiotic stress mitigation for improved plant nutrition. *Ecologies, 5*(3), 375–401. https://doi.org/10.3390/ecologies5030024

Adhikari, K., Timilsina, A., & Chen, H. (2024). The effects of size and surface-coating of CuO-nanoparticles on extractable Cu and enzyme activities in soil. *Soil & Environmental Health, 2*(1), 100065. https://doi.org/10.1016/j.seh.2024.100065

Afzal, S., & Singh, N. K. (2022). Effect of zinc and iron oxide nanoparticles on plant physiology, seed quality and microbial community structure in a rice-soil-microbial ecosystem. *Environmental Pollution, 314*, 120224. https://doi.org/10.1016/j.envpol.2022.120224

Agri, U., Chaudhary, P., Sharma, A., & Kukreti, B. (2022). Physiological response of maize plants and its rhizospheric microbiome under the influence of potential bioinoculants and nanochitosan. *Plant and Soil, 474*(1–2), 451–468. https://doi.org/10.1007/s11104-022-05351-2

Ahmed, T., Shou, L., Guo, J., Noman, M., Qi, Y., Yao, Y., Masood, H. A., Rizwan, M., Ali, M. A., Ali, H. M., Li, B., & Qi, X. (2024). Modulation of rhizosphere microbial community and metabolites by bio-functionalized nanoscale silicon oxide alleviates cadmium-induced phytotoxicity in bayberry plants. *Science of the Total Environment, 933*, 173068. https://doi.org/10.1016/j.scitotenv.2024.173068

Al-Gahtany, S. A., Ghobashy, M. M., Meganid, A. S., Alshangiti, D. M., Alkhursani, S. A., & Madani, M. (2024). pH-tailored monodispersed nanopolystyrene: Environmental impact on PGPR and nitrogen cycling in agricultural soils. *Green Chemistry Letters and Reviews, 17*(1). https://doi.org/10.1080/17518253.2024.2302043

Al-Juthery, H. W., Lahmod, N. R., & Al-Taee, R. A. (2021). Intelligent, nano-fertilizers: A new technology for improvement nutrient use efficiency (article review). In *IOP Conference Series: Earth and Environmental Science* (Vol. 735(1), p. 012086). IOP Publishing. https://doi.org/10.1088/1755-1315/735/1/012086

Al-Selwey, W. A., Alsadon, A. A., Ibrahim, A. A., Labis, J. P., & Seleiman, M. F. (2023). Effects of zinc oxide and silicon dioxide nanoparticles on physiological, yield, and water use efficiency traits of potato grown under water deficit. *Plants, 12*(1), 218. https://doi.org/10.3390/plants12010218

Arya, S. S., Tanwar, N., & Lenka, S. K. (2021). Prospects of nano- and peptide-carriers to deliver CRISPR cargos in plants to edit across and beyond central dogma. *Nanotechnology For Environmental Engineering, 6*(2), 22. https://doi.org/10.1007/s41204-021-00118-z

Ashraf, H., Anjum, T., Riaz, S., Ahmad, I. S., Irudayaraj, J., Javed, S., et al. (2021). Inhibition mechanism of green-synthesized copper oxide nanoparticles from Cassia fistula towards Fusarium oxysporum by boosting growth and defense response in tomatoes. *Environmental Science: Nano, 8*(6), 1729–1748.

Avila-Arias, H., Nies, L. F., Gray, M. B., & Turco, R. F. (2019). Impacts of molybdenum-, nickel-, and lithium- oxide nanomaterials on soil activity and microbial community structure. *Science of the Total Environment, 652*, 202–211. https://doi.org/10.1016/j.scitotenv.2018.10.189

Awais, M., Naqvi, S. M. Z. A., Zhang, H., Li, L., Zhang, W., Awwad, F. A., Ismail, E. A. A., Khan, M. I., Raghavan, V., & Hu, J. (2023). AI and machine learning for soil analysis: An assessment of sustainable agricultural practices. *Bioresources and Bioprocessing, 10*(1), 90. https://doi.org/10.1186/s40643-023-00710-y

Baig, N., Kammakakam, I., & Falath, W. (2021). Nanomaterials: A review of synthesis methods, properties, recent progress, and challenges. *Materials Advances, 2*(6), 1821–1871. https://doi.org/10.1039/D0MA00807A

Bayda, S., Adeel, M., Tuccinardi, T., Cordani, M., & Rizzolio, F. (2019). The history of nanoscience and nanotechnology: From chemical–physical applications to nanomedicine. *Molecules, 25*(1), 112. https://doi.org/10.3390/molecules25010112

Benelli, G. (2018). Mode of action of nanoparticles against insects. *Environmental Science and Pollution Research, 25*(13), 12329–12341. https://doi.org/10.1007/s11356-018-1850-4

Bhaskar, R., Mishra, A. K., & Mohanty, B. (2017). Neonatal exposure to endocrine disrupting chemicals impairs learning behaviour by disrupting hippocampal organization in male Swiss Albino mice. *Basic & Clinical Pharmacology & Toxicology, 121*(1), 44–52. https://doi.org/10.1111/bcpt.12767

Bhaskar, R., Pandey, S. P., Kumar, U., Kim, H., Jayakodi, S. K., Gupta, M. K., & Han, S. S. (2024). Nanobionics for sustainable crop production: Recent development to regulate plant growth and protection strategies from pests. *OpenNano, 15*, 100198. https://doi.org/10.1016/j.onano.2023.100198

Bhattacharjya, S., Adhikari, T., Sahu, A., & Patra, A. K. (2021). Ecotoxicological effect of TiO2 nano particles on different soil enzymes and microbial community. *Ecotoxicology, 30*(4), 719–732. https://doi.org/10.1007/s10646-021-02398-2

Celestina, C., Wood, J. L., Manson, J. B., Wang, X., Sale, P. W., Tang, C., & Franks, A. E. (2019). Microbial communities in top-and subsoil of repacked soil columns respond differently to amendments but their diversity is negatively correlated with plant productivity. *Scientific Reports, 9*(1), 8890.

Chakraborty, U., Chakraborty, B., Dey, P., & Chakraborty, A. P. (2015). Role of microorganisms in alleviation of abiotic stresses for sustainable agriculture. In *Abiotic stresses in crop plants* (pp. 232–253). Cabi.

Chandi, K., Udomkun, P., Boonupara, T., & Kaewlom, P. (2024). Enhancing soil health, microbial count, and hydrophilic methomyl and hydrophobic lambda-cyhalothrin remediation with biochar and nano-biochar. *Scientific Reports, 14*(1), 19551. https://doi.org/10.1038/s41598-024-70515-2

Chaudhary, P., Sharma, A., Chaudhary, A., Khati, P., Gangola, S., & Maithani, D. (2021). Illumina based high throughput analysis of microbial diversity of maize rhizosphere treated with nanocompounds and Bacillus sp. *Applied Soil Ecology, 159*, 103836. https://doi.org/10.1016/j.apsoil.2020.103836

Chen, Y., Wu, N., Mao, H., Zhou, J., Su, Y., Zhang, Z., et al. (2019). Different toxicities of nanoscale titanium dioxide particles in the roots and leaves of wheat seedlings. *RSC Advances, 9*(33), 19243–19252. https://doi.org/10.1039/C9RA02984B

Chen, S., Dong, C., Gao, Y., Li, Y., & Shi, Y. (2023). Effects of nanocarbon and nano-calcium carbonate on soil enzyme activities and soil microbial community in wheat (*Triticum aestivum* L.) rhizosphere soil. *Journal of Plant Nutrition and Soil Science*. https://doi.org/10.1002/jpln.202300146

Chen, S., Teng, Y., Luo, Y., Kuramae, E., & Ren, W. (2024). Threats to the soil microbiome from nanomaterials: A global meta and machine-learning analysis. *Soil Biology and Biochemistry, 188*, 109248. https://doi.org/10.1016/j.soilbio.2023.109248

Ciurli, A., Giagnoni, L., Pastorelli, R., Sega, D., Zamboni, A., Renella, G., & Varanini, Z. (2022). A novel P nanofertilizer has no impacts on soil microbial communities and soil microbial activity. *Applied Soil Ecology, 178*, 104570. https://doi.org/10.1016/j.apsoil.2022.104570

Das, N., & Chandran, P. (2011). Microbial degradation of petroleum hydrocarbon contaminants: An overview. *Biotechnology Research International, 2011*(1), 941810.

Das, P., Davis, K., Penton, C. R., Westerhoff, P., & Bi, Y. (2022). Impacts of graphitic nanofertilizers on nitrogen cycling in a sandy, agricultural soil. *Journal of Nanoparticle Research, 24*(6), 120. https://doi.org/10.1007/s11051-022-05500-9

Dhaigude, V., Chaudhary, A., Gore, P., Kaur, J., Ghosh, A., Aravindharajan, S. T. M., Bhandari, G., Upadhayay, V. K., & Abou Fayssal, S. (2024). *Microbial inoculants for improved soil nutrient accessibility and maintenance of soil fertility* (pp. 281–301). https://doi.org/10.1007/978-981-97-0633-4_12

Ding, M., Dai, H., He, Y., Liang, T., Zhai, Z., Zhang, S., et al. (2024). Continuous cropping system altered soil microbial communities and nutrient cycles. *Frontiers in Microbiology, 15*, 1374550.

Ďúranová, H., Kšiňan, S., Kuželová, L., Šimora, V., Ďurišová, Ľ., Olexíková, L., et al. (2024). Nanoparticle-plant interactions: Physico-chemical characteristics, application strategies, and transmission electron microscopy-based ultrastructural insights, with a focus on stereological research. *Chemosphere*, 142772. https://doi.org/10.1016/j.chemosphere.2024.142772

Elsawy, H., El-shahawy, A., Ibrahim, M., El-Halim, A. E.-H. A., Talha, N., Sedky, A., Alfwuaires, M., Alabbad, H., Almeri, N., & Mahmoud, E. (2022). Properties of recycled nanomaterials and their effect on biological activity and yield of canola in degraded soils. *Agriculture, 12*(12), 2096. https://doi.org/10.3390/agriculture12122096

Enamala, M. K., Kolapalli, B., Divya Sruthi, P., Sarkar, S., Kuppam, C., & Chavali, M. (2019). Applications of nanomaterials and future prospects for nanobionics. In *Plant nanobionics: Volume 2, Approaches in nanoparticles, biosynthesis, and toxicity* (pp. 177–197). https://doi.org/10.1007/978-3-030-16379-2_6

Faizan, M., Karabulut, F., Alam, P., Yusuf, M., Tonny, S. H., Adil, M. F., Sehar, S., Ahmed, S. M., & Hayat, S. (2023). Nanobionics: A sustainable agricultural approach towards understanding plant response to heavy metals, drought, and salt stress. *Nanomaterials, 13*(6), 974. https://doi.org/10.3390/nano13060974

Fetyan, N. A. H., Essa, T. A., Salem, T. M., Taha, A. A., Elgobashy, S. F., Tharwat, N. A., & Elsakhawy, T. (2024). Promising eco-friendly nanoparticles for managing bottom Rot disease in Lettuce (Lactuca sativa var. longifolia). *Microbiology Research, 15*(1), 196–212. https://doi.org/10.3390/microbiolres15010014

Fouad, F. A., Youssef, D. G., Shahat, F. M., & Abd El-Ghany, M. N. (2022). Role of microorganisms in biodegradation of pollutants. In *Handbook of biodegradable materials* (pp. 1–40). Springer International Publishing.

Ge, Y., Schimel, J. P., & Holden, P. A. (2011). Evidence for negative effects of TiO_2 and ZnO nanoparticles on soil bacterial communities. *Environmental Science & Technology, 45*(4), 1659–1664. https://doi.org/10.1021/es103040t

Ghannam, R. B., & Techtmann, S. M. (2021). Machine learning applications in microbial ecology, human microbiome studies, and environmental monitoring. *Computational and Structural Biotechnology Journal, 19*, 1092–1107. https://doi.org/10.1016/j.csbj.2021.01.028

Gil-Díaz, M., Álvarez-Aparicio, J., Alonso, J., Mancho, C., Lobo, M. C., González, J., & García-Gonzalo, P. (2024). Soil properties determine the impact of nZVI on Lactuca sativa L and its rhizosphere. *Environmental Pollution, 341*, 122683. https://doi.org/10.1016/j.envpol.2023.122683

Gorczyca, A., Przemieniecki, S. W., & Oćwieja, M. (2024). Comparative effect of silver nanoparticles on maize rhizoplane microbiome in initial phase of plants growth. *International Agrophysics, 38*(2), 155–164. https://doi.org/10.31545/intagr/184863

Grün, A.-L., Straskraba, S., Schulz, S., Schloter, M., & Emmerling, C. (2018). Long-term effects of environmentally relevant concentrations of silver nanoparticles on microbial biomass, enzyme activity, and functional genes involved in the nitrogen cycle of loamy soil. *Journal of Environmental Sciences, 69*, 12–22. https://doi.org/10.1016/j.jes.2018.04.013

Gupta, S. D., Agarwal, A., & Pradhan, S. (2018). Phytostimulatory effect of silver nanoparticles (AgNPs) on rice seedling growth: An insight from antioxidative enzyme activities and gene expression patterns. *Ecotoxicology and Environmental Safety, 161*, 624–633. https://doi.org/10.1016/j.ecoenv.2018.06.023

Hafeez, R., Guo, J., Ahmed, T., Jiang, H., Raza, M., Shahid, M., Ibrahim, E., Wang, Y., Wang, J., Yan, C., An, Q., White, J. C., & Li, B. (2024). Bio-formulated chitosan nanoparticles enhance disease resistance against rice blast by physiomorphic, transcriptional, and microbiome modulation of rice (Oryza sativa L.). *Carbohydrate Polymers, 334*, 122023. https://doi.org/10.1016/j.carbpol.2024.122023

Hartmann, M., & Six, J. (2023). Soil structure and microbiome functions in agroecosystems. *Nature Reviews Earth & Environment, 4*(1), 4–18.

Hassanzadeh, P., Atyabi, F., & Dinarvand, R. (2022). Nanobionics: From plant empowering to the infectious disease treatment. *Journal of Controlled Release, 349*, 890–901. https://doi.org/10.1016/j.jconrel.2022.07.028

He, S., Feng, Y., Ren, H., Zhang, Y., Gu, N., & Lin, X. (2011). The impact of iron oxide magnetic nanoparticles on the soil bacterial community. *Journal of Soils and Sediments, 11*(8), 1408–1417. https://doi.org/10.1007/s11368-011-0415-7

Huang, H., Chen, J., Liu, S., & Pu, S. (2022). Impact of ZnO nanoparticles on soil lead bioavailability and microbial properties. *Science of the Total Environment, 806*, 150299. https://doi.org/10.1016/j.scitotenv.2021.150299

Huang, J., Qian, X., Li, X., Hu, Q., Cao, C., & Yan, C. (2023a). A comparative study on responses of soil under three typical nanoparticles exposure: Enzyme activities and microbial community structure. *Water, Air, & Soil Pollution, 234*(4), 229. https://doi.org/10.1007/s11270-023-06218-7

Huang, Q., Zhu, W., Gao, X., Liu, X., Zhang, Z., & Xing, B. (2023b). Nanoparticles-mediated ion channels manipulation: From their membrane interactions to bioapplications. *Advanced Drug Delivery Reviews, 195*, 114763. https://doi.org/10.1016/j.addr.2023.114763

Jaafar, Z., Quelennec, B., Moreau, C., Lourdin, D., Maigret, J. E., Pontoire, B., et al. (2020). Plant cell wall inspired xyloglucan/cellulose nanocrystals aerogels produced by freeze-casting. *Carbohydrate Polymers, 247*, 116642. https://doi.org/10.1016/j.carbpol.2020.116642

Jaggi, V., Upadhayay, V. K., Joshi, S., Dasila, H., & Sahgal, M. (2023). Rhizosphere engineering for sustainable agriculture. In *Advanced microbial technology for sustainable agriculture and environment* (pp. 119–136). Elsevier. https://doi.org/10.1016/B978-0-323-95090-9.00004-2

Jayaraman, S., Naorem, A. K., Lal, R., Dalal, R. C., Sinha, N. K., Patra, A. K., & Chaudhari, S. K. (2021). Disease-suppressive soils—beyond food production: A critical review. *Journal of Soil Science and Plant Nutrition, 21*, 1437–1465.

Jiao, L., Yue, L., Le, X., Cao, X., Chen, F., Wang, C., Zhang, X., Zou, H., & Wang, Z. (2024). Dynamic microbial regulation of triiron tetrairon phosphate nanomaterials in the tomato rhizosphere. *Environmental Science: Nano, 11*(3), 1157–1169. https://doi.org/10.1039/D3EN00797A

Kačergius, A., Sivojienė, D., Gudiukaitė, R., Bakšienė, E., Masevičienė, A., & Žičkienė, L. (2023). Comparison of the structure of soil microbial communities of different ecosystems using the microbiome sequencing approach. *Soil Systems, 7*(3), 70.

Kannaujia, R., Singh, P., Prasad, V., & Pandey, V. (2022). Evaluating impacts of biogenic silver nanoparticles and ethylenediurea on wheat (Triticum aestivum L.) against ozone-induced damages. *Environmental Research, 203*, 111857. https://doi.org/10.1016/j.envres.2021.111857

Karigar, C. S., & Rao, S. S. (2011). Role of microbial enzymes in the bioremediation of pollutants: A review. *Enzyme Research, 2011*(1), 805187.

Karimi, E., & Mohseni Fard, E. (2017). *Nanomaterial effects on soil microorganisms* (pp. 137–200). https://doi.org/10.1007/978-3-319-46835-8_5

Karny, A., Zinger, A., Kajal, A., Shainsky-Roitman, J., & Schroeder, A. (2018). Therapeutic nanoparticles penetrate leaves and deliver nutrients to agricultural crops. *Scientific Reports, 8*(1), 1–10. https://doi.org/10.1038/s41598-018-25197-y

Karunakaran, G., Suriyaprabha, R., Manivasakan, P., Rajendran, V., & Kannan, N. (2014). Influence of nano and bulk SiO2 and Al2O3 particles on PGPR and soil nutrient contents. *Current Nanoscience, 10*(4), 604–612. https://doi.org/10.3390/mi11010092

Kaviya, N., Upadhayay, V. K., Singh, J., Khan, A., Panwar, M., & Singh, A. V. (2019). Role of microorganisms in soil genesis and functions. In *Mycorrhizosphere and pedogenesis* (pp. 25–52). Springer Singapore. https://doi.org/10.1007/978-981-13-6480-8_2

Khan, A., Singh, J., Upadhayay, V. K., Singh, A. V., & Shah, S. (2019). Microbial biofortification: A green technology through plant growth promoting microorganisms. In *Sustainable green technologies for environmental management* (pp. 255–269). Springer Singapore. https://doi.org/10.1007/978-981-13-2772-8_13

Khan, A., Upadhayay, V. K., Panwar, M., & Singh, A. V. (2020). *Soil microbiota: A key bioagent for revitalization of soil health in Hilly Regions* (pp. 183–200). https://doi.org/10.1007/978-981-15-1902-4_10

Khan, A., Singh, A. V., Gautam, S. S., Agarwal, A., Punetha, A., Upadhayay, V. K., Kukreti, B., Bundela, V., Jugran, A. K., & Goel, R. (2023). Microbial bioformulation: A microbial assisted biostimulating fertilization technique for sustainable agriculture. *Frontiers in Plant Science, 14*. https://doi.org/10.3389/fpls.2023.1270039

Khan, A., Singh, A. V., Kukreti, B., Pandey, D. T., Upadhayay, V. K., Kumar, R., & Goel, R. (2024). Deciphering the impact of cold-adapted bioinoculants on rhizosphere dynamics, biofortification, and yield of kidney bean across varied altitudinal zones. *Science of the Total Environment, 927*, 172204. https://doi.org/10.1016/j.scitotenv.2024.172204

Khatri, K., & Rathore, M. S. (2018). Plant nanobionics and its applications for developing plants with improved photosynthetic capacity. *Photosynthesis: From Its Evolution to Future Improvements in Photosynthetic Efficiency using Nanomaterials, 95*, 1016.

Kolenčík, M., Ernst, D., Urík, M., Ďurišová, Ľ., Bujdoš, M., Šebesta, M., et al. (2020). Foliar application of low concentrations of titanium dioxide and zinc oxide nanoparticles to the common sunflower under field conditions. *Nanomaterials, 10*(8), 1619. https://doi.org/10.3390/nano10081619

Kondal, R., Kalia, A., Krejcar, O., Kuca, K., Sharma, S. P., Luthra, K., Dheri, G. S., Vikal, Y., Taggar, M. S., Abd-Elsalam, K. A., & Gomes, C. L. (2021). Chitosan-urea nanocomposite for improved fertilizer applications: The effect on the soil enzymatic activities and microflora dynamics in N cycle of potatoes (Solanum tuberosum L.). *Polymers, 13*(17), 2887. https://doi.org/10.3390/polym13172887

Kordrostami, M., Mafakheri, M., & Al-Khayri, J. M. (2021). Contributions of nano biosensors in managing environmental plant stress under climatic changing era. In J. M. Al-Khayri, M. I. Ansari, & A. K. Singh (Eds.), *Nanobiotechnology*. Springer. https://doi.org/10.1007/978-3-030-73606-4_6

Koshila Ravi, R., Anusuya, S., Balachandar, M., & Muthukumar, T. (2019). Microbial interactions in soil formation and nutrient cycling. *Mycorrhizosphere and pedogenesis*, 363–382.

Kumari, A., Rana, V., Yadav, S. K., & Kumar, V. (2023a). Nanotechnology as a powerful tool in plant sciences: Recent developments, challenges and perspectives. *Plant Nano Biology*, 100046. https://doi.org/10.1016/j.plana.2023.100046

Kumari, R., Suman, K., Karmakar, S., Mishra, V., Lakra, S. G., Saurav, G. K., & Mahto, B. K. (2023b). Regulation and safety measures for nanotechnology-based agri-products. *Frontiers in Genome Editing, 5*. https://doi.org/10.3389/fgeed.2023.1200987

Labouyrie, M., Ballabio, C., Romero, F., Panagos, P., Jones, A., Schmid, M. W., et al. (2023). Patterns in soil microbial diversity across Europe. *Nature Communications, 14*(1), 3311.

Landa, P. (2021). Positive effects of metallic nanoparticles on plants: Overview of involved mechanisms. *Plant Physiology and Biochemistry, 161*, 12–24. https://doi.org/10.1016/j.plaphy.2021.01.039

Lastochkina, O., Aliniaeifard, S., SeifiKalhor, M., Bosacchi, M., Maslennikova, D., & Lubyanova, A. (2022). Novel approaches for sustainable horticultural crop production: Advances and prospects. *Horticulturae, 8*(10), 910. https://doi.org/10.3390/horticulturae8100910

Lew, T. T. S., Koman, V. B., Gordiichuk, P., Park, M., & Strano, M. S. (2020a). The emergence of plant nanobionics and living plants as technology. *Advanced Materials Technologies, 5*(3), 1900657. https://doi.org/10.1002/admt.201900657

Lew, T. T. S., Koman, V. B., Silmore, K. S., Seo, J. S., Gordiichuk, P., Kwak, S. Y., et al. (2020b). Real-time detection of wound-induced H2O2 signalling waves in plants with optical nanosensors. *Nature Plants, 6*(4), 404–415.

Lew, T. T. S., Sarojam, R., Jang, I. C., Park, B. S., Naqvi, N. I., Wong, M. H., et al. (2020c). Species-independent analytical tools for next-generation agriculture. *Nature Plants, 6*(12), 1408–1417. https://doi.org/10.1038/s41477-020-00808-7

Lew, T. T. S., Park, M., Cui, J., & Strano, M. S. (2021). Plant nanobionic sensors for arsenic detection. *Advanced Materials, 33*(1), 2005683. https://doi.org/10.1002/adma.202005683

Li, Y., Zhao, H. P., & Zhu, L. (2021). Remediation of soil contaminated with organic compounds by nanoscale zero-valent iron: A review. *Science of the Total Environment, 760*, 143413. https://doi.org/10.1016/j.scitotenv.2020.143413

Li, X., Ghanizadeh, H., Han, Z., Wang, Q., Li, F., Qiu, Y., et al. (2023). Metabolic profile and physiological mechanisms underlying the promoting effects of TiO2NPs on the photosynthesis and growth of tomato. *Scientia Horticulturae, 322*, 112394. https://doi.org/10.1016/j.scienta.2023.112394

Li, H., Guo, Y., Liang, A., Xu, X., Hao, Y., Shang, H., Li, C., Cai, Z., Han, L., Zhao, J., Jia, W., White, J. C., Ma, C., & Xing, B. (2024). Commonly used engineered nanomaterials improve soil health via suppressing soil-borne *Fusarium* and positively altering soil microbiome. *ACS ES&T Engineering*, *4*(4), 915–927. https://doi.org/10.1021/acsestengg.3c00501.

Liu, W., Yao, J., Cai, M., Chai, H., Zhang, C., Sun, J., Chandankere, R., & Masakorala, K. (2014). Synthesis of a novel nanopesticide and its potential toxic effect on soil microbial activity. *Journal of Nanoparticle Research, 16*(11), 2677. https://doi.org/10.1007/s11051-014-2677-7

Liu, Y., Yue, L., Wang, Z., & Xing, B. (2019). Processes and mechanisms of photosynthesis augmented by engineered nanomaterials. *Environmental Chemistry, 16*(6), 430–445. https://doi.org/10.1071/EN19046

Maithani, D., Sharma, A., Dasila, H., Tiwari, A., & Upadhayay, V. K. (2023). Nanotechnology for crop improvement and sustainable agriculture. In *Advances in nanotechnology for smart agriculture* (pp. 323–339). CRC Press. https://doi.org/10.1201/9781003345565-16

Maliszewska, I. (2016). Effects of the biogenic gold nanoparticles on microbial community structure and activities. *Annals of Microbiology, 66*(2), 785–794. https://doi.org/10.1007/s13213-015-1158-2

Mao, L., Song, Q., Li, M., Liu, X., Shi, Z., Chen, F., et al. (2023). Decreasing photosystem antenna size by inhibiting chlorophyll synthesis: A double-edged sword for photosynthetic efficiency. *Crop and Environment, 2*(1), 46–58. https://doi.org/10.1016/j.crope.2023.02.006

McGee, C. F., Storey, S., Clipson, N., & Doyle, E. (2017). Soil microbial community responses to contamination with silver, aluminium oxide and silicon dioxide nanoparticles. *Ecotoxicology, 26*(3), 449–458. https://doi.org/10.1007/s10646-017-1776-5

Meier, M. J., Dodge, A. E., Samarajeewa, A. D., & Beaudette, L. A. (2020). Soil exposed to silver nanoparticles reveals significant changes in community structure and altered microbial transcriptional profiles. *Environmental Pollution, 258*, 113816. https://doi.org/10.1016/j.envpol.2019.113816

Mekuye, B., & Abera, B. (2023). Nanomaterials: An overview of synthesis, classification, characterization, and applications. *Nano Select, 4*(8), 486–501. https://doi.org/10.1002/nano.202300038

Mishra, J., Singh, R., & Arora, N. K. (2017). Alleviation of heavy metal stress in plants and remediation of soil by rhizosphere microorganisms. *Frontiers in Microbiology, 8*, 1706.

Moll, J., Klingenfuss, F., Widmer, F., Gogos, A., Bucheli, T. D., Hartmann, M., & van der Heijden, M. G. A. (2017). Effects of titanium dioxide nanoparticles on soil microbial communities and wheat biomass. *Soil Biology and Biochemistry, 111*, 85–93. https://doi.org/10.1016/j.soilbio.2017.03.019

Montes de Oca-Vásquez, G., Solano-Campos, F., Vega-Baudrit, J. R., López-Mondéjar, R., Odriozola, I., Vera, A., Moreno, J. L., & Bastida, F. (2020). Environmentally relevant concentrations of silver nanoparticles diminish soil microbial biomass but do not alter enzyme activities or microbial diversity. *Journal of Hazardous Materials, 391*, 122224. https://doi.org/10.1016/j.jhazmat.2020.122224

Mony, C., Kaur, P., Rookes, J. E., Callahan, D. L., Eswaran, S. V., Yang, W., & Manna, P. K. (2022). Nanomaterials for enhancing photosynthesis: Interaction with plant photosystems and scope of nanobionics in agriculture. *Environmental Science: Nano, 9*(10), 3659–3683. https://doi.org/10.1039/D2EN00451H

Mubeen, B., Hasnain, A., Wang, J., Zheng, H., Naqvi, S. A. H., Prasad, R., ur Rehman, A., Sohail, M. A., Hassan, M. Z., Farhan, M., Khan, M. A., & Moustafa, M. (2023). Current progress and open challenges for combined toxic effects of Manufactured Nano-Sized Objects (MNO's) on soil biota and microbial community. *Coatings, 13*(1), 212. https://doi.org/10.3390/coatings13010212

Nafees, M., Ali, S., Rizwan, M., Aziz, A., Adrees, M., Hussain, S. M., Ali, Q., & Junaid, M. (2020). *Effect of nanoparticles on plant growth and physiology and on soil microbes* (pp. 65–85). https://doi.org/10.1007/978-3-030-34544-0_5

Naylor, D., Sadler, N., Bhattacharjee, A., Graham, E. B., Anderton, C. R., McClure, R., et al. (2020). Soil microbiomes under climate change and implications for carbon cycling. *Annual Review of Environment and Resources, 45*(1), 29–59.

Nepal, J., Xin, X., Maltais-Landry, G., Ahmad, W., Pereira, J., Santra, S., Wright, A. L., Ogram, A., Stofella, P. J., & He, Z. (2023). Carbon nanomaterials are a superior soil amendment for sandy soils than biochar based on impacts on lettuce growth, physiology and soil biochemical quality. *NanoImpact, 31*, 100480. https://doi.org/10.1016/j.impact.2023.100480

Nogueira, V., Lopes, I., Rocha-Santos, T., Santos, A. L., Rasteiro, G. M., Antunes, F., Gonçalves, F., Soares, A. M. V. M., Cunha, A., Almeida, A., Gomes, N. N. C. M., & Pereira, R. (2012). Impact of organic and inorganic nanomaterials in the soil microbial community structure. *Science of the Total Environment, 424*, 344–350. https://doi.org/10.1016/j.scitotenv.2012.02.041

Ogunkunle, C. O., Oyedeji, S., Okoro, H. K., & Adimula, V. (2021). Interaction of nanoparticles with soil. In *Nanomaterials for soil remediation* (pp. 101–132). Elsevier. https://doi.org/10.1016/B978-0-12-822891-3.00006-2

Pariona, N., Paraguay-Delgado, F., Basurto-Cereceda, S., Morales-Mendoza, J. E., Hermida-Montero, L. A., & Mtz-Enriquez, A. I. (2020). Shape-dependent antifungal activity of ZnO particles against phytopathogenic fungi. *Applied Nanoscience, 10*, 435–443. https://doi.org/10.1007/s13204-019-01127-w

Parveen, A., & Siddiqui, Z. A. (2022). Foliar spray and seed priming of titanium dioxide nanoparticles and their impact on the growth of tomato, defense enzymes and some bacterial and fungal diseases. *Archives of Phytopathology and Plant Protection, 55*(5), 527–548. https://doi.org/10.1080/03235408.2022.2035535

Philippot, L., Spor, A., Hénault, C., Bru, D., Bizouard, F., Jones, C. M., et al. (2013). Loss in microbial diversity affects nitrogen cycling in soil. *The ISME Journal, 7*(8), 1609–1619.

Pomerantseva, E., Bonaccorso, F., Feng, X., Cui, Y., & Gogotsi, Y. (2019). Energy storage: The future enabled by nanomaterials. *Science, 366*(6468), eaan8285. https://doi.org/10.1126/science.aan8285

Prajapati, M. K., Sen, A., & Upadhayay, V. K. (2024). *Microbial inoculants in amelioration of stress by regulation of signalling cscades in plant* (pp. 375–410). https://doi.org/10.1007/978-981-97-0633-4_16

Prywes, N., Phillips, N. R., Tuck, O. T., Valentin-Alvarado, L. E., & Savage, D. F. (2023). Rubisco function, evolution, and engineering. *Annual Review of Biochemistry, 92*(1), 385–410. https://doi.org/10.1146/annurev-biochem-040320-101244

Przemieniecki, S. W., Ruraż, K., Kosewska, O., Oćwieja, M., & Gorczyca, A. (2024). The impact of various forms of silver nanoparticles on the rhizosphere of wheat (Triticum aestivum L.) – Shifts in microbiome structure and predicted microbial metabolic functions. *Science of the Total Environment, 914*, 169824. https://doi.org/10.1016/j.scitotenv.2023.169824

Pullagurala, V. L. R., Adisa, I. O., Rawat, S., White, J. C., Zuverza-Mena, N., Hernandez-Viezcas, J. A., & Gardea-Torresdey, J. L. (2019). Fate of engineered nanomaterials in agroenvironments and impacts on agroecosystems. In *Exposure to engineered nanomaterials in the environment* (pp. 105–142). Elsevier. https://doi.org/10.1016/B978-0-12-814835-8.00004-2

Qu, K., Guo, F., Liu, X., Lin, Y., & Zou, Q. (2019). Application of machine learning in microbiology. *Frontiers in Microbiology, 10*. https://doi.org/10.3389/fmicb.2019.00827

Rajput, R. S., Jyoti Singh, J. S., Prachi Singh, P. S., Chitara, M. K., Ram, R. M., Sandhya Mishra, S. M., & Singh, H. B. (2018). Nanotechnology: Comprehensive understanding of interaction, toxicity and the fate of biosynthesized nanoparticles in the agroecosystem. In *Emerging trends in agri-nanotechnology: Fundamental and applied aspects* (pp. 135–163). CAB International. https://doi.org/10.1079/9781786391445.0135

Rajput, V., Minkina, T., Mazarji, M., Shende, S., Sushkova, S., Mandzhieva, S., & Jatav, H. (2020). Accumulation of nanoparticles in the soil-plant systems and their effects on human health. *Annals of Agricultural Sciences, 65*(2), 137–143. https://doi.org/10.1016/j.aoas.2020.08.001

Rajput, V. D., Singh, A., Singh, V. K., Minkina, T. M., & Sushkova, S. (2021). Impact of nanoparticles on soil resource. In *Nanomaterials for soil remediation* (pp. 65–85). Elsevier. https://doi.org/10.1016/B978-0-12-822891-3.00004-9

Rajput, V. D., Kumari, A., Upadhyay, S. K., Minkina, T., Mandzhieva, S., Ranjan, A., Sushkova, S., Burachevskaya, M., Rajput, P., Konstantinova, E., Singh, J., & Verma, K. K. (2023). Can nanomaterials improve the soil microbiome and crop productivity? *Agriculture, 13*(2), 231. https://doi.org/10.3390/agriculture13020231

Rani, N., Kumari, K., & Hooda, V. (2024). Impact evaluation of bare and chitosan complexed silver nanoparticles on the growth of Sorghum bicolor (L.) Moench and soil bacterial diversity. *Plant Physiology Reports, 29*(2), 316–331. https://doi.org/10.1007/s40502-023-00774-0

Ranjan, A., Rajput, V. D., Kumari, A., Mandzhieva, S. S., Sushkova, S., Prazdnova, E. V., Zargar, S. M., Raza, A., Minkina, T., & Chung, G. (2022). Nanobionics in crop production: an emerging approach to modulate plant functionalities. *Plants, 11*(5), 692. https://doi.org/10.3390/plants11050692

Rehman, A., ur Rahman, S., Li, P., Shah, I. H., Manzoor, M. A., Azam, M., Cao, J., Malik, M. S., Jeridi, M., Ahmad, N., Alabbosh, K. F., Liu, Q., Khalid, M., & Niu, Q. (2024). Modulating plant-soil microcosm with green synthesized ZnONPs in arsenic contaminated soil. *Journal of Hazardous Materials, 470*, 134130. https://doi.org/10.1016/j.jhazmat.2024.134130

Ren, Y., Wang, G., Su, Y., Li, J., Zhang, H., Ma, G., & Han, J. (2024). Effect of CeO2, TiO2 and SiO2 nanoparticles on the growth and quality of model medicinal plant Salvia miltiorrhiza by acting on soil microenvironment. *Ecotoxicology and Environmental Safety, 280*, 116552. https://doi.org/10.1016/j.ecoenv.2024.116552

Rodrigues, D. F., Jaisi, D. P., & Elimelech, M. (2013). Toxicity of functionalized single-walled carbon nanotubes on soil microbial communities: Implications for nutrient cycling in soil. *Environmental Science & Technology, 47*(1), 625–633. https://doi.org/10.1021/es304002q

Roy, A., Sharma, A., Yadav, S., Jule, L. T., & Krishnaraj, R. (2021). Nanomaterials for remediation of environmental pollutants. *Bioinorganic Chemistry and Applications, 2021*(1), 1764647. https://doi.org/10.1155/2021/1764647

Saha, S., Bansal, S., & Khanuja, M. (2022). Classification of nanomaterials and their physical and chemical nature. In *Nano-enabled agrochemicals in agriculture* (pp. 7–34). Academic Press. https://doi.org/10.1016/B978-0-323-91009-5.00001-X

Sahu, N., Vasu, D., Sahu, A., Lal, N., & Singh, S. K. (2017). Strength of microbes in nutrient cycling: A key to soil health. In *Agriculturally important microbes for sustainable agriculture: Volume I: Plant-soil-microbe nexus* (pp. 69–86).

Sahu, P. K., Singh, D. P., Prabha, R., Meena, K. K., & Abhilash, P. C. (2019). Connecting microbial capabilities with the soil and plant health: Options for agricultural sustainability. *Ecological Indicators, 105*, 601–612.

Saleh, T. A. (2020). Nanomaterials: Classification, properties, and environmental toxicities. *Environmental Technology & Innovation, 20*, 101067. https://doi.org/10.1016/j.eti.2020.101067

Samarajeewa, A. D., Velicogna, J. R., Princz, J. I., Subasinghe, R. M., Scroggins, R. P., & Beaudette, L. A. (2017). Effect of silver nano-particles on soil microbial growth, activity and community diversity in a sandy loam soil. *Environmental Pollution, 220*, 504–513. https://doi.org/10.1016/j.envpol.2016.09.094

Shaban, A., Telegdi, J., & Vastag, G. (2022). Self-assembling of nanobionics: From theory to applications. In *Fundamentals of bionanomaterials* (pp. 111–138). Elsevier. https://doi.org/10.1016/B978-0-12-824147-9.00005-4

Shah, V., Collins, D., Walker, V. K., & Shah, S. (2014). The impact of engineered cobalt, iron, nickel and silver nanoparticles on soil bacterial diversity under field conditions. *Environmental Research Letters, 9*(2), 024001. https://doi.org/10.1088/1748-9326/9/2/024001

Shah, G. M., Ali, H., Ahmad, I., Kamran, M., Hammad, M., Shah, G. A., Bakhat, H. F., Waqar, A., Guo, J., Dong, R., & Rashid, M. I. (2022). Nano agrochemical zinc oxide influences microbial activity, carbon, and nitrogen cycling of applied manures in the soil-plant system. *Environmental Pollution, 293*, 118559. https://doi.org/10.1016/j.envpol.2021.118559

Shende, S. S., Rajput, V. D., Ranjan, A., Mandzhieva, S. S., Sushkova, S. N., & Minkina, T. M. (2024). Nanotoxicity to soil ecosystems. In *Nanotoxicology for agricultural and environmental applications* (pp. 109–134). Elsevier. https://doi.org/10.1016/B978-0-443-15570-3.00013-2

Simonin, M., Richaume, A., Guyonnet, J. P., Dubost, A., Martins, J. M. F., & Pommier, T. (2016). Titanium dioxide nanoparticles strongly impact soil microbial function by affecting archaeal nitrifiers. *Scientific Reports, 6*(1), 33643. https://doi.org/10.1038/srep33643

Simonin, M., Cantarel, A. A. M., Crouzet, A., Gervaix, J., Martins, J. M. F., & Richaume, A. (2018). Negative effects of copper oxide nanoparticles on carbon and nitrogen cycle microbial activities in contrasting agricultural soils and in presence of plants. *Frontiers in Microbiology, 9*. https://doi.org/10.3389/fmicb.2018.03102

Singh, A., Tomar, B., Margaryan, G., Rajput, P., Minkina, T., Mandzhieva, S., et al. (2024). Emerging technologies for sustainable soil management and precision farming. *Nanotechnology Applications and Innovations for Improved Soil Health*, 210–235. 979-8-3693-1471-5. ch010.f01.

Singha, K. M., Singha, L. P., Acharya, D., & Pandey, P. (2024). Inhibitory effects of silver silicate (AgSiO2) nanoparticles on uncultivable bacterial phyla present in soil. *Environmental Sustainability, 7*(2), 209–219. https://doi.org/10.1007/s42398-024-00314-y

Sonali, J. M. I., Gayathri, K. V., Kumar, P. S., Rangasamy, G., Poorva, C. S., Kumar, S. U., Doss, C. G. P., & Gnanasambandan, R. (2024). Effect of biogenic selenium nanocomposite on Okra and metagenomic analysis of bacterial community associated with rhizosphere and phyllosphere. *Waste and Biomass Valorization, 15*(3), 1819–1831. https://doi.org/10.1007/s12649-023-02215-3

Song, P., Liu, J., Ma, W., & Gao, X. (2024). Remediation of cadmium-contaminated soil by biochar-loaded nano-zero-valent iron and its microbial community responses. *Journal of Environmental Chemical Engineering, 12*(2), 112311. https://doi.org/10.1016/j.jece.2024.112311

Sri Sindhura, K., Prasad, T. N. V. K., Panner Selvam, P., & Hussain, O. M. (2014). Synthesis, characterization and evaluation of effect of phytogenic zinc nanoparticles on soil exo-enzymes. *Applied Nanoscience, 4*(7), 819–827. https://doi.org/10.1007/s13204-013-0263-4.

Srivastav, A., Shukla, A., Singhal, R. K., Srivastav, S., Ganjewala, D., & Shrivastava, M. (2023). Soil and plant enzymes responses to zinc oxide nanoparticles in submerged rice (*Oryza sativa* L.) ecosystem. *Trends in Sciences, 20*(9), 5558. https://doi.org/10.48048/tis.2023.5558

Talukder, P., Sinha, B., Biswas, S., Ghosh, A., Banerjee, A., & Paul, S. (2024). A study on the prospect of converting C3 Plants into C4 plants. *Biocatalysis and Agricultural Biotechnology*, 103191. https://doi.org/10.1016/j.bcab.2024.103191

Tang, Q., Xu, Z., Hong, A., Zhang, X., Kah, M., Li, L., & Wang, Y. (2021). Response of soil enzyme activity and bacterial community to copper hydroxide nanofertilizer and its ionic analogue under single versus repeated applications. *Science of the Total Environment, 796*, 148974. https://doi.org/10.1016/j.scitotenv.2021.148974

Teleanu, D. M., Negut, I., Grumezescu, V., Grumezescu, A. M., & Teleanu, R. I. (2019). Nanomaterials for drug delivery to the central nervous system. *Nanomaterials, 9*(3), 371. https://doi.org/10.3390/nano9030371

Tighe-Neira, R., Carmora, E., Recio, G., Nunes-Nesi, A., Reyes-Diaz, M., Alberdi, M., et al. (2018). Metallic nanoparticles influence the structure and function of the photosynthetic apparatus in plants. *Plant Physiology and Biochemistry, 130*, 408–417. https://doi.org/10.1016/j.plaphy.2018.07.024

Tripathi, G. D., Javed, Z., & Dashora, K. (2024). Toxicity of copper oxide nanoparticles on agriculturally important soil rhizobacteria Bacillus megaterium. *Emerging Contaminants, 10*(1), 100280. https://doi.org/10.1016/j.emcon.2023.100280

Trivedi, B., Chaudhary, A., Abou Fayssal, S., Upadhayay, V. K., & Kumar, G. (2024). *Commercialized microbial inoculants for crop improvement* (pp. 229–246). https://doi.org/10.1007/978-981-97-0633-4_10

Upadhayay, V. K., Singh, J., Khan, A., Lohani, S., & Singh, A. V. (2019). Mycorrhizal mediated micronutrients transportation in food based plants: A biofortification strategy. In *Mycorrhizosphere and pedogenesis* (pp. 1–24). Springer Singapore. https://doi.org/10.1007/978-981-13-6480-8_1

Upadhayay, V. K., Singh, A. V., & Khan, A. (2022a). Cross talk between zinc-solubilizing bacteria and plants: A short tale of bacterial-assisted zinc biofortification. *Frontiers in Soil Science, 1*. https://doi.org/10.3389/fsoil.2021.788170

Upadhayay, V. K., Singh, A. V., Khan, A., Singh, J., Pareek, N., & Raghav, A. (2022b). FE-SEM/EDX based zinc mobilization analysis of Burkholderia cepacia and Pantoea rodasii and their functional annotation in crop productivity, soil quality, and zinc biofortification of Paddy. *Frontiers in Microbiology, 13*. https://doi.org/10.3389/fmicb.2022.852192

Upadhayay, V. K., Chitara, M. K., Mishra, D., Jha, M. N., Jaiswal, A., Kumari, G., Ghosh, S., Patel, V. K., Naitam, M. G., Singh, A. K., Pareek, N., Taj, G., Maithani, D., Kumar, A., Dasila, H., & Sharma, A. (2023a). Synergistic impact of nanomaterials and plant probiotics in agriculture: A tale of two-way strategy for long-term sustainability. *Frontiers in Microbiology, 14*. https://doi.org/10.3389/fmicb.2023.1133968

Upadhayay, V. K., Maithani, D., Dasila, H., Taj, G., & Singh, A. V. (2023b). Microbial services for mitigation of biotic and abiotic stresses in plants. In *Advanced microbial techniques in agriculture, environment, and health management* (pp. 67–81). Elsevier. https://doi.org/10.1016/B978-0-323-91643-1.00003-X

Upadhayay, V. K., Naik, Y. D., Ranjan, N., Kastury, C., Shekhar, S., Kumar, S., & Jaggi, V. (2023c). Prolific nicrobial agents as key products for sustainable agriculture. In *Microbial bioactive compounds* (pp. 181–205). Springer Nature. https://doi.org/10.1007/978-3-031-40082-7_10

Upadhayay, V. K., Naitam, M. G., Ghosh, S., Patel, V. K., Sharma, A., Chandra, A. K., & Taj, G. (2023d). Techniques for integrating nanotechnology in agriculture. In *Advances in nanotechnology for smart agriculture* (pp. 19–40). CRC Press. https://doi.org/10.1201/9781003345565-2

Upadhayay, V. K., Chitara, M. K., Sharma, A., Chaudhary, P., Jaiswal, A., Kumari, G., & Mishra, D. (2024a). Role of nanomaterials on soil microbial community and functionality: Nanomaterial impact - Unraveling the dynamics of soil microbial communities and functions. In V. Rajput (Ed.), *Nanotechnology applications and innovations for improved soil health* (pp. 333–353). IGI Global. https://doi.org/10.4018/979-8-3693-1471-5.ch017

Upadhayay, V. K., de los Santos Villalobos, S., Aravindharajan, S. T. M., Kukreti, B., Chitara, M. K., Jaggi, V., Sharma, A., & Singh, A. V. (2024b). *Microbial advancement in agriculture* (pp. 95–125). https://doi.org/10.1007/978-981-97-0633-4_5

Ur Rahim, H., Qaswar, M., Uddin, M., Giannini, C., Herrera, M. L., & Rea, G. (2021). Nano-enable materials promoting sustainability and resilience in modern agriculture. *Nanomaterials, 11*(8), 2068. https://doi.org/10.3390/nano11082068

van der Meer, J. R. (2006). Environmental pollution promotes selection of microbial degradation pathways. *Frontiers in Ecology and the Environment, 4*(1), 35–42.

Wang, Q., Liao, Y., Zhao, W., Yi, T., Jiang, Y., Zhu, G., Sun, Y., Wang, Q., Huang, L., Chen, F., Zhang, P., & Rui, Y. (2024). Potassium-based nanomaterials significantly enhance nutrient utilization efficiency and promote high crop yields. *Environmental Science: Nano, 11*(7), 2906–2922. https://doi.org/10.1039/D4EN00148F

Wei, L., Wang, S., Zuo, Q., Liang, S., Shen, S., & Zhao, C. (2016). Nano-hydroxyapatite alleviates the detrimental effects of heavy metals on plant growth and soil microbes in e-waste-contaminated soil. *Environmental Science: Processes & Impacts, 18*(6), 760–767. https://doi.org/10.1039/C6EM00121A

Weng, Y., Bai, X., Kang, M., Huang, Y., Ji, Y., Wang, H., & Hua, Z. (2024). Comparative analysis of chemically and green synthesized titanium dioxide nanoparticles for the regulation of photosynthesis in Lactuca sativa L. *Environmental Science: Nano, 11*(1), 161–174. https://doi.org/10.1039/D3EN00589E

Wijesooriya, S. N., Madanayake, N. H., & Adassooriya, N. M. (2023). Impact of nanomaterials on beneficial soil micro-organisms. In *Nanotechnology in agriculture and agroecosystems* (pp. 367–385). Elsevier. https://doi.org/10.1016/B978-0-323-99446-0.00006-4

Xu, C., Peng, C., Sun, L., Zhang, S., Huang, H., Chen, Y., & Shi, J. (2015). Distinctive effects of TiO 2 and CuO nanoparticles on soil microbes and their community structures in flooded paddy soil. *Soil Biology and Biochemistry, 86*, 24–33. https://doi.org/10.1016/j.soilbio.2015.03.011

Xu, N., Kang, J., Ye, Y., Zhang, Q., Ke, M., Wang, Y., Zhang, Z., Lu, T., Peijnenburg, W. J. G. M., Penuelas, J., Bao, G., & Qian, H. (2022a). Machine learning predicts ecological risks of nanoparticles to soil microbial communities. *Environmental Pollution, 307*, 119528. https://doi.org/10.1016/j.envpol.2022.119528

Xu, Z., Long, X., Jia, Y., Zhao, D., & Pan, X. (2022b). Occurrence, transportand toxicity of nanomaterials in soil ecosystems: A review. *Environmental Chemistry Letters, 20*(6), 3943–3969. https://doi.org/10.1007/s10311-022-01507-9

Yangen, F., Xiuxiu, Z., Hanyue, W., Yueyue, T., Qinzeng, X., & Lixia, Z. (2019). Effects of light intensity on metabolism of light-harvesting pigment and photosynthetic system in Camellia sinensis L. cultivar 'Huangjinya'. *Environmental and Experimental Botany, 166*, 103796. https://doi.org/10.1016/j.envexpbot.2019.06.009

Ye, D., Wu, L., Li, X., Atoba, T. O., Wu, W., & Weng, H. (2023). A synthetic review of various dimensions of non-destructive plant stress phenotyping. *Plants, 12*(8), 1698. https://doi.org/10.3390/plants12081698

Yousefi, S., Kartoolinejad, D., & Naghdi, R. (2017). Effects of priming with multi-walled carbon nanotubes on seed physiological characteristics of Hopbush (Dodonaeaviscosa L.) under drought stress. *International Journal of Environmental Studies, 74*(4), 528–539. https://doi.org/10.1080/00207233.2017.1325627

Zahra, Z., Habib, Z., Chung, S., & Badshah, M. A. (2020). Exposure route of TiO2 NPs from industrial applications to wastewater treatment and their impacts on the agro-environment. *Nanomaterials, 10*(8), 1469. https://doi.org/10.3390/nano10081469

Zand, A. D., Mikaeili Tabrizi, A., & Vaezi Heir, A. (2020). Application of titanium dioxide nanoparticles to promote phytoremediation of Cd-polluted soil: Contribution of PGPR inoculation. *Bioremediation Journal, 24*(2–3), 171–189. https://doi.org/10.1080/10889868.2020.1799929

Zhang, H., Huang, M., Zhang, W., Gardea-Torresdey, J. L., White, J. C., Ji, R., & Zhao, L. (2020a). Silver nanoparticles alter soil microbial community compositions and metabolite profiles in unplanted and cucumber-planted soils. *Environmental Science & Technology, 54*(6), 3334–3342. https://doi.org/10.1021/acs.est.9b07562

Zhang, W., Jia, X., Chen, S., Wang, J., Ji, R., & Zhao, L. (2020b). Response of soil microbial communities to engineered nanomaterials in presence of maize (Zea mays L.) plants. *Environmental Pollution, 267*, 115608. https://doi.org/10.1016/j.envpol.2020.115608

Zhang, J., Xu, M., Zou, X., & Chen, J. (2022). Structural and functional characteristics of soil microbial community in a Pinus massoniana forest at different elevations. *PeerJ, 10*, e13504.

Zhang, H., Zhao, X., Bai, J., Tang, M., Du, W., Lv, Z., Siddique, K. H. M., & Mao, H. (2024a). Effect of *ZnO* nanoparticle application on crop safety and soil environment: A case study of potato planting. *Environmental Science: Nano, 11*(1), 351–362. https://doi.org/10.1039/D3EN00680H

Zhang, M.-X., Zhao, L.-Y., He, Y.-Y., Hu, J.-P., Hu, G.-W., Zhu, Y., Khan, A., Xiong, Y.-C., & Zhang, J.-L. (2024b). Potential roles of iron nanomaterials in enhancing growth and nitrogen fixation and modulating rhizomicrobiome in alfalfa (Medicago sativa L.). *Bioresource Technology, 391*, 129987. https://doi.org/10.1016/j.biortech.2023.129987

Zhang, Z., Lin, J., Owens, G., & Chen, Z. (2024c). Deciphering silver nanoparticles perturbation effects and risks for soil enzymes worldwide: Insights from machine learning and soil property integration. *Journal of Hazardous Materials, 469*, 134052. https://doi.org/10.1016/j.jhazmat.2024.134052

Zhao, F., Xin, X., Cao, Y., Su, D., Ji, P., Zhu, Z., & He, Z. (2021). Use of carbon nanoparticles to improve soil fertility, crop growth and nutrient uptake by Corn (Zea mays L.). *Nanomaterials, 11*(10), 2717. https://doi.org/10.3390/nano11102717

Zhuang, Q., Zhang, Y., Liu, Q., Sun, Y., Sharma, S., Tang, S., Dhankher, O. P., & Yuan, H. (2024). Effects of sulfur nanoparticles on rhizosphere microbial community changes in oilseed rape plantation soil under mercury stress. *International Journal of Phytoremediation, 26*(10), 1545–1555. https://doi.org/10.1080/15226514.2024.2335207

Chapter 4
Role of Nanobionics as Fertilizer in Horticulture

Sagar Mondal, Swati Priya, Khusboo Pandey, Seema Ramniwas, Jutishna Bora, Smita Lata, Ajay Kumar Mahalka, Abeer Hashem, Elsayed Fathi Abd Allah, and Sumira Malik

Abstract Nano bionics, a cutting-edge interdisciplinary field, has emerged as a promising approach in horticulture, particularly in the realm of fertilisation. The integration of nanotechnology into plant biology offers innovative solutions to enhance crop productivity, nutrient efficiency, and environmental sustainability. Nano bionics involves the use of nanoscale materials, such as nanoparticles and nanocomposites, that can be engineered to deliver nutrients more effectively to plants. These materials have unique properties, including increased surface area and reactivity, which enable them to interact with plant tissues at the cellular and molecular levels. As fertilizers, nano bionic materials can enhance nutrient uptake, reduce leaching losses, and promote plant growth under stress conditions, such as drought and high temperatures. This approach not only improves the efficiency of nutrient delivery but also minimises the environmental impact associated with conventional fertilizers. The potential of nano bionics in horticulture extends beyond nutrient management, offering prospects for plant protection, growth regulation, and stress

Sagar Mondal and Swati Priya contributed equally with all other contributors.

S. Mondal · S. Priya · K. Pandey · J. Bora · S. Lata · A. K. Mahalka · S. Malik (✉)
Amity Institute of Biotechnology, Amity University Jharkhand, Ranchi, India
e-mail: smalik@rnc.amity.edu

S. Ramniwas
University Center for Research & Development (UCRD) Chandigarh University, Mohali, Punjab, India

A. Hashem
Department of Botany and Microbiology, College of Science, King Saud University, Riyadh, Saudi Arabia

E. F. A. Allah
Department of Plant Production, College of Food and Agricultural Sciences, King Saud University, Riyadh, Saudi Arabia

V. D. Rajput et al. (eds.), *Contemporary Suitability of Nanobionics in Agriculture*, Nanotechnology in the Life Sciences, https://doi.org/10.1007/978-3-031-82779-2_4

resistance. This abstract explores the role of nano bionics as a novel fertilizer strategy in horticulture, highlighting its benefits, challenges, and future prospects for sustainable agriculture.

Keywords Nano bionics · Sustainable agriculture · Nano fertilizer · Horticulture · Fruit crops

4.1 Introduction

Nanotechnology exhibits profound potential in enhancing agricultural output, therefore contributing to future food security. Achieving both quality control and minimising the spoiling of harvested vegetables and fruits is a highly demanding endeavour in horticulture. A wide range of nanomaterials have demonstrated significant promise in augmenting productivity, prolonging shelf life, mitigating postharvest damage, and enhancing the quality of crops used in agriculture. The application of antimicrobial nanomaterials as nanofilms on harvested items and/or packaging materials is appropriate for the preservation and transportation of fruits and vegetables. Furthermore, nanomaterials enhance the liveliness of each cut flower. Nanofertilizers are selectively applied, gradually released, and very effective in enhancing the vertical development, pollination, and reproductive capacity in flowers, leading to higher crop production and enhanced quality of fruit trees as well as vegetables. Formulated nanopesticides are selectively targeted, environmentally friendly, and astonishingly effective. Nanosensors provide real-time monitoring of vegetative development, plant diseases, and insect infestation in agricultural crops in natural field settings. These innovative sensors are required to accurately detect soil moisture, humidity, crop pest population, pesticide residues, and determine nutrient needs. The objective of this review was to present a current overview of the latest progress in nanomaterials and their possible applications in improving productivity, product quality, insect protection, and minimising postharvest losses in horticultural crops. This work establishes that nanotechnology has the potential to develop advanced methods for enhancing the yield and quality of horticulture crops, therefore ensuring adequate nutrition and food for the continuously growing global population as in Fig. 4.1.

The field of nanotechnology is a developing technology with a historical background spanning more than 2000 years. Furthermore, it has been thriving as a sophisticated instrument in several scientific fields such as chemistry, physics, medicine, materials science, aviation, pharmaceutical, agriculture, food, and even in gardening. In the field of agriculture, horticulture refers to the systematic cultivation and administration of vegetables, fruits, flowers, including ornamental plants. Undoubtedly, guaranteeing the food and health needs of the continuously growing global population is a very difficult and demanding undertaking in the dynamic global environment. Thus, enhancing production and reducing post-harvest losses by cutting-edge technological methods like biotechnology and nanotechnology are

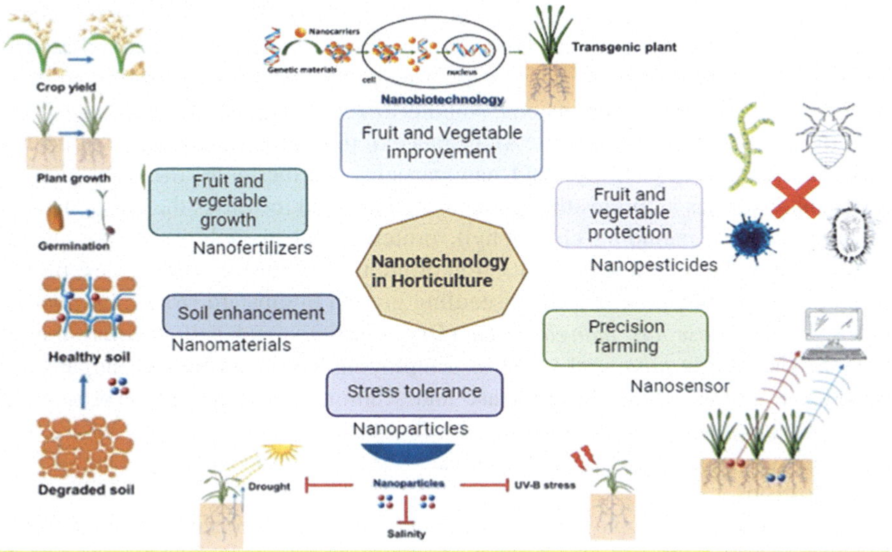

Fig. 4.1 Nanotechnology in horticulture

regarded as the most effective practices for tackling this difficult problem. The application of nanomaterials in horticulture is crucial for enhancing production, improving product quality, and minimising post-harvest losses of vegetables and fruit. Microbiological deterioration and physiological processes are responsible for the loss of up to 30% of horticulture agricultural products in underdeveloped nations. The implementation of nanofilm as well as nano-packaging containing antimicrobial nanomaterials can effectively reduce post-harvest losses to 5–10%, resulting in substantial preservation of nutritious food products. Minimising these expenses cannot just enhance the financial earnings of farmers but also guarantee superior quality and nutritional value of the food items. The commercially available synthetic fertilizers and pesticides pose significant risks to both the environment along with human health, and are also prohibitively costly. An extensive amount of these chemical inputs disappears by leaching, volatilisation, evaporation, and rainwater. Conversely, the novel compositions of nanopesticides and nanofertilizers exhibit exceptional efficiency, specificity towards defined targets, and safety.

4.2 Development and Growth of Horticultural Crops Influenced by Nanomaterials

A nanoparticle is a solid colloidal particle composed of macromolecular components. The nanoparticles absorb, encapsulate, dissolve, or entrap active components such as bioactive substances or therapeutic molecules. Nanofertilizers are economically viable and ecologically benign substances that enhance the nutritional efficiency of plants, thereby leading to higher crop production overall. Nanofertilizers deliver nutrients to crop plants through three methods: (i) the nutrient can be encapsulated by nanoparticles in the kind of nanoporous materials or nanotubes; (ii) encased by a thin protective layer of polymer; and (iii) supplied as emulsion or particles with nanoscale dimensions. Nanofertilizers are gradually, selectively, and effectively delivered to the plants. An instance of this is the enhancement of peanut (*Arachis hypogaea*) yield by ZnO nanoparticles. Furthermore, the utilisation of SiO2 nanoparticles amplifies the biomass of plants and increases the concentration of biomolecules, including chlorophyll, proteins, and phenols, in maize grains. Carbon nanotubes at a low concentration promote root development in hexaploidy wheat, seed germination as well as seedling growth in mustard (*Brassica juncea*), black gram (*Phaseolus mungo*), rice (*Oryza sativa*), and cell proliferation in tobacco (*Nicotiana tabacum*) by 16%. Employing TiO2 and SiO2 nanoparticles boosts the activity of nitrate reductase and seemingly accelerates the process of seed germination as well as seedling development in soya bean. Similar to field crops, the use of nanomaterials also enhances the development and growth of horticulture crops.

Nanofertilizers are employed in horticulture to enhance vegetative growth, pollination, and viability of flowers, leading to higher crop output and enhanced product quality for trees that bear fruit. Applying nano-Ca as an external supplement to blueberries grown under salt stress conditions leads to enhanced vegetative growth and higher chlorophyll content in the leaves. Furthermore, the application of nano-boron spray on the foliage of mango trees has a beneficial impact on augmenting the total production and chemical characteristics of the fruits. This effect is likely associated with the augmentation of chlorophyll levels and the presence of vital nutrient elements such as nitrogen (N), phosphorus (P), potassium (K), manganese (Mn), magnesium (Mg), boron (B), zinc (Zn), and iron (Fe) in the leaves. The application of nano-zinc to mango trees results in enhanced fruit weight, fruit number and yield, leaf chlorophyll and carotene levels, as well as concentrations of several essential nutrients such as nitrogen, phosphorus, potassium, and zinc. Applying nano-boron and nano-zinc fertilizers enhances fruit quality, boosts fruit yield, optimises the ratio of total soluble sugars (TSS) to maturity index, and increases total sugars as well as total phenols in pomegranates (Davarpanah et al., 2016).

4.2.1 Nanofertilizers

A fertilizer is a chemical or natural product, excluding liming materials, that is added to the soil or applied to plant tissues to contribute to one or more vital nutritional elements to enhance the nourishment of crop plants. A wide variety of fertilizers are available, derived from either organic or synthetic sources. In contemporary agriculture, the use of fertilizers is essential for achieving greater crop output. An inherent limitation of fertilisation in crops is the significant loss of spent fertilizers through several means, which eventually leads to environmental pollution and escalates production costs. The utilisation of nanofertilizers has emerged as a significant recent advancement in mitigating the environmental loss caused by applied fertilizers. Nanofertilizers are being developed by integrating plant nutrients within nanomaterials, coating nutrient molecules with a thin layer of nanomaterials, and generating aerosolised nanosized emulsions. The categories of nanofertilizers and nanobiofertilizers include both natural and synthetic substances, respectively, which effectively enhance bioavailability and overall soil fertility in comparison to conventional fertilizers. Notwithstanding, the most crucial features of nanofertilizers are (i) the particles must be at least 100 nm in size; (ii) the bulk size should be around 100 nm; and (iii) the nanoproduct should be both environmentally benign and long-lasting. An additional characteristic of a nanofertilizer is its ability to maintain its nanoscale dimensions and organised structures when it comes into contact with soil particles through the roots of agricultural plants. All the reactive properties of nanofertilizers are determined entirely by the morphology and dimensions of the nanoparticles. The most critical aspects of nanofertilizers are the enhancement of nutrient usage efficiency, control of active components, and minimisation of residual influence on soil biodiversity. Nanofertilizers are primarily classified as macronutrient and micronutrient nanofertilizers based on their distribution, nutrient balance, and the proportion needed by the plant (Fig. 4.2). Further the application of nano fertilizers for sustainable fruit production is explained in Fig. 4.3.

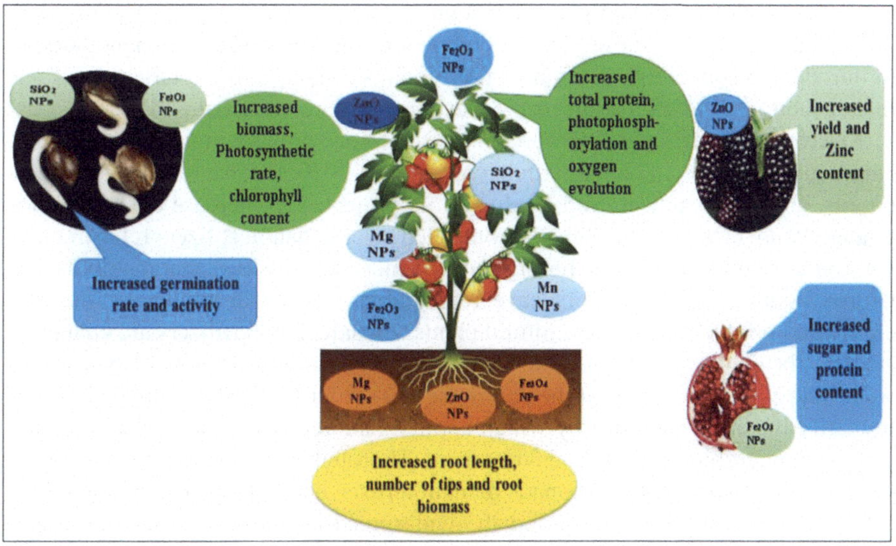

Fig. 4.2 Schematic summary of enhancements in seed germination, plant growth, and biomass or yield generation with the use of several types of nano-fertilizers. (Adapted from Rana et al. (2021))

Fig. 4.3 Application of nanofertilizers for sustainable fruit production

4.2.2 Macronutrient Nanofertilizers

Macronutrients refer to the vital nutrient elements that are necessary in substantially higher amounts for optimal plant development and productivity. To optimise their utilisation, a couple of these nutrient components are often linked to nanoparticles to introduce a suitable proportion of nutritional substances to the target crops and reduce their amount with additional advantages (Sidorowicz et al., 2019).

Macronutrient nanofertilizers are composed of one or more nutrient components enclosed within specific nanoparticle carriers. By 2020, the use of nitrogen, phosphorus, and potassium in crop production is projected to increase to 265 million tonnes (Wang et al., 2016). Slow or controlled-release nanofertilizers refer to nitrogen source nanofertilizers, including zeolites, mesoporous silica nanomaterials, and hydroxyapatite. Nanofertilizers have remarkable efficacy in enhancing the efficiency and productivity of agricultural plants. Biosafe nanofertilizers are characterised by the inclusion of phosphorus (P) as a constituent. It is a suspension of water-phosphorite particles at the nanoscale (60–120 nm). Further study is required to clarify the exact methods by which applied nanofertilizers interact with plants, soil, plant microbiome, and the environment. Thorough investigation is necessary to evaluate the environmental impact of nanofertiser application and guarantee its safety. The economic viability and the widespread availability of nanofertilizers enable broader use of these innovative agricultural chemicals.

4.2.3 *Micronutrient Nanofertilizers*

Micronutrients are associated with the production of proteins, carbohydrates, along with the control of hormones (such as auxins), therefore assuring the protection of plants against infections and pests (Broadley et al., 2007). This study reports on the efficacy of zinc-based nanofertilizers in several horticultural crop plants, including garden pea, cucumber, spinach, tomato, eggplants, chilli, coriander, and onion.

By using Zn nanoparticles, a substantial enhancement in crop productivity was shown in wheat, maize, rice, sugarcane, potato, sunflower, and *Brassica* (Monreal et al., 2016). Utilising iron (Fe) source nanofertilizers enhances crop productivity in several horticultural plant species including cucumber, lettuce, and garden pea. Another crucial micronutrient, copper (Cu), is needed for the growth and the development of plants. The biological development of several horticultural crop plants, including lettuce and tomato, is significantly improved by the application of Cu nanoparticles at suitable concentrations. Conversely, high levels of copper have detrimental effects on many agricultural crops. Manganese (Mn) functions crucially in the physiological and metabolic processes of plants. Furthermore, it enhances plants' ability to withstand various environmental pressures by regulating the activity of numerous enzymes as a cooperation factor. The application of nano-Mn fertilizer enhances the development and productivity of agricultural plants in cases when the soil lacks this essential component. Accurate and appropriate dosages of all micronutrients are essential for the development and productivity of horticulture plants. Thus, it is imperative to do soil analysis before suggesting the use of any nanofertilizers as a micronutrient source.

4.3 Nano-Plant Growth Stimulator

The utilisation of nanomaterials (NMs) in horticulture has significantly grown in recent decades. The structure and components of the NMs are utilised in the control of pests and diseases to significantly improve the development and productivity of crops (Servin & White, 2016). However, scientific literature has also shown both advantageous and harmful impacts of using nanomaterials on plants. While the precise mechanisms remain incompletely elucidated, available data indicate that increased quantities of nanoparticles have harmful effects on plants (Table 4.1). The influence and application of the nanoparticles are indicated in Table 4.2.

4.4 Nanopesticide

Agricultural pests are the primary constraints on agricultural productivity and must be effectively managed. Pest management by conventional methods entails the application of substantial amounts of chemical pesticides, leading to environmental

Table 4.1 Effects of nanoparticles on plant growth and development

Nanoparticles	Dose (mg/L)	Crop	Effect on plant growth and development	Reference
Cerium oxide (CeO_2)	125–4000	*Cucumis sativus*	Adverse effects inside the molecular and metabolic systems of plants	Rana et al. (2021)
Titanium oxide (TiO_2)	1000–2000	*Spinacia oleracea L.*	Promotes the processes of growth and photosynthesis	
Carbon nanotubes (MWCNT)	10–40	*Solanum lycopersicum*	Accelerates germination and growth rate while suppressing root elongation in tomato plants	
Carbon nanotubes (MWCNT)	10–40	*Allium cepa* and *Cucumis sativus*	Augments the elongation of the root	
Carbon nanotubes (MWCNT)	0, 500, 1000 or 5000	*Cucurbita pepo, Solanum lycopersicum, Zea mays, Glycine max*	Biomass in maize and soybean is reduced by 500 mg/kg, however the growth of tomato as well as zucchini remains unaffected	
Iron oxide (Fe_3O_4)	0.67	*Lactuca sativa, Spinacia oleracea L., Raphanus sativus, Cucumis sativus, Solanum lycopersicum,* Peppers	Retards seed germination	
Zinc oxide (ZnO)	100–1000	*Pisum sativum*	While seed germination is unaffected, nodulation along with root length are impacted	
Silver (Ag)	800	*Vicia faba L.*	Reduced germination	
Silver (Ag)	0, 125, 250, 500	*Raphanus sativus*	Germination unaffected	

issues and raising production costs (Sharon et al., 2010). Dilution of pesticides with nanotreated water has the potential to significantly enhance their efficacy and minimise the amount of chemicals required. Nanopesticides exhibit superior efficacy beyond traditional pesticides in the management of pests. Furthermore, its application results in a 50% reduction in cost when compared to traditional pesticides.

Nanopesticides are generally environmentally friendly, and most of their formulations are designed to selectively target specific areas and release them in a controlled manner. Nanopesticides provide distinct characteristics that optimise the use of pesticides and significantly reduce residual levels and environmental contamination. Highly polymeric nanomicrocapsule formulations exhibit delayed release and

Table 4.2 Application and effects of various nanoparticles

Nanoparticles	Size of nanoparticles in (NM)	Effect of nanoparticle	Application of nanoparticle	Reference
Au (Gold)	6–50	Quick identification of organophosphrous pesticide	Enhancing nutrient uptake, promoting root growth, etc.	Zhai et al. (2022)
Ag-NP (silver)	8–70	The method used for detecting chlorpyrifos (CPS) pesticides	Preserves the quality of horticultural goods and lowers post-harvest losses	Sahu et al. (2022)
MXene-based NM	60–100	They are useful in pesticide detection	The environmental impact of pesticides is minimised and effective pest control is ensured	Siwal et al. (2022)
Zn-NP (Zinc)	30–150	The elimination of undesirable microorganisms	Advantageous in the management of plant pathogens, including bacteria and fungi, due to their antimicrobial properties	Zhao et al. (2022)
Fe3+ (Ferric)	25	Reduces the need for agricultural pesticides	Employed as carriers to facilitate the controlled and efficient delivery of agrochemicals or nutrients to plants	Lin et al. (2021)
C (carbon)	10–45	Widespread pesticide use has a detrimental effect on the ecosystem and aquatic life	Identify nutrient deficiencies, environmental stressors, or pathogens at an early stage	Rashed et al. (2022)

protective properties due to their formulation using light-sensitive, humidity-sensitive, temperature-sensitive, enzyme-sensitive, and soil pH-sensitive components. Novel nanopesticide formulations enhance the adherence of droplets to plant surfaces, therefore enhancing the dispersion as well as bioactivity of the active components in pesticide formulations. Hence, nanopesticides have superior effectiveness in comparison to traditional pesticide formulations. Nanopesticides have greater effectiveness due to their compact dimensions, ability to be wetted, enhanced ductility of pesticide droplets, and adsorption towards the intended target. By employing nanoencapsulation, insecticidal properties can also be enhanced. This technique involves the encapsulation of the nanosized active pesticide component by a thin protective layer. This method significantly enhances efficiency and minimises the chemical usage and associated environmental contamination. "Halloysite" clay nanotubes serve as a pesticide transporter, significantly reducing the quantity of insecticides needed to control target pests (Kumar et al., 2019).

4.4.1 Nanoinsecticides

Nanoencapsulation facilitates the incremental release of a chemical compound to the targeted host for pest insect management via several processes including biodegradation, dissolution, osmotic pressure, and diffusion at a precise pH level (Vidhyalakshmi et al., 2009). Nanoparticles containing garlic oil efficaciously manage *Tribolium castaneum* disease (Barik et al., 2008). Nano-encapsulated insecticides are often designed to specifically target a certain insect, therefore reducing the required product amount in comparison to traditional pesticides. Significant mortality of *Rhyzopertha dominica* along with *Sarocladium oryzae* insects has been observed in wheat following a 3-day treatment with nanostructured alumina (Stadler et al., 2010). The pesticide ethiprole has been enclosed within nanospheres composed of poly (lactic) acid and polycaprolactone. The analysis of nanoformulation indicates that nanospheres do not provide a controlled release (CR) of the active components of agrochemicals. Instead, due to the small size of nanospheres, they enhance the penetration of the active ingredient into the plant cell in comparison to the suspension without nanoformulation (Boehm et al., 2003). A study done on Egyptian cotton leaf-worm larvae (*Spodoptera littoralis*) had shown that the nanoparticles of novaluron achieved the same level of toxicity as the conventional commercial formulation. An investigation was conducted to assess the efficacy of nanoformulations, including a commercially available one, in managing dominant soybean pests such as stem fly (*Melanagromyza sojae*) and whitefly (*Bemisia tabaci*). Most imidacloprid CR formulations exhibit superior activity in controlling specific insect pests compared to non-CR industrial formulations. Furthermore, the use of some CR formulations resulted in greater agricultural productivity in comparison to the non-CR pesticide formulations and the untreated control environment. The imidacloprid and carbofuran CR formulations demonstrated enhanced control of the leafhopper *Amrasca biguttula* Ishida and the aphid *Aphis gossypii* on potato crops, in comparison to traditional non-CR pesticide formulations.

Efficient management of fruit fly populations using pheromones is environmentally friendly. Nano-pheromones provide a straightforward and efficient sampling technique for capturing insect pests in guava and mango orchards. A nanogel composed of methyl eugenol and pheromone, integrated with a low-molecular-weight organic gelator such as all-trans tri (p-phenylene vinylene) bis-aldoxime, exhibits enhanced stability in ambient conditions, reduces evaporation, and ensures the optimal release of the active component of the pheromone. Furthermore, this nanopheromone offers a seamless sampling method for efficiently capturing pests in guava and mango crops. Crucially, the employment of pheromone nano-gel-based technology leads to effective management of the well-known fruit insect pest, *Bactrocera dorsalis*, which badly affects guava and mango crops (Bhagat et al., 2013).

4.4.2 Nanofungicides

Various chemically varied fungicides, including dicarboximides, cupric salt, dithiocarbamates, dinitrophenol, triazoles, thiabendazole, organotin compounds, as well as thiocarbamates, are used to control fungal illnesses. The target specificity (TS) of most of these fungicides is inadequate. The implementation of nanotechnology can readily guarantee the thermal stability of these chemical fungicides. The incorporation of nanofungicides is contingent upon their formulation. In a recent work conducted by Kumar et al. (2016), a nanoform of a commonly used fungicide was synthesised. This nanoform consists of 25% trifloxystrobin and 50% tebuconazole and serves as an efficient control agent against the broad-host-range fungus, *Macrophomina phaseolina*. The developed nanoform exhibited a prolonged reaction with enhanced inhibitory effects, including changes in hyphal shape, lysis of hyphal cells, and development of aberrant sclerotia in *M. phaseolina*. Antonoglou et al. (2018) developed a cost-effective Cu-Zn bimetallic nanoparticle for the selective regulation of *Saccharomyces cerevisiae*. The metallic nanofungicide has a limited impact on the photosystem II of tomato leaves but exhibits higher efficacy in inhibiting the growth of *S. cerevisiae*.

4.4.3 Nanonematicides

Management of nematode pests using chemical nematicides is a formidable challenge in agricultural output. Within agriculture, the often employed nematicides include aldicarb, fenamiphos, fosthiazate, and oxamyl compounds (Ntalli & Caboni, 2012). The nematicides in question pose a significant threat to both human health and the environment due to their fast release of hazardous active components. Innovative nanotechnological methods appear to be successful in tackling these problems. An exemplary demonstration of this method is the implementation of silver nanoparticles to regulate *Maloidogyne incognita* and *M. graminis* (Cromwell et al., 2014). The efficacy of green silver nanoparticles in nematicide for suppressing *M. javanica* in eggplant has been shown to be superior to that of conventional chemical nematicides.

4.5 Enhancement of Shelf Life of Horticultural Crops by Nanomaterials

Unfortunately, the perishable nature of most fruits and vegetables poses a threat to their acceptable shelf life spans when stored under typical settings. While there are various conventional preservation methods, they are all costly and either marginally effective in prolonging the shelf life or restricted by an unwanted residue. The regulatory properties of nanoparticles make nanotechnology-based exploitation of shelf life extension strategies capable of mitigating the limitations of traditional approaches.

4.5.1 Nanofilms/Coatings

Nanofilms serve as carriers of metal nanoparticles (NPs) as antibacterial agents, implemented as protective barriers. These innovative materials reduce the rate of respiration, control the breakdown and change in colour, maintain a desirable storage quality, and extend the shelf life of horticulture crops. Edible nanofilms are suggested as a feasible method for compensating for the properties of horticulture goods during storage and shelf life. Dubey et al. (2019) recently engineered a nanocomposite edible film consisting of glycerol, aloe vera gel, and ZnO-NPs solutions. Zinc oxide nanoparticles (ZnO-NPs) solutions have been shown to improve the shelf life and quality of mango fruits for up to 9 days at room temperature. This is achieved by reducing mass loss, suppressing soluble solids, increasing titratable acidity as well as ascorbic acid concentration in the fruit, regulating pH, increasing thickness, and enhancing transmittance.

4.5.2 Nanopackaging

Overall, the components used in wrapping and processing play a significant influence in improving the longevity and quality of food items (Grumezescu, 2016). Given the lack of adequate expertise and technology in packing, a substantial quantity of food is wasted annually worldwide. Among the cutting-edge food processing and packaging technologies, nano-based technologies have already been shown to be the most efficient. Nano packaging and processing techniques involve the use of nanomaterials to enhance and maintain the exceptional quality and longevity of food. Metal nanoparticles such as Ag, Cu, TiO2, and MgO (Islam et al., 2018), edible antibacterial nanocomposite films, and gas scavengers are examples of active food nano-packaging agents. For instance, the use of Ag-NPs nanoparticles improved the longevity of fruits and vegetables by intercepting ethylene gas. The rapid deterioration of fruits and vegetables is attributed to oxygen. Therefore, the use of gaseous scavenger-NPs in packaging and processing materials can effectively inhibit oxidative processes, thus improving the quality and storage longevity of fruits and vegetables. The study revealed that nanocrystalline-TiO2 functions as an oxygen scavenger in the context of food packaging and processing (Kuswandi, 2017). Research conducted by Kasaai (2015) revealed that including nano-Si as a surface coating in food packaging enhances the quality and storage capacity of horticulture crops. Nanoparticles (Ag-NPs) used for packaging materials can reduce microbial activity, thus extending the shelf life of food products (Sidorowicz et al., 2021). In addition, Ag-NPs, ZnO-NPs, TiO2-NPs, MgO-NPs, CuO, chitosan nanoparticles, carbon nanotubes, quantum dots, and numerous other similar nanoparticles have significant commercial value in enhancing the quality and longevity of stored fruits, vegetables, other perishable foods. This is due to their capacity to exhibit antimicrobial properties.

4.5.3 Nanomaterials for Enhancing the Shelf Life of Fruits

Extending the shelf life of horticulture commodities, particularly fruits and vegetables, is the goal of research on nano-based technologies to increase their stability and enhance their quality. Organic and synthetic coating nanoparticles (NPs) including chitosan, silicon (Si), titanium dioxide (TiO2), along with their derivative composites are employed as coating materials for perishable fruits. Specifically, these nanomaterials enhance the longevity of Chinese bayberry (Wang et al., 2010), loquat (Nasrollahzadeh et al., 2019), and strawberry (Li et al., 2017) by adjusting their mechanical and antibacterial characteristics. The subsequent sub-sections provide descriptions of some promising nanoparticles (NPs) and their feasible impacts on enhancing the shelf life of agricultural goods.

4.5.3.1 Silver (Ag) Nanoparticles

Empirical evidence has extensively shown that several categories of silver nanoparticles (NPs) exert diverse effects on the longevity of fruits. Economically prosperous nations such as China, India, Brazil, and the United States have widely employed silver nanoparticles (Ag NPs) to enhance the longevity of perishable agricultural products. In India, the second-largest global producer of fresh produce, many nanoparticles (NPs) derived from silver nanostructures are employed to enhance the long-term preservation capacity and freshness of fruits and vegetables. For instance, silver nanoparticles are effective in extending the shelf life of lime (*Citrus aurantifolium*) and apple (*Malus domestica*) (Gudadhe et al., 2014). A separate investigation carried out by Chowdappa et al. (2014) observed that chitosan-based Ag-NPs composites significantly reduce post-harvest losses in fresh mangoes and enhance quality due to their anti-fungal characteristics. Kumar et al. (2018) found that metal nanoparticles (Ag-NPs), in combination with biocompatible coating ingredients such as chitosan and glycol, extended the shelf life of red grapes to a maximum of 14 days. The incorporation of silver nanostructures into the coating of organic polymers enhanced the mechanical characteristics of the coated films (Kumar et al., 2018). In addition to India, China is extensively using Ag-NPs to enhance the preservation and quality of vegetables and fruits (Sidorowicz et al., 2021). By virtue of the notable antibacterial properties of Ag-NPs, the longevity of cherry tomatoes can be extended by a maximum of 15 days. In Brazil, the use of biologically produced silver nanoparticles (Ag-NPs) loaded with *Myxobacteria virescens* extract is employed to enhance the shelf life of fresh apples (Bhople et al., 2016). These bioactive silver nanoparticles (Ag-NPs) are incorporated and synthesised directly on the paper used for fruit packing. The synthesis of silver nanoparticles (Ag-NPs) on paper has significant antibacterial properties and extends the shelf life of fresh apples by 2 weeks (Bhople et al., 2016).

4.5.3.2 Hexanal

Hexanal ($C_6H_{12}O_6$) is a highly volatile aromatic molecule that contributes to taste development throughout the process of ripening. The inhibition of enzymatic activity on the fruit surface inhibits the synthesis of ethylene during fruit ripening, thereby decreasing the occurrence of post-harvest illnesses (Seethapathy et al., 2016). Hence, hexanal finds commercial application as pre-harvest sprays or dip treatments to prolong the peak freshness of fruits by a duration of 2–3 weeks, while maintaining their quality. Empirical evidence has demonstrated that applying a 0.02% hexanal solution for 10 min (2 L of solution kg/fruit) and storing mangoes in dry sheds reduced the rate of ethylene evolution, physiological weight loss, and soluble solids. Additionally, it increased the chlorophyll content and delayed the ripening process in mangoes by a minimum of 2 days under ambient conditions (25 ± 0.8 °C and 60 ± 10% relative humidity) (Jincy et al., 2017).

4.6 Nanosensors in Precision Horticulture

A nanosensor is a device that accurately transmits data and evidence regarding the behaviour and properties of nanoparticles (NPs) at the nanoscale level to the macroscopic level. Nanosensors are essential for accurately monitoring the real-time progress of field crops, including their growth and the occurrence of pests and diseases. Nanoelectromechanical materials suitable for sensor fabrication include metal nanotubes, nanowires, nanofibers, nanocomposites, nanorods, nanostructured polymers, as well as various carbon allotropes such as carbon nanotubes, graphene, and fullerenes (Márquez & Morant, 2015). Continuous monitoring in real-time can effectively reduce the excessive use of pesticides and fertilizers in crop production, therefore contributing to the mitigation of environmental pollution and the reduction of production costs. The utilisation of nanosensors transforms traditional agriculture into smart agriculture, thus enhancing energy efficiency and promoting sustainable farming practices. Smart agricultural practices in horticultural crop production include the use of nanoformulation-based fertilizers or pesticide delivery systems to enhance the deposition and wetting of nutrients (Hoque et al., 2021). Additionally, nanodetectors are employed to detect pesticide or fertilizer residues. Furthermore, remote-sensing-based monitoring systems are used to track disease incidence and crop growth. Within the field of horticulture, nanosensors find application in the identification of soil moisture content, pesticide residues, nutrient needs, and crop pest detection.

4.7 Concluding Remarks and Prospect of Nanotechnology

Nanotechnology is progressing as a cutting-edge technology in contemporary agriculture to enable sustainable crop cultivation. Furthermore, it has significant potential in the field of horticulture, where distinct types of nanomaterials are employed to enhance the output and quality of agricultural products, as well as minimise the deterioration of fruits and vegetables after harvest. Nanotechnology harnesses the unique properties of nanoparticles and their distribution techniques to enhance the yield of horticulture crops. Their role is to mitigate the excessive use of chemical fertilizers and pesticides. Nanomaterials are characterised by their simplicity, cost-effectiveness, and eco-friendliness, enabling their production in a minimal timeframe and with less effort, without producing any detrimental effects on the environment. Nanotechnology enhances the quality and prolongs the shelf life of both unprocessed and processed horticultural agricultural fruits, vegetables, and cut flowers. Nanosensors facilitate the monitoring of soil moisture, detection of pesticide residue, determination of nutritional needs, and diagnosis of agricultural pests. Conversely, the inappropriate use of nanomaterials will yield detrimental effects on both crop plants and the environment. Therefore, it is necessary to provide research-based and appropriate use of nanomaterials in order to decrease post-harvest spoiling of horticulture crops and enhance the quality of food output. Finally, it has been shown that the extensive use of nanotechnology will greatly enhance growth, boost yields, and decrease production costs and post-harvest losses by preserving the exceptional quality and storage longevity of fresh and processed vegetables and fruits. An essential requirement for guaranteeing food and nutritional security for the growing global population in a changing climate scenario is the availability of practical nanoparticles and thorough safety evaluations of their practical use. The wider implementation of nanotechnology will result in a horticulture system that is sensitive to climatic change, minimises losses after harvesting, and enhances the overall quality of the crops.

References

Antonoglou, O., Moustaka, J., Adamakis, I.-D. S., Sperdouli, I., Pantazaki, A. A., Moustakas, M., & Dendrinou-Samara, C. (2018). Nanobrass CuZn nanoparticles as foliar spray nonphytotoxic fungicides. *ACS Applied Materials & Interfaces, 10*(5), 4450–4461.

Barik, T., Sahu, B., & Swain, V. (2008). Nanosilica—from medicine to pest control. *Parasitology Research, 103*, 253–258.

Bhagat, D., Samanta, S. K., & Bhattacharya, S. (2013). Efficient management of fruit pests by pheromone nanogels. *Scientific Reports, 3*(1), 1–8.

Bhople, S., Gaikwad, S., Deshmukh, S., Bonde, S., Gade, A., Sen, S., Brezinska, A., Dahm, H., & Rai, M. (2016). Myxobacteria-mediated synthesis of silver nanoparticles and their impregnation in wrapping paper used for enhancing shelf life of apples. *IET Nanobiotechnology, 10*(6), 389–394.

Boehm, A., Martinon, I., Zerrouk, R., Rump, E., & Fessi, H. (2003). Nanoprecipitation technique for the encapsulation of agrochemical active ingredients. *Journal of Microencapsulation, 20*(4), 433–441.

Broadley, M. R., White, P. J., Hammond, J. P., Zelko, I., & Lux, A. (2007). Zinc in plants. *New Phytologist, 173*(4), 677–702.

Chowdappa, P., Gowda, S., Chethana, C., & Madhura, S. (2014). Antifungal activity of chitosan-silver nanoparticle composite against Colletotrichum gloeosporioides associated with mango anthracnose. *African Journal of Microbiology Research, 8*(17), 1803–1812.

Cromwell, W., Yang, J., Starr, J., & Jo, Y.-K. (2014). Nematicidal effects of silver nanoparticles on root-knot nematode in bermudagrass. *Journal of Nematology, 46*(3), 261.

Davarpanah, S., Tehranifar, A., Davarynejad, G., Abadía, J., & Khorasani, R. (2016). Effects of foliar applications of zinc and boron nano-fertilizers on pomegranate (Punica granatum cv. Ardestani) fruit yield and quality. *Scientia Horticulturae, 210*, 57–64.

Dubey, P. K., Shukla, R. N., Srivastava, G., Mishra, A. A., & Pandey, A. (2019). Study on quality parameters and storage stability of mango coated with developed nanocomposite edible film. *International Journal of Current Microbiology and Applied Sciences, 8*(4), 2899–2935.

Grumezescu, A. M. (2016). *Nanotechnology in the agri-food industry volume 3: Emulsions*. Nikki Levy.

Gudadhe, J. A., Yadav, A., Gade, A., Marcato, P. D., Durán, N., & Rai, M. (2014). Preparation of an agar-silver nanoparticles (A-AgNp) film for increasing the shelf-life of fruits. *IET Nanobiotechnology, 8*(4), 190–195.

Hoque, M. I. U., Chowdhury, A.-N., Islam, M. T., Firoz, S. H., Luba, U., Alowasheeir, A., Rahman, M. M., Rehman, A. U., Ahmad, S. H. A., & Holze, R. (2021). Fabrication of highly and poorly oxidized silver oxide/silver/tin (IV) oxide nanocomposites and their comparative anti-pathogenic properties towards hazardous food pathogens. *Journal of Hazardous Materials, 408*, 124896.

Islam, M. S., Akter, N., Rahman, M. M., Shi, C., Islam, M. T., Zeng, H., & Azam, M. S. (2018). Mussel-inspired immobilization of silver nanoparticles toward antimicrobial cellulose paper. *ACS Sustainable Chemistry & Engineering, 6*(7), 9178–9188.

Jincy, M., Djanaguiraman, M., Jeyakumar, P., Subramanian, K., Jayasankar, S., & Paliyath, G. (2017). Inhibition of phospholipase D enzyme activity through hexanal leads to delayed mango (Mangifera indica L.) fruit ripening through changes in oxidants and antioxidant enzymes activity. *Scientia Horticulturae, 218*, 316–325.

Kasaai, M. R. (2015). Nanosized particles of silica and its derivatives for applications in various branches of food and nutrition sectors. *Journal of Nanotechnology, 2015*(1), 852394.

Kumar, G. D., Natarajan, N., & Nakkeeran, S. (2016). Antifungal activity of nanofungicide trifloxystrobin 25%+ Tebuconazole 50% against Macrophomina phaseolina. *African Journal of Microbiology Research, 10*(4), 100–105.

Kumar, S., Shukla, A., Baul, P. P., Mitra, A., & Halder, D. (2018). Biodegradable hybrid nanocomposites of chitosan/gelatin and silver nanoparticles for active food packaging applications. *Food Packaging and Shelf Life, 16*, 178–184.

Kumar, D., Malik, N., & Sengar, R. S. (2019). Emerging technologies to enrich agricultural and horticultural crop quality and production. *Annals of Horticulture, 12*(1), 55–61.

Kuswandi, B. (2017). Environmental friendly food nano-packaging. *Environmental Chemistry Letters, 15*(2), 205–221.

Li, D., Ye, Q., Jiang, L., & Luo, Z. (2017). Effects of nano-TiO2-LDPE packaging on postharvest quality and antioxidant capacity of strawberry (Fragaria ananassa Duch.) stored at refrigeration temperature. *Journal of the Science of Food and Agriculture, 97*(4), 1116–1123.

Lin, X., Yu, Q., Yang, W., He, C., Zhou, Y., Duan, N., & Wu, S. (2021). Double-enzymes-mediated fluorescent assay for sensitive determination of organophosphorus pesticides based on the quenching of upconversion nanoparticles by Fe3+. *Food Chemistry, 345*, 128809.

Márquez, F., & Morant, C. (2015). Nanomaterials for sensor applications. *Soft Nanoscience Letters, 5*(01), 1.

Monreal, C., DeRosa, M., Mallubhotla, S., Bindraban, P., & Dimkpa, C. (2016). Nanotechnologies for increasing the crop use efficiency of fertilizer-micronutrients. *Biology and Fertility of Soils, 52*, 423–437.

Nasrollahzadeh, M., Issaabadi, Z., Sajjadi, M., Sajadi, S. M., & Atarod, M. (2019). Types of nanostructures. *Interface Science and Technology, 28*, 29–80.

Ntalli, N. G., & Caboni, P. (2012). Botanical nematicides: a review. *Journal of Agricultural and Food Chemistry, 60*(40), 9929–9940.

Rana, R. A., Siddiqui, M. N., Skalicky, M., Brestic, M., Hossain, A., Kayesh, E., Popov, M., Hejnak, V., Gupta, D. R., & Mahmud, N. U. (2021). Prospects of nanotechnology in improving the productivity and quality of horticultural crops. *Horticulturae, 7*(10), 332.

Rashed, M. A., Faisal, M., Alsareii, S., Alsaiari, M., Jalalah, M., & Harraz, F. A. (2022). Highly sensitive and selective electrochemical sensor for detecting imidacloprid pesticide using novel silver nanoparticles/mesoporous carbon/hematite ore ternary nanocomposite. *Journal of Environmental Chemical Engineering, 10*(5), 108364.

Sahu, B., Kurrey, R., Khalkho, B. R., & Deb, M. K. (2022). α-Cyclodextrin functionalized silver nanoparticles as colorimetric sensor for micro extraction and trace level detection of chlorpyrifos pesticide in fruits and vegetables. *Colloids and Surfaces A: Physicochemical and Engineering Aspects, 654*, 129947.

Seethapathy, P., Gurudevan, T., Subramanian, K. S., & Kuppusamy, P. (2016). Bacterial antagonists and hexanal-induced systemic resistance of mango fruits against Lasiodiplodia theobromae causing stem-end rot. *Journal of Plant Interactions, 11*(1), 158–166.

Servin, A. D., & White, J. C. (2016). Nanotechnology in agriculture: next steps for understanding engineered nanoparticle exposure and risk. *Nano Impact, 1*, 9–12.

Sharon, M., Choudhary, A. K., & Kumar, R. (2010). Nanotechnology in agricultural diseases and food safety. *Journal of Phytology, 2*(4).

Sidorowicz, A., Maqbool, Q., & Nazar, M. (2019). *Future of nanofertilizer* (pp. 143–152). Crop Production & Protection.

Sidorowicz, A., Szymański, T., & Rybka, J. D. (2021). Photodegradation of biohazardous dye brilliant blue R using organometallic silver nanoparticles synthesized through a green chemistry method. *Biology, 10*(8), 784.

Siwal, S. S., Sheoran, K., Mishra, K., Kaur, H., Saini, A. K., Saini, V., Vo, D.-V. N., Nezhad, H. Y., & Thakur, V. K. (2022). Novel synthesis methods and applications of MXene-based nanomaterials (MBNs) for hazardous pollutants degradation: Future perspectives. *Chemosphere, 293*, 133542.

Stadler, T., Buteler, M., & Weaver, D. K. (2010). Novel use of nanostructured alumina as an insecticide. *Pest Management Science: Formerly Pesticide Science, 66*(6), 577–579.

Vidhyalakshmi, R., Bhakyaraj, R., & Subhasree, R. (2009). Encapsulation "the future of probiotics"—A review. *Advances in Biology Research, 3*(3–4), 96–103.

Wang, K., Jin, P., Shang, H., Li, H., Xu, F., Hu, Q., & Zheng, Y. (2010). A combination of hot air treatment and nano-packing reduces fruit decay and maintains quality in postharvest Chinese bayberries. *Journal of the Science of Food and Agriculture, 90*(14), 2427–2432.

Wang, P., Lombi, E., Zhao, F.-J., & Kopittke, P. M. (2016). Nanotechnology: a new opportunity in plant sciences. *Trends in Plant Science, 21*(8), 699–712.

Zhai, R., Chen, G., Liu, G., Huang, X., Xu, X., Li, L., Zhang, Y., Wang, J., Jin, M., & Xu, D. (2022). Enzyme inhibition methods based on Au nanomaterials for rapid detection of organophosphorus pesticides in agricultural and environmental samples: A review. *Journal of Advanced Research, 37*, 61–74.

Zhao, W., Liu, Y., Zhang, P., Zhou, P., Wu, Z., Lou, B., Jiang, Y., Shakoor, N., Li, M., & Li, Y. (2022). Engineered Zn-based nano-pesticides as an opportunity for treatment of phytopathogens in agriculture. *NanoImpact, 28*, 100420.

Chapter 5
Nanomaterials Based Agrochemicals Delivery: An Advanced Approach for Sustainable Agriculture

Rashmi C. Kulkarni, Chitra Katti, Raviraj M. Kulkarni, and Yunus Shukor

Abstract The agrochemical sector has increasingly become interested in nanomaterials (NMs) due to their unique qualities, including their small particle size at the nanoscale level, solubility, surface structure, and chemical composition aspects. Nanotechnology in agriculture offers advantages over traditional agrochemical products by effectively addressing inherent limitations such as low bioavailability, risks of organic solvent contamination, and rapid degradation through photolysis processes. Nanoparticles have been extensively studied for their applications in agriculture, mainly as pesticides and fertilizer delivery systems. In this chapter, we will focus on the development of these nanomaterials over time. Furthermore, this chapter explores the emerging functions of nanomaterials (NMs) in different domains of agrochemicals, including nucleic acid-derived pesticides, biopesticides, plant growth regulators (PGRs), and pheromones. In addition, we examine the challenges associated with the proliferation of nanomaterials in the agricultural chemical industry and their increasing industrialization. The utilization of nanotechnology in the development of agricultural chemical distribution systems has a significant potential to enhance agricultural production and sustainability. Improving the stability and dispersion of components in these systems enables accurate application of agricultural pesticides, leading to lowered pollution levels, lower residue, and decreased labor costs in farming techniques. This approach increases farm output and improves food security by optimizing the effectiveness of agricultural pesticides and ensuring the long-term sustainability of farming practices.

R. C. Kulkarni · C. Katti · R. M. Kulkarni (✉)
Department of Chemistry, Centre for Nanoscience and Nanotechnology, KLS Gogte Institute of Technology, Visvesvaraya Technological University, Belagavi, Karnataka, India
e-mail: ravirajmk@git.edu

Y. Shukor
Department of Biochemistry, Faculty of Biotechnology and Biomolecular Sciences, Universiti Putra Malaysia (UPM), Serdang, Selangor, Malaysia

V. D. Rajput et al. (eds.), *Contemporary Suitability of Nanobionics in Agriculture*, Nanotechnology in the Life Sciences, https://doi.org/10.1007/978-3-031-82779-2_5

Keywords Nanomaterials · Nanotechnology · Agrochemicals · Nanoformulations · Sustainable agriculture

5.1 Introduction

Agriculture is the most significant sector, which provides essential ingredients for the feed and food segments (Gupta et al., 2022a, b). The growing world population and depletion of natural resources require efficient, profitable, and environmentally sustainable agricultural expansion. This transformation will significantly impact the achievement of numerous objectives in the upcoming year (Mukhopadhyay, 2014). Safeguarding food security for the worldwide population is a significant challenge that requires significant attention and action. By 2050, the world's dietary intake is estimated to rise from 59% to 98% (Duro et al., 2020). The use of nano-modified stimulants in farming can expedite the advancement of sustainable agriculture. Nanotechnology research has the potential to boost soil quality and food security by improving various facets of food security (Ashraf et al., 2021). Since agriculture is the cornerstone of sustainable development, fostering its expansion is essential to eradicating poverty and malnutrition. Therefore, we should significantly shift our focus toward agricultural production.

Through precision farming, nanotechnology is making a big splash in the agricultural industry. It can be strategically utilized to align with the broader framework of sustainable agriculture, especially by introducing novel, environmentally advantageous, and resourceful technologies (Duhan et al., 2017; Fraceto et al., 2016; Raliya et al., 2017). The implementation of nanotechnology in the agricultural economy has become increasingly viable with the advent of nanotechnology in medications and pharmaceuticals (Robles-García et al., 2016; Vega-Vásquez et al., 2020). The production of novel products and technologies and a rise in applications in the agriculture industry have been made possible by integrating the field of nanotechnology with many scientific domains (Fig. 5.1). The existing agricultural system may be drastically improved by nanotechnology through the preservation of crop inputs, livestock, animal husbandry, the aquaculture sector, and fisheries. Using nanoscale products to maximize crop production (yields) while minimizing agricultural inputs (such as fertilizers, herbicides, and agrochemicals) with continuous monitoring systems is the fundamental objective of agri-nanotechnology (Pramanik et al., 2020; Scott & Chen, 2012). Nanotechnology offers the ability to enhance crop yield and quality at every stage of agriculture, ranging from seed storage to seed priming, fertigation, the germination process, and post-harvest activities. By managing water quality and controlling nutrients, nanotechnology contributes significantly to increased production in agriculture and facilitates sustainable development. Nano agrochemical technology (NT) describes products and formulas designed and manufactured for agricultural purposes. These specialized

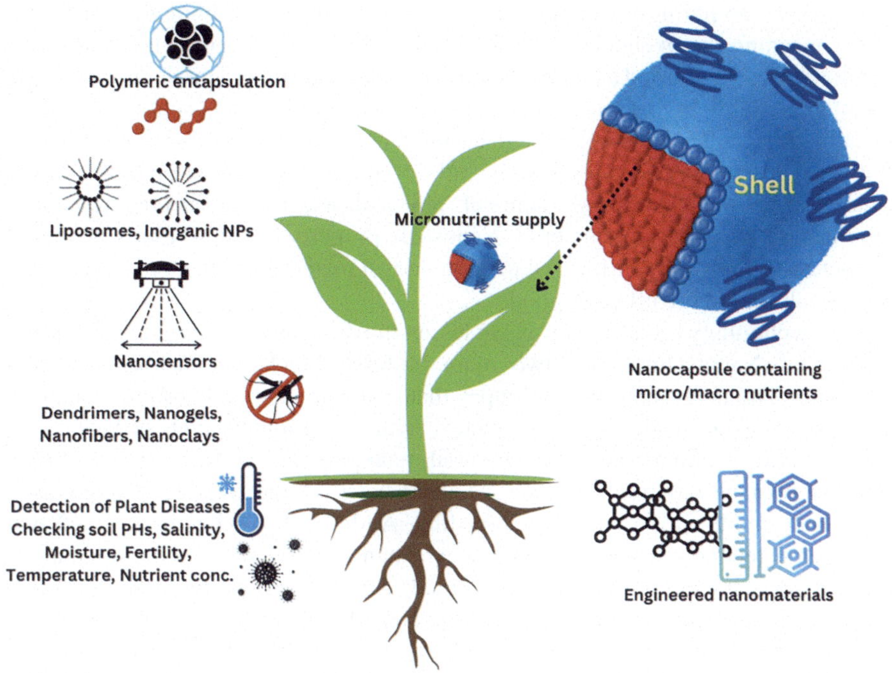

Fig. 5.1 Role of nanomaterials as agrochemicals

agricultural concoctions called nanopesticides (NP) are distinguished by their elongated shape and consist of elements with sizes ranging up to 100 nanometers. Promoting the use of environmentally friendly nanopesticides that offer precise targeting abilities and improved control over their application could revolutionize the agriculture industry (Forini et al., 2022). Recent studies have demonstrated a rise in efficiency levels by 31.5%, surpassing the conventional standards set for traditional pesticides (D. Wang et al., 2022). Therefore, persistent chemicals (NPC) address the shortcomings of pesticides by enhancing their ability to be taken in by living organisms while reducing susceptibility to degradation from light exposure and diminishing the adverse impact of organic solvent contamination on the environment. The illustration in Fig. 5.1 showcases the advantages and key characteristics linked to using NPCs, which represents NPC's strengths (Kannan et al., 2023).

5.2 Fundamentals of Nanotechnology

Nanotechnology is defined as "directing morphological and the modification of size at nanometer level ranging from 1–100 nm through the synthesizing, designing, characterizing, along with the use of assemblies, tools, as well as systems" (Yadollahi et al., 2009). To put things in perspective, a nanometer is one billionth (10^{-9}) of a meter, meaning that technology can be used at this scale. Nanoscience and nanotechnologies are regarded as methods in research and development fields today. It is essential to highlight that when examining the characteristics and actions of materials at the molecular levels up to the macromolecular level, it is crucial to recognize that their properties may vary significantly compared to what is seen at a larger scale. The resultant products' chemical, physical, and biological properties vary significantly from the bulk material. Nanoscale property research results in altered characteristics that may be utilized to create novel materials with altered structures, better tools, and more efficient goods.

Nanotechnology has offered ways to lessen the drawbacks of current agricultural techniques. Nanotechnology's two main building blocks are nanodevices and nanotools. The many nanomaterial types that make up nanotools include nanoparticles made of metal and oxides of metals with exceptionally high surface areas, carbon-based nanomaterials, and mineral nanoparticles (Lowry et al., 2019). Conversely, nano-based technological devices have enhanced the technology used for genetic modification and plant breeding. Additionally, it has been noted that agricultural techniques based on engineered nanomaterials (ENMs) can adapt to changing climatic circumstances (Mittal et al., 2020). Nutrient distribution is more targeted in the events where ENMs can be coated with antibodies or aptamers as molecular recognition components. Reducing the adverse effects on the environment enhances agricultural yields and promotes sustainable utilization of nutrients. By improving the targeted administration of agrochemicals and their effective use and oversight in the fields, these ENMs have solved the problems associated with conventional and contemporary agricultural techniques. By creating nanoscaled agrochemicals, ENMs have demonstrated encouraging sustainable uses in agricultural operations (Gilbertson et al., 2020). ENMs can intensify agricultural sustainability by (i) refining soil texture and lowering the usage of nitrogen fertilizers; (ii) applying Cu and Zn nanoparticle surface coating to seedlings has increased the rate of seed germination compared with conventional methods using high-embodied energy polyethylene glycol (PEG), potassium nitrate, and potassium chloride; (iii) effective, targeted agrichemical absorption and foliar spraying of ENMs, which include nano ZnO, nano Zn, and nano TiO2, have also increased crop production. ENM-based foliar treatment can improve crop productivity by boosting the chlorophyll amount, refining light absorption efficacy, scavenging the reactive oxygen species, fostering an efficient electron transport system, and preventing the impact of abiotic stresses on the photosynthesis rate. Research has shown that high photosynthetic rates can occasionally reduce nutrients in food. Consequently, spraying ENMs containing micronutrients like iron, zinc, and silica helps keep up the high nutrient content of food. Furthermore, altering the surface chemistry of ENMs makes them more adhereable to leaves, which boosts the mesophyll and cuticle cells' capacity to absorb nutrients. Moreover, these ENMs have enhanced sensing tools that gauge crop health with a significant spatial and temporal resolution (Giraldo et al., 2019).

5.2.1 Advantages of Nanotechnology in Agrochemical Delivery

(a) *Enhanced Solubility and Stability:* Under the protection of NMs, active ingredients (AIs) that degrade rapidly due to photolysis can be stabilized. Despite its widespread use, the effective bio-pesticide avermectin (AV) is photosensitive and has a short half-life of only 6 hours (Pang et al., 2020). Zhu et al. found that grafting AV onto a long PEG-CMC chain significantly improved the grafted AV's stability compared to the control. Furthermore, after 72 hours of illumination, the rate of AV degradation was less than 50% (Zhu et al., 2020). Using NMs as an AI's outer layer of protection is a further strategy. After enclosing AV in hollow mesoporous silica, Kaziem et al. (2018) attached cyclodextrin to the particles' outermost layer. These nanoparticles effectively protected the AV within, as only about 9% of it was photolyzed after 24 hours of exposure to UV light. Certain unstable pesticides, such as quinalphos, fosthiazate, and phoxim, would probably also be preserved by utilizing NMs, much like AV. AIs can also be shielded by a shell of other NMs, such as lignin and carbon-based materials (Bin Liu et al., 2021; Zhou et al., 2018).

(b) *Targeted Delivery and Controlled Release:* AIs can be released selectively and under control using NMs at the ideal working concentration. This approach lessens the possibility of resistance brought on by AI concentrations that are extremely high or too low, along with the loss of AI resources and accidental damage to unintended third parties (Wu et al., 2019). These substances are often nanoscale polymers produced by specific processes, which include urea–formaldehyde resin (Heidary et al., 2021), polyurethane (F. Zheng et al., 2020), and polyurea (Luo et al., 2020). By using in situ deposition, Xiang et al. created a magnetic nanocarrier made of Fe_3O_4 and diatomite. This formulation included wrapping both the herbicide (glyphosate) and an insecticide (cypermethrin), followed by a coating of chitosan that could be dissolved to regulate the pesticide's release in acidic environments. Fe_3O_4's magnetic qualities aid in removing nanocarriers from water and soil, making material recycling easier (Xiang et al., 2017).

(c) *NMs' Small Size and High Specific Surface Area:* As carriers, NMs make AIs more soluble while shielding them from breakdown and volatilization (Campos et al., 2019). Moreover, the use of pesticides is influenced by the size of their particles (L. Zheng et al., 2018). Because of their large specific surface area and small dimensions, NMs have the potential to bind more pesticide-specific AIs to the target per unit area than larger particles. The area where the AIs interact with the target is widened, and their bioavailability is increased (X. Zhao et al., 2017). By coating the Lambda-Cyhalothrin with polylactic acid (PLA), Liu et al. could construct a range of drug delivery systems with a particle size that could be adjusted between 0.68 μm and 4.6 μm. The researchers noted that good water dispersion and evident slow-release properties benefited their microencapsulated drug delivery systems. The 0.68 μm microcapsules, meanwhile, showed good biological activity and thermal and UV stability (Baoxia Liu et al., 2016). Similar to this, Wang et al. investigated the creation of various AV nano-delivery system particle sizes using PLA and assessed the systems' performance. The nano-delivery system's AV release rate could be fine-tuned, thanks to particle size variations, and the findings indicated that reducing particle size increased biological activity (Wang et al., 2018).

(d) *Reduction in Agrochemical Usage and Environmental Impact:* By utilizing the material's isolating action on AIs and organisms, which helps to expand the pesticide's applicability, NMs could lower the harmful effects of pesticides on nontargets. Pyraclostrobin has demonstrated superior efficacy in managing rice blast disease. This drug causes physiological modifications in crops, enhancing yields by strengthening the crops' resistance to the disease (Kabir et al., 2021). However, the application of this molecule to rice crops is limited since it is very hazardous to aquatic species (H. Li et al., 2021). Utilizing an interfacial polymerization reaction involving amines and isocyanates, Li et al. created pyraclostrobin-loaded polyurea nano microcapsules in this specific case. The particle sizes were almost spherical, with an 815 and 867 nm dispersion. Coating pyraclostrobin will lessen the harmful effects of pesticides on aquatic species, making it possible to spray pyraclostrobin onto rice while enhancing environmental safety (M. Li et al., 2018). A core-shell structured, magnetically responsive, and controlled-release herbicide nanocomposite was created by Chi et al. with palygorskite, ferroferrie oxide, glyphosate, and amino silicon oil. A magnetic field could speed up the release of glyphosate, and a partial application of amino silicon oil may effectively limit loss by preventing the herbicide from coming into complete contact with the environment. According to a biosafety experiment conducted on zebrafish, this delivery strategy has superior biosafety than glyphosate (Chi et al., 2021).

5.3 Applications of Nanotechnology in Agrochemical Delivery

This section discusses several nano-agrochemicals that have been used in agriculture, including fertilizers, pesticides, sensing materials, and enhanced products, as well as potential uses for advancing sustainability in the environment (Fig. 5.2).

5.3.1 Nanofertilizers

Besides reducing the harm to the environment caused by using fertilizers in agriculture practices, nanofertilizers show promise in enhancing crop nourishment effectiveness and cost efficiency. Nanofertilizers are developed through cutting-edge

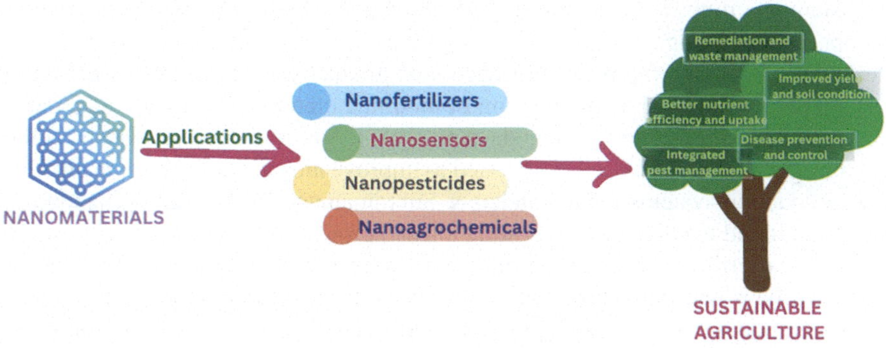

Fig. 5.2 Applications of nano-agrochemicals for sustainable agriculture

technology advancements; they hold the potential to address the escalating food shortage crisis and promote eco-farming methods as well. A viable approach to tackling issues related to eutrophication and water contamination involves adopting nanofertilizers over fertilizer applications since they release nutrients into the soil gradually and in a controlled manner. When TiO2 nanoparticles were used on corn plants, they showed a growth enhancement effect; however, using bulk TiO2 did not seem to have any noticeable effect on the plant's growth rate or health status in any way at all. Several unique characteristics of nanofertilizers involve rapid absorption rates into the plant system and contribute to higher yield and photosynthesis activity in the plants; additionally, these fertilizers also lead to a visible increase in leaf surface area (Council, 2009).

Applying nanofertilizers can improve plant and soil health utilization while also reducing the need for frequent fertilizer use and minimizing the harmful effects of excessive fertilization (Naderi & Abedi, 2012). In scenarios of agricultural advancement, zinc oxide nanoparticles may offer positive results as nanofertilizers for tomato cultivation. In trials using pots, tomato plants treated with a spray containing 20 mg mL^{-1} of zinc oxide nanoparticles showed better growth and increased biomass compared to untreated plants (de la Rosa et al., 2013; Panwar, 2012).

Hopefully, nanotechnology-based fertilizers will surpass options in the future as a multidisciplinary strategy is essential for advancing nanofertilizer innovation. The nutrients in nanofertilizers can be delivered to plants via emulsions or through nanoencapsulation techniques (DeRosa et al., 2010). An option to enhance absorption effectiveness is utilizing slow-release or controlled-release (CR) fertilizers produced using nanotechnology. To enhance crop production in agricultural fields and address the challenge of water retention in soils with low moisture levels, engineered nanoparticles are employed to improve nutrient availability in the root zone or rhizosphere. The controlled release of nutrients from fertilizer capsules can be achieved by coating and binding nanostructured materials on a scale. Furthermore, the growth and productivity of grain crops have been positively impacted by the application of nanocomposites containing elements such as nitrogen, phosphorus, potassium, micronutrients, mannose, and amino acids. Moreover; plant growth regulators are now being administered through the use of zinc aluminum layered double hydroxide nanocomposites. Nitrogen was gradually released over a period at varying pH levels using a combination of modified urea hydroxyapatite nanoparticles and *Gliricidia sepium* (Kottegoda et al., 2011). A new method involving zeolite and nitrogen fertilizer has been suggested to improve nitrogen utilization in agricultural production systems (Manikandan & Subramanian, 2014). Carbon nanotubes have been found to enhance the germination and growth rates of tomato seeds when applied as a fertilizer. The carbon nanotubes were found to penetrate the layer of seeds and help with absorbing water per research (Khodakovskaya et al., 2009). Three techniques exist for packaging fertilizers in nanoparticles: (1) the nutrient can be stored in substances; (2) it can be wrapped with a thin polymer layer; or (3) it can be delivered as nanoscale particles or emulsions. There have been studies discussing the benefits of nanofertilizers in agriculture (El-Ramady, 2014). As shown in Table 5.1, NMs have been used in various fertilizers.

5.3.2 Nanopesticides

Protection against insects is one of the most crucial tactics for raising crop output. Using pesticides to control weeds and pests is a standard practice in agriculture. Conventionally applied pesticides and crop protection methods are ineffective in promoting sustainable development and environmental safety (Davydov et al., 2018; Lamichhane et al., 2016). Pesticide use on crops has increased significantly globally, from 2,285,881 metric tons in 1990 to 4,113,591 tonnes in 2017. This is a 79% increase. Adverse health effects on humans, loss of soil biodiversity, pesticide accumulation in the environment, disease resistance, and targeting beneficial and nonthreatening species are just a few of the many problems caused by the excessive

Table 5.1 Nanomaterials applied to fertilizers

Fertilizer type	Active ingredients	Materials
Bio-fertilizer	PGPR and endophytic bacteria	Graphite and silica composites
Trace element fertilizer	Iron	Iron oxide nanoparticles
	Zinc	Nano Zn
	Selenium	Nano Se
Major element fertilizer	Nitrogen	Nano nitrogen chelate
	Phosphorus and potassium compound	Zeolite
Medium element fertilizer	Calcium	Nano-Ca_2SO_4
Organic fertilizer	Vermi compost	Nano Vermi compost organic matter

use of pesticides in agriculture (Mahmood et al., 2016). Insecticides, weedkillers, fungicides, and bactericides are all types of nano pesticides that are used because of their efficient delivery systems, low cost, wide availability, and ease of application (Baker et al., 2019; Yadav & Srivastava, 2015). The best way to increase the pesticide value of the active ingredient bonded by protective coverings is by nanoencapsulation (Pascoli et al., 2019). Because they are safe, biocompatible, and environmentally friendly, nanocarriers aid in improving the dissolution of active ingredients and provide superior defense against degradation and volatilization. They have incredible qualities like enhanced stability and permeability, regulated and gradual release, and resistance to premature degradation. Pesticide manufacturing expenses and usage can be significantly reduced with the help of pesticide delivery systems developed using nanoscale engineering. Pesticide development and application for environmentally friendly agricultural practices would benefit immensely from this (Owens & Owens, 2010). Because of their insecticidal, bactericidal, and fungicidal qualities, various formulations, including polymer NPs, inorganic NPs (silica, copper, silver, alumina, titanium dioxide), solid–lipid NPs, carbon-based substances (MWCNTs, graphene, etc.), nano gels, and nanoemulsions, have been shown to be effective nanopesticides (Fig. 5.3). By using polymer NPs, metal NPs, as well as nano alumina-silicates exactly as nanocarriers for herbicides and pesticides, it has been possible to increase plant growth (Shah et al., 2019). At present, there is interest in using nanoemulsions/nanodispersions (that consist of an oil phase and are colloidal liquids with a droplet size in the range of 20–200 nm that are optically isotropic and kinetically stable, surfactant, as well as water) as possible vehicles for the targeted delivery of pesticides (Nile et al., 2020). As shown in Table 5.2, NMs have been applied in various pesticides.

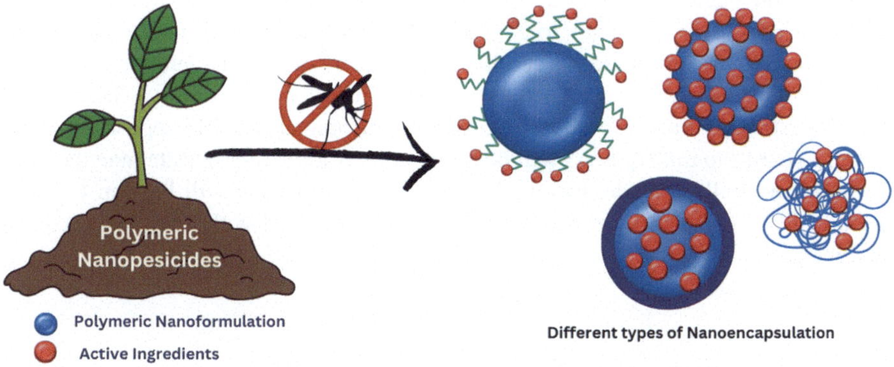

Fig. 5.3 Effective pest management with nanopesticides: polymers, clay, lipids, and metal-organic nanomaterials are nanocarriers for pesticides

Table 5.2 Nanomaterials applied to pesticides

Pesticide type	Active ingredients	Materials
Bacterial	*Bacillus thuringiensis*	Nanotubular sodium titanate
Fungal	*Beauveria bassiana*	Carbon fibers and silica nanoparticles
Plant-based	Essential oils	Nanosilver coated with chitosan
Agricultural antibiotics	Avermectin	PEG-CMC
	Validamycin and thiafluzamide	Polylactic acid
Biochemical	Gibberllic acid	Layered double hydroxide

5.3.3 *Nanoweedicides/Herbicides*

Researchers have developed nanoweedicides and nanoherbicides to combat specific weeds and herbs that invade crop habitats and deplete crop nutrients. Using encapsulated nanocarriers (NMs) for weed management can lessen the risk of phytotoxicity of herbicides among crops (Kumar et al., 2017). The use of nanoactive herbicides can increase yields while decreasing the need for synthetic herbicides. Polymers like poly(epsilon-caprolactone), chitosan, and alginate have shown great effectiveness when encapsulating herbicides, especially atrazine, ametrine, simazine, and paraquat (Abigail & Chidambaram, 2017). A recent study on Indian mustard plants revealed the herbicidal effects of atrazine nanocapsules and nanoformulations when sprayed on leaves (Bombo et al., 2019). They proved successful against hairy beggarticks (*Bidens pilosa* L.) and thin amaranth (*Amaranthus viridis* L.). Soil sorption and degradation of herbicide active ingredients (AI) can be prevented by loading them into nanocomposites for controlled release (Abigail, 2020). Various support materials, including NPs, polymeric microparticles, and clays, have recently been studied for use in herbicide formulations (Tosco et al., 2020). In conclusion, by reducing the need for synthetic chemicals along with the associated hazards, nanopesticides play a crucial role in the efficient and durable control of pests. Understanding nanopesticides' cellular mechanism of action and their harmful effects on target organisms is necessary for their broader applicability. The dangers of nanopesticides must be confirmed before a safe and sustainable approach can be chosen. Thanks to this, farmers will be better able to adopt appropriate management methods and will be made aware of them. To find out what will happen to them ecologically, an extensive evaluation of the risks involved with their use needs to be done.

5.4 Mechanisms of Action

Nanoparticle (NP) interactions with plants have been the subject of much research. Unfortunately, there is a lack of centralized data on NP penetration, translocation, transformation rates, and their effects on crops. Hence, immediate and expressive consideration is required to improve the systematic explanation of NPs' importance in agriculture. This review attempts to fill the empirical void in this area by offering a comprehensive perspective on how NPs affect crops' overall performance. We examine how NPs are absorbed, transported, transformed, and affect agricultural productivity while shielding crops from different types of stress. Further, we provide a synopsis of the relevant experimental contexts and significant findings concerning the translocation, uptake, transformation, and physiological impacts of NPs on plants.

5.5 Nanoparticle Uptake and Translocation in Plants

Even though a number of findings on the absorption and movement of NPs have been reviewed in recent years, it is still unclear how NPs are transported and absorbed in various plants. The three stages of the NPs–plant interaction are as follows: NPs are absorbed and transformed on the surface of a leaf, root, or stem; they are then taken up by the epidermis, cuticle, and stomata; finally, the xylem and phloem carry them and transform them further inside the plant. NPs can be incorporated into plants by a variety of methods, including root application, foliar spraying, trunk feeding, and seed treatment (Wise et al., 2014). Of these, foliar and root exposure is simple, making them good for delivering nanoparticles. Analyzing NPs penetration is crucial to understanding the NPs–plant interaction. The plant species and the size, stability, chemical conformation, and concentration of the NPs influence the uptake and translocation (Lv et al., 2019). The NPs sprayed foliarly enter the plants through the stomata or leaf epidermis. Following absorption, they use apoplastic or symplastic routes to enter vascular tissue. Vascular tissue (xylem and phloem) carries the NPs upward and downward from their principal storage sites in vacuoles and cell walls to diverse plant parts. Abiotic variables in the atmosphere, such as light, humidity, temperature, and NP characteristics, such as shape, size, and charge transfer, influence nanoparticle penetration and accumulation (NPs). For example, the cucumber plant uptakes CeO2 NPs measuring 8 ± 1 nm (Hong et al., 2014), and the lettuce plant uptakes Cu(OH)2 NPs ranging from approximately 50 to 100 nm (L. Zhao et al., 2016). Foliar spray NP supplementation is more effective and eco-friendly than root or soil exposure (Fan et al., 2020).

5.5.1 Role of Nanocarriers in Agrochemical Delivery

Smart nanocarriers are being developed at the moment to ensure that plant growth regulators, insecticides, and fertilizers are absorbed efficiently. Soil rich in organic matter limits the distribution of these clever nanocarriers to the region immediately surrounding the roots of crop plants. Soil fertilizer release was improved by encasing potassium nitrate in graphene oxide. Consequently, both the stability and cost of fertilizer will rise (Shang et al., 2019). Applying ENMs to soil systems also increases the rhizosphere roots' soil–microbiome interaction, boosting crop yield. When metallic ENMs are introduced to soil, they change the carbon, phosphorus, and nitrogen cycles. These ENMs facilitate the colonization of plant roots by nitrogen-fixing bacteria or endophytes, which are beneficial to plants. Also, ZnO, TiO2, MWCNTs, and FeO nanoparticles have improved the growth of several crops, including peanuts, soybeans, mung beans, wheat, tomatoes, and potatoes (Shojaei et al., 2019). Consequently, this method has become more consistent while reducing efforts to mitigate atmospheric deformations.

5.6 Technological Innovations and Techniques

5.6.1 Nano-encapsulation and Nano-formulation Techniques

Conventional crop conservation efforts have many advantages, but nanoparticles provide a large-scale opportunity for even more improvement. Weeds in crops are difficult, if not impossible, to eradicate, and over 90% of pesticides wind up in the environment. Almost every pesticide has some unintended side effect (Zaytseva & Neumann, 2016). Not only does this increase the value of plant species synthesis but it also destroys the agricultural system and no longer works. Targeting specific active compounds with nanoparticles at low concentrations can enhance agricultural productivity and pest control. In light of this, new technologies were developed for plant conservation, which was crucial for agricultural research. The technology known as pesticide nanoencapsulation, or nano-formulation, was created and is essential to plant conservation (Fig. 5.4). Encapsulation was applied to pesticides,

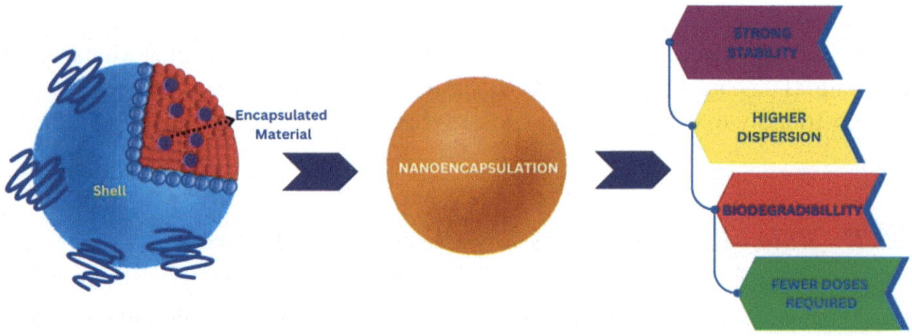

Fig. 5.4 Nano-based smart pesticide encapsulation

which should have a lower component concentration that serves as a lively pesticide (Ul Haq & Ijaz, 2019).

A pesticide's active ingredient is covered with various materials ranging in size in the nano range by a developed technology known as nano-encapsulation. Agrochemicals are the encapsulated core materials; the material used for encapsulation is called the external phase and the material encased is called the internal phase. In other words, progressively increasing the concentration of pesticides or herbicides causes an increase in the inhibition of the target microorganisms, decreases the pesticide solubility within water, and negatively impacts organisms. The advancement of nano-formulation technology would aid in mitigating the shortcomings identified earlier. For example, Petosa et al. (2017) suggest that controlling the pesticides' migration ability increases crop output. This nano-formulation shows high elution over time because it combines the pyrethroid-bifenthrin (nCAP4-BIF) with polymeric nanocapsules. Based on these findings, $nCAP_4$ could potentially serve as a pesticide transfer vector for pyrethroids in crop conservation efforts.

5.6.2 *Smart Delivery Systems and Precision Agriculture*

Precision agriculture, or PA, is a farming method that maximizes crop yields and streamlines resource management by utilizing advanced technologies and data-driven decision-making tools. Precision agriculture could undergo a revolution, thanks to nanotechnology, which can provide growers and farmers with new tools and methods for increasing crop yield. PA maximizes crop yields and resource management by applying cutting-edge technologies with the help of data-driven decision-making tools. These two topics converge on the use of nanotechnology to improve PA practices. With global information systems (GIS) and remote sensing technologies, PA helps farmers monitor and control field variability. These technologies may provide prescription maps for the variable-rate input application, which lowers input costs and has a positive environmental impact (Finch et al., 2002). By employing real-time sensor data to monitor crop growth, soil moisture, and various other environmental conditions, PA enables farmers to take the appropriate action at the appropriate time and location, improving agricultural yields and minimizing waste (Dhanaraju et al., 2022). Cutting input costs and optimizing resource use can greatly boost productivity. Furthermore, by combining farmers' knowledge and perspectives, PA helps improve agricultural management decision-making by gradually increasing control. An integral part of PA, variable rate technology (VRT) allows for the precise dosing of inputs such as fertilizers, herbicides, and insecticides in response to alterations in soil type and crop health. Farmers can maximize crop development, minimize waste, and increase yields by supplying inputs only where they are needed. Nanoparticles (NPs) can further enhance VRT efficiency by transporting these inputs.

5.7 Role of Sensors and Nanobiosensors

By controlling agricultural soil and water, as well as identifying contaminants like excess fertilizer, pesticide residues, pathogens of plants (bacteria, fungi, as well as viruses), heavy metals, soil nutrients, and moisture content, nanosensors can significantly increase crop yields (Srivastava et al., 2018). Nanosensors have several advantages over traditional sensors, including a large surface area to volume ratio, real-time detection, small size, exceptional stability, selectivity, and sensitivity (Deep & Kumar, 2019). Tehe next generation of detection devices, known as nano biosensors, combine a biological sensing element with a transducer to pick up on physicochemical signals. This signal can be electrochemical, optical, or mechanical and can be produced even at very low concentrations of the analyte or parameter of choice. Numerous nanobiosensors have been developed using various types of nanomaterials, including graphene oxide (GO), carbon nanotubes (CNTs), magnetic nanoparticles (NPs), gold nanoparticles (AuNPs), and quantum dots (QDs). NMs are useful for identifying soil nutrients, including nitrate, urea, ammonium, nitrite, and urease, and these, when present in excess, can have toxicological effects on soils (Ahmed et al., 2017). Furthermore, NM-based sensors have measured the organic matter phosphates, total carbon content, and more for soil analysis (Shang et al., 2019).

Pathogens and toxins in food and agricultural systems have also been identified using nanobiosensors. The capacity to detect contagious plant diseases prior to the emergence of symptoms makes these sensors crucial to "smart" agriculture (Dar et al., 2020). For instance, optical nanobiosensors were developed to detect water food-borne infections, and research into plant disease prevention has made use of metal oxide NPs, quantum dots (QDs), and polymeric NPs that detect environmental microbial pathogens (Du et al., 2022). Additionally, bacteria including *Salmonella enterica serovar Typhimurium, Staphylococcus aureus, Salmonella enterica, Escherichia coli,* and *Listeria monocytogenes* have been optically detected using metal nanoparticles (NPs) like AgNPs, AuNPs, and magnetic NPs. Crop growth is negatively impacted when hazardous metal ions are present in the environment in excess of critical thresholds. AuNPs, graphene, AgNPs, CNTs, QDs, and their nanocomposites are among the different types of nanomaterials (NMs) that have been utilized in surface-enhanced Raman scattering (SERS), colorimetric, and fluorescent, along with electrochemical sensors to identify different toxic heavy metal ions (like Hg^{2+}, Pb^{2+}, and Cd^{3+}) in soil and water samples (Sun et al., 2020).

5.8 Future Perspectives, Challenges, and Limitations

Several pressing concerns must be addressed during ongoing research into NMs for agrochemical applications, including production costs, evaluation criteria, public concern issues, and safety assessments. Before it is decided that nanoparticles are safe for usage, their effects must be critically evaluated (Rai et al., 2015).

5.8.1 Environmental Safety Assessment

Concerns have been raised regarding the possible safety hazards NMs pose to both people and nontarget creatures. According to Juárez-Maldonado et al., the direct interaction between NMs and microbial cells is also highly complicated and may result in particular alterations in the physiology and gene expression of microbial cells, which may promote the growth of any cell (Juárez-Maldonado et al., 2021). These dangers could lessen the variety and quantity of particular microbial groups in soil. To improve nutrient capture, targeting, uptake, delivery, and long-term impacts on soil microbial communities, Zhang et al. proposed combining nanoinformatics methods with existing nutrient cycling and crop productivity models. This would lead to the development of environmentally predictive nanoscale agrochemicals with optimal safety and functional features (Padrilah et al., 2024; Zhang et al., 2021). Kah et al. state that extensive field testing is required to determine the effectiveness and environmental effects of nano-agrochemicals (Kah et al., 2018). Further research and analysis of nano-agrochemicals will be conducted to fairly evaluate the benefits and potential hazards of currently available goods. The commercial production of nano-agrochemicals derived from NMs and nanotechnology is affected by many elements, including product economics, environmental concerns, assessment criteria, usage suggestions, and policy concerns. Because NMs and nanotechnology can enhance the precision, efficacy, and targeting of agrochemicals, their application to agriculture is presumed to yield enormous advantages. In addition, reasonable laws and widespread use of new technologies can both speed up their development.

5.9 Conclusion

Sustainable agriculture is becoming increasingly valuable for the sake of both the environment and the safety of our food supply. Agriculture should be innovative and modernized so it can keep up with the world's population growth and the demand for food without compromising environmental sustainability. Using few agrochemicals while reaching high yields and low production costs is fundamental to sustainable agriculture. The creation and use of nano-agrochemicals in agriculture have restored scientists' and researchers' conviction in the dependability of nanotechnology. Sustained release and nanoscale delivery technologies are the most advanced tools for modern agriculture. The agro-food sector can reap the benefits of nanotechnology through the use of novel approaches such as the dispensing of micronutrients and nanofertilizers, the development of agrochemical-encapsulated nanocarrier systems, and the distribution of nanopesticides, which encompass both environmentally friendly and biological pesticides. Waste management, water and soil remediation, agro-product processing and storage, plant disease detection, and growth monitoring with nanosensors have all been huge boons to the agricultural workforce. Thorough evaluations of all of these nanoproducts are necessary before their use. Most of the research in this area is still in the lab, and there is very little documentation of the rules and regulations that govern it. Commercial workers and farmers alike are wary of investing in this sector because of the high product costs and low return on investment. Neither technological nor commercial viability has allowed these solutions to keep up with the ever-increasing demands for food production and distribution. Because there are few working examples of commercial products, this problem needs fixing immediately. Additionally, the effects of NMs on human health, safety, and the environment are poorly understood. It is essential to assess the physical and chemical properties of the soil around the crops before applying NPs to reduce plant hazards. Conducting human and eco-toxicological research can help us understand the complex relationship between the agro environment, nanoscale agricultural chemicals, exposure levels, and humans. An in-depth analysis is necessary to assess the agroecological toxicity and application of NMs.

References

Abigail, E. A., & Chidambaram, R. (2017). Nanotechnology in herbicide resistance. In *Nanostructured materials: Fabrication to applications* (pp. 207–212). IntechOpen.

Abigail, M. E. A. (2020). Nanocomposites: New trends for sensing and controlled-release of herbicides. In *Multifunctional hybrid nanomaterials for sustainable agri-food and ecosystems* (pp. 255–269). Elsevier.

Ahmed, M., Rauf, M., Mukhtar, Z., & Saeed, N. A. (2017). Excessive use of nitrogenous fertilizers: An unawareness causing serious threats to environment and human health. *Environmental Science and Pollution Research, 24*, 26983–26987.

Ashraf, S. A., Siddiqui, A. J., Abd Elmoneim, O. E., Khan, M. I., Patel, M., Alreshidi, M., Moin, A., Singh, R., Snoussi, M., & Adnan, M. (2021). Innovations in nanoscience for the sustainable development of food and agriculture with implications on health and environment. *Science of the Total Environment, 768*, 144990.

Baker, S., Satish, S., Prasad, N., & Chouhan, R. S. (2019). Nano-agromaterials: Influence on plant growth and crop protection. In *Industrial applications of nanomaterials* (pp. 341–363). Elsevier.

Bombo, A. B., Pereira, A. E. S., Lusa, M. G., de Medeiros Oliveira, E., de Oliveira, J. L., Campos, E. V. R., de Jesus, M. B., Oliveira, H. C., Fraceto, L. F., & Mayer, J. L. S. (2019). A mechanistic view of interactions of a nanoherbicide with target organism. *Journal of Agricultural and Food Chemistry, 67*(16), 4453–4462.

Campos, E. V. R., Proença, P. L. F., Oliveira, J. L., Bakshi, M., Abhilash, P. C., & Fraceto, L. F. (2019). Use of botanical insecticides for sustainable agriculture: Future perspectives. *Ecological Indicators, 105*, 483–495.

Chi, Y., Chen, C., Zhang, G., Ye, Z., Su, X., Ren, X., & Wu, Z. (2021). Fabrication of magnetic-responsive controlled-release herbicide by a palygorskite-based nanocomposite. *Colloids and Surfaces B: Biointerfaces, 208*, 112115.

Council, I. N. I. (2009). *First nano-organic iron chelated fertilizer invented in Iran [webpage on the Internet]*. Iran Nanotechnology Initiative Council.

Dar, F. A., Qazi, G., & Pirzadah, T. B. (2020). Nano-biosensors: NextGen diagnostic tools in agriculture. In *Nanobiotechnology in ariculture: An approach towards sustainability* (pp. 129–144).

Davydov, R., Sokolov, M., Hogland, W., Glinushkin, A., & Markaryan, A. (2018). The application of pesticides and mineral fertilizers in agriculture. *MATEC Web of Conferences, 245*, 11003.

de la Rosa, G., López-Moreno, M. L., de Haro, D., Botez, C. E., Peralta-Videa, J. R., & Gardea-Torresdey, J. L. (2013). Effects of ZnO nanoparticles in alfalfa, tomato, and cucumber at the germination stage: Root development and X-ray absorption spectroscopy studies. *Pure and Applied Chemistry 4th, 85*(12), 2161–2174.

Deep, A., & Kumar, S. (2019). *Advances in nanosensors for biological and environmental analysis*. Elsevier.

DeRosa, M. C., Monreal, C., Schnitzer, M., Walsh, R., & Sultan, Y. (2010). Nanotechnology in fertilizers. *Nature Nanotechnology, 5*(2), 91.

Dhanaraju, M., Chenniappan, P., Ramalingam, K., Pazhanivelan, S., & Kaliaperumal, R. (2022). Smart farming: Internet of Things (IoT)-based sustainable agriculture. *Agriculture, 12*(10), 1745.

Du, H., Li, Z., Wang, Y., Yang, Q., & Wu, W. (2022). Nanomaterial-based optical biosensors for the detection of foodborne bacteria. *Food Reviews International, 38*(4), 655–684.

Duhan, J. S., Kumar, R., Kumar, N., Kaur, P., Nehra, K., & Duhan, S. (2017). Nanotechnology: The new perspective in precision agriculture. *Biotechnology Reports, 15*, 11–23.

Duro, J. A., Lauk, C., Kastner, T., Erb, K.-H., & Haberl, H. (2020). Global inequalities in food consumption, cropland demand and land-use efficiency: A decomposition analysis. *Global Environmental Change, 64*, 102124.

El-Ramady, H. R. (2014). Integrated nutrient management and postharvest of crops. *Sustainable Agriculture Reviews, 13*, 163–274.

Fan, X., Cao, X., Zhou, H., Hao, L., Dong, W., He, C., Xu, M., Wu, H., Wang, L., & Chang, Z. (2020). Carbon dioxide fertilization effect on plant growth under soil water stress associates with changes in stomatal traits, leaf photosynthesis, and foliar nitrogen of bell pepper (Capsicum annuum L.). *Environmental and Experimental Botany, 179*, 104203.

Finch, H. J. S., Samuel, A. M., & Lane, G. P. F. (2002). *Lockhart & Wiseman's crop husbandry: Including grassland.*

Forini, M. M. L., Pontes, M. S., Antunes, D. R., de Lima, P. H. C., Santos, J. S., Santiago, E. F., & Grillo, R. (2022). Nano-enabled weed management in agriculture: From strategic design to enhanced herbicidal activity. *Plant Nano Biology, 1*, 100008.

Fraceto, L. F., Grillo, R., de Medeiros, G. A., Scognamiglio, V., Rea, G., & Bartolucci, C. (2016). Nanotechnology in agriculture: Which innovation potential does it have? *Frontiers in Environmental Science, 4*, 186737.

Gilbertson, L. M., Pourzahedi, L., Laughton, S., Gao, X., Zimmerman, J. B., Theis, T. L., Westerhoff, P., & Lowry, G. V. (2020). Guiding the design space for nanotechnology to advance sustainable crop production. *Nature Nanotechnology, 15*(9), 801–810.

Giraldo, J. P., Wu, H., Newkirk, G. M., & Kruss, S. (2019). Nanobiotechnology approaches for engineering smart plant sensors. *Nature Nanotechnology, 14*(6), 541–553.

Gupta, A., Bano, A., Rai, S., Mishra, R., Singh, M., Sharma, S., & Pathak, N. (2022a). Mechanistic insights of plant-microbe interaction towards drought and salinity stress in plants for enhancing the agriculture productivity. *Plant Stress, 4*, 100073.

Gupta, A., Dubey, P., Kumar, M., Roy, A., Sharma, D., Khan, M. M., Bajpai, A. B., Shukla, R. P., Pathak, N., & Hasanuzzaman, M. (2022b). Consequences of arsenic contamination on plants and mycoremediation-mediated arsenic stress tolerance for sustainable agriculture. *Plants, 11*(23), 3220.

Heidary, M., Karimzadeh, J., Jafari, S., Negahban, M., & Shakarami, J. (2021). Aphicidal activity of urea–formaldehyde nanocapsules loaded with the Thymus daenensis Celak essential oil on Brevicoryne brassicae L. *International Journal of Tropical Insect Science*, 1–12.

Hong, J., Peralta-Videa, J. R., Rico, C., Sahi, S., Viveros, M. N., Bartonjo, J., Zhao, L., & Gardea-Torresdey, J. L. (2014). Evidence of translocation and physiological impacts of foliar applied CeO2 nanoparticles on cucumber (Cucumis sativus) plants. *Environmental Science & Technology, 48*(8), 4376–4385.

Juárez-Maldonado, A., Tortella, G., Rubilar, O., Fincheira, P., & Benavides-Mendoza, A. (2021). Biostimulation and toxicity: The magnitude of the impact of nanomaterials in microorganisms and plants. *Journal of Advanced Research, 31*, 113–126.

Kabir, M. H., Tisha, F. S., Nayan, H. R., Islam, M. A., Kashem, M. A., Uddin, M. I., Islam, M. R., & Meah, M. B. (2021). Determining an effective and economic fungicide spray schedule for reducing blast of wheat. *International Journal of Agricultural Research, Innovation and Technology (IJARIT), 11*(1), 10–16.

Kah, M., Kookana, R. S., Gogos, A., & Bucheli, T. D. (2018). A critical evaluation of nanopesticides and nanofertilizers against their conventional analogues. *Nature Nanotechnology, 13*(8), 677–684.

Kannan, M., Bojan, N., Swaminathan, J., Zicarelli, G., Hemalatha, D., Zhang, Y., Ramesh, M., & Faggio, C. (2023). Nanopesticides in agricultural pest management and their environmental risks: A review. *International journal of Environmental Science and Technology, 20*(9), 10507–10532.

Kaziem, A. E., Gao, Y., Zhang, Y., Qin, X., Xiao, Y., Zhang, Y., You, H., Li, J., & He, S. (2018). α-Amylase triggered carriers based on cyclodextrin anchored hollow mesoporous silica for enhancing insecticidal activity of avermectin against Plutella xylostella. *Journal of Hazardous Materials, 359*, 213–221.

Khodakovskaya, M., Dervishi, E., Mahmood, M., Xu, Y., Li, Z., Watanabe, F., & Biris, A. S. (2009). Carbon nanotubes are able to penetrate plant seed coat and dramatically affect seed germination and plant growth. *ACS Nano, 3*(10), 3221–3227.

Kottegoda, N., Munaweera, I., Madusanka, N., & Karunaratne, V. (2011). A green slow-release fertilizer composition based on urea-modified hydroxyapatite nanoparticles encapsulated wood. *Current Science*, 73–78.

Kumar, S., Bhanjana, G., Sharma, A., Dilbaghi, N., Sidhu, M. C., & Kim, K.-H. (2017). Development of nanoformulation approaches for the control of weeds. *Science of the Total Environment, 586*, 1272–1278.

Lamichhane, J. R., Dachbrodt-Saaydeh, S., Kudsk, P., & Messéan, A. (2016). Toward a reduced reliance on conventional pesticides in European agriculture. *Plant Disease, 100*(1), 10–24.

Li, H., Jing, T., Li, T., Huang, X., Gao, Y., Zhu, J., Lin, J., Zhang, P., Li, B., & Mu, W. (2021). Ecotoxicological effects of pyraclostrobin on tilapia (Oreochromis niloticus) via various exposure routes. *Environmental Pollution, 285*, 117188.

Li, M., Xu, W., Hu, D., & Song, B. (2018). Preparation and application of pyraclostrobin microcapsule formulations. *Colloids and Surfaces A: Physicochemical and Engineering Aspects, 553*, 578–585.

Liu, B., Wang, Y., Yang, F., Wang, X., Shen, H., Cui, H., & Wu, D. (2016). Construction of a controlled-release delivery system for pesticides using biodegradable PLA-based microcapsules. *Colloids and Surfaces B: Biointerfaces, 144*, 38–45.

Liu, B., Zhang, J., Chen, C., Wang, D., Tian, G., Zhang, G., Cai, D., & Wu, Z. (2021). Infrared-light-responsive controlled-release pesticide using hollow carbon microspheres@ polyethylene glycol/α-cyclodextrin gel. *Journal of Agricultural and Food Chemistry, 69*(25), 6981–6988.

Lowry, G. V., Avellan, A., & Gilbertson, L. M. (2019). Opportunities and challenges for nanotechnology in the agri-tech revolution. *Nature Nanotechnology, 14*(6), 517–522.

Luo, J., Zhang, D., Jing, T., Liu, G., Cao, H., Li, B., Hou, Y., & Liu, F. (2020). Pyraclostrobin loaded lignin-modified nanocapsules: Delivery efficiency enhancement in soil improved control efficacy on tomato Fusarium crown and root rot. *Chemical Engineering Journal, 394*, 124854.

Lv, J., Christie, P., & Zhang, S. (2019). Uptake, translocation, and transformation of metal-based nanoparticles in plants: Recent advances and methodological challenges. *Environmental Science: Nano, 6*(1), 41–59.

Mahmood, I., Imadi, S. R., Shazadi, K., Gul, A., & Hakeem, K. R. (2016). Effects of pesticides on environment. In *Plant, soil and microbes: Volume 1: Implications in crop science* (pp. 253–269).

Manikandan, A., & Subramanian, K. S. (2014). Fabrication and characterisation of nanoporous zeolite based N fertilizer. *African Journal of Agricultural Research, 9*(2), 276–284.

Mittal, D., Kaur, G., Singh, P., Yadav, K., & Ali, S. A. (2020). Nanoparticle-based sustainable agriculture and food science: Recent advances and future outlook. *Frontiers in Nanotechnology, 2*, 579954.

Mukhopadhyay, S. S. (2014). Nanotechnology in agriculture: Prospects and constraints. In *Nanotechnology, science and applications* (pp. 63–71).

Naderi, M. R., & Abedi, A. (2012). Application of nanotechnology in agriculture and refinement of environmental pollutants. *Journal of Nanotechnology, 11*(1), 18–26.

Nile, S. H., Baskar, V., Selvaraj, D., Nile, A., Xiao, J., & Kai, G. (2020). Nanotechnologies in food science: Applications, recent trends, and future perspectives. *Nano-Micro Letters, 12*, 1–34.

Owens, S. A., & Owens, D. A. A. (2010). *International center for technology assessment.*

Padrilah, S. N., Samsudin, N. I. P., Shukor, M. Y. A., & Masdor, N. A. (2024). Nanoemulsion strategies in controlling fungal contamination and toxin production on grain corn using essential oils. *Green Chemistry Letters and Reviews, 17*(1), 2315138.

Pang, Y., Qin, Z., Wang, S., Yi, C., Zhou, M., Lou, H., & Qiu, X. (2020). Preparation and application performance of lignin-polyurea composite microcapsule with controlled release of avermectin. *Colloid and Polymer Science, 298*, 1001–1012.

Panwar, J. (2012). Positive effect of zinc oxide nanoparticles on tomato plants: A step towards developing nano-fertilizers. In *International conference on environmental research and technology (ICERT).*

Pascoli, M., Jacques, M. T., Agarrayua, D. A., Avila, D. S., Lima, R., & Fraceto, L. F. (2019). Neem oil based nanopesticide as an environmentally-friendly formulation for applications in sustainable agriculture: An ecotoxicological perspective. *Science of the Total Environment, 677*, 57–67.

Petosa, A. R., Rajput, F., Selvam, O., Öhl, C., & Tufenkji, N. (2017). Assessing the transport potential of polymeric nanocapsules developed for crop protection. *Water Research, 111*, 10–17.

Pramanik, P., Krishnan, P., Maity, A., Mridha, N., Mukherjee, A., & Rai, V. (2020). Application of nanotechnology in agriculture. *Environmental Nanotechnology, 4*, 317–348.

Rai, M., Ribeiro, C., Mattoso, L., & Duran, N. (2015). *Nanotechnologies in food and agriculture* (Vol. 33). Springer.

Raliya, R., Saharan, V., Dimkpa, C., & Biswas, P. (2017). Nanofertilizer for precision and sustainable agriculture: Current state and future perspectives. *Journal of Agricultural and Food Chemistry, 66*(26), 6487–6503.

Robles-García, M. A., Rodríguez-Félix, F., Marquez-Rios, E., Aguilar, J. A., Barrera-Rodríguez, A., Aguilar, J., Ruiz-Cruz, S., & Del-Toro-Sánchez, C. L. (2016). Applications of nanotechnology in the agriculture, food, and pharmaceuticals. *Journal of Nanoscience and Nanotechnology, 16*(8), 8188–8207.

Scott, N., & Chen, H. (2012). Nanoscale science and engineering for agriculture and food systems. *Industrial Biotechnology, 8*(6), 340–343.

Shah, T., Xu, J., Zou, X., Cheng, Y., Zhang, X., Hussain, Q., & Gill, R. A. (2019). Impact of nanomaterials on plant economic yield and next generation. In *Advances in phytonanotechnology* (pp. 203–214). Elsevier.

Shang, Y., Hasan, M. K., Ahammed, G. J., Li, M., Yin, H., & Zhou, J. (2019). Applications of nanotechnology in plant growth and crop protection: A review. *Molecules, 24*(14), 2558.

Shojaei, T. R., Salleh, M. A. M., Tabatabaei, M., Mobli, H., Aghbashlo, M., Rashid, S. A., & Tan, T. (2019). Applications of nanotechnology and carbon nanoparticles in agriculture. In *Synthesis, technology and applications of carbon nanomaterials* (pp. 247–277). Elsevier.

Srivastava, A. K., Dev, A., & Karmakar, S. (2018). Nanosensors and nanobiosensors in food and agriculture. *Environmental Chemistry Letters, 16*, 161–182.

Sun, J., Lu, Y., He, L., Pang, J., Yang, F., & Liu, Y. (2020). Colorimetric sensor array based on gold nanoparticles: Design principles and recent advances. *TrAC Trends in Analytical Chemistry, 122*, 115754.

Tosco, T., Granetto, M., Re, L., Audino, A., Serpella, L., Fogliatto, S., & Vidotto, F. (2020). A novel nano-formulation to reduce the environmental dispersion of herbicides. *EGU General Assembly Conference Abstracts, 18817*.

Ul Haq, I., & Ijaz, S. (2019). Use of metallic nanoparticles and nanoformulations as nanofungicides for sustainable disease management in plants. In *Nanobiotechnology in bioformulations* (pp. 289–316).

Vega-Vásquez, P., Mosier, N. S., & Irudayaraj, J. (2020). Nanoscale drug delivery systems: From medicine to agriculture. *Frontiers in Bioengineering and Biotechnology, 8*, 79.

Wang, A., Wang, Y., Sun, C., Wang, C., Cui, B., Zhao, X., Zeng, Z., Yao, J., Yang, D., & Liu, G. (2018). Fabrication, characterization, and biological activity of avermectin nano-delivery systems with different particle sizes. *Nanoscale Research Letters, 13*, 1–7.

Wang, D., Saleh, N. B., Byro, A., Zepp, R., Sahle-Demessie, E., Luxton, T. P., Ho, K. T., Burgess, R. M., Flury, M., & White, J. C. (2022). Nano-enabled pesticides for sustainable agriculture and global food security. *Nature Nanotechnology, 17*(4), 347–360.

Wise, J. C., VanWoerkom, A. H., Acimovic, S. G., Sundin, G. W., Cregg, B. M., & Vandervoort, C. (2014). Trunk injection: A discriminating delivering system for horticulture crop IPM. *Entomology, Ornithology & Herpetology, 3*(2), 1.

Wu, D., Qin, M., Xu, D., Wang, L., Liu, C., Ren, J., Zhou, G., Chen, C., Yang, F., & Li, Y. (2019). A bioinspired platform for effective delivery of protein therapeutics to the central nervous system. *Advanced Materials, 31*(18), 1807557.

Xiang, Y., Zhang, G., Chi, Y., Cai, D., & Wu, Z. (2017). Fabrication of a controllable nanopesticide system with magnetic collectability. *Chemical Engineering Journal, 328*, 320–330.

Yadav, A. S., & Srivastava, D. S. (2015). Application of nano-technology in weed management: A review. *Research & Reviews: Journal of Crop Science and Technology, 4*(2), 21–23.

Yadollahi, A., Arzani, K., & Khoshghalb, H. (2009). The role of nanotechnology in horticultural crops postharvest management. *Southeast Asia Symposium on Quality and Safety of Fresh and Fresh-Cut Produce, 875*, 49–56.

Zaytseva, O., & Neumann, G. (2016). Carbon nanomaterials: Production, impact on plant development, agricultural and environmental applications. *Chemical and Biological Technologies in Agriculture, 3*, 1–26.

Zhang, P., Guo, Z., Ullah, S., Melagraki, G., Afantitis, A., & Lynch, I. (2021). Nanotechnology and artificial intelligence to enable sustainable and precision agriculture. *Nature Plants, 7*(7), 864–876.

Zhao, L., Ortiz, C., Adeleye, A. S., Hu, Q., Zhou, H., Huang, Y., & Keller, A. A. (2016). Metabolomics to detect response of lettuce (Lactuca sativa) to Cu (OH) 2 nanopesticides: Oxidative stress response and detoxification mechanisms. *Environmental Science & Technology, 50*(17), 9697–9707.

Zhao, X., Cui, H., Wang, Y., Sun, C., Cui, B., & Zeng, Z. (2017). Development strategies and prospects of nano-based smart pesticide formulation. *Journal of Agricultural and Food Chemistry, 66*(26), 6504–6512.

Zheng, F., Li, Y., Zhang, Z., Jia, J., Hu, P., Zhang, C., & Xu, H. (2020). Novel strategy with an eco-friendly polyurethane system to improve rainfastness of tea saponin for highly efficient rice blast control. *Journal of Cleaner Production, 264*, 121685.

Zheng, L., Cao, C., Cao, L., Chen, Z., Huang, Q., & Song, B. (2018). Bounce behavior and regulation of pesticide solution droplets on rice leaf surfaces. *Journal of Agricultural and Food Chemistry, 66*(44), 11560–11568.

Zhou, M., Xiong, Z., Yang, D., Pang, Y., Wang, D., & Qiu, X. (2018). Preparation of slow release nanopesticide microspheres from benzoyl lignin. *Holzforschung, 72*(7), 599–607.

Zhu, H., Shen, Y., Cui, J., Li, N., Wang, C., Cui, B., Sun, C., Zhao, X., Wang, C., & Gao, F. (2020). Avermectin loaded carboxymethyl cellulose nanoparticles with stimuli-responsive and controlled release properties. *Industrial Crops and Products, 152*, 112497.

Chapter 6
Prospecting the Significance of Nanobionics in Crop Improvement

Divya Upadhyay, Kartikeya Shukla, Arti Mishra, and Smriti Shukla

Abstract Nanobionics, an interdisciplinary field combining nanotechnology and biology, holds immense promise for revolutionizing crop improvement strategies. By integrating nanomaterials with biological systems, nanobionics offers novel approaches to enhance crop productivity, resilience to environmental stressors, and nutrient uptake efficiency. This review explores the significance of nanobionics in crop improvement, elucidating its potential applications in plant growth regulation, disease management, and environmental monitoring. Additionally, challenges and future prospects in the field are discussed, highlighting the need for further research and development to realize the full potential of nanobionics in sustainable agriculture. The synergy between nanostructured materials and plant cell organelles, particularly chloroplasts, presents a promising avenue for enhancing organelle functions and performance. Plant nanobionics-based technologies offer solutions to pressing agricultural challenges such as nutrient deficiencies, stagnant yields, climate change, and water scarcity. Innovations in nanostructured materials, including carbon-based and metal-based nanoparticles (NPs) such as iron (Fe), Copper (Cu), Zinc (Zn), Silver (Ag), and cerium (Ce), hold the potential for delivering nutrients and controlling pests with precision and environmental sustainability. This review explores recent advancements in nanostructured materials for nutrient transportation, emphasizing the role of nanobiotechnology in augmenting crop production

D. Upadhyay · K. Shukla
Amity Institute of Environmental Sciences, Amity University, Noida, Uttar Pradesh, India

A. Mishra
Department of Botany, Hansraj College, University of Delhi, Delhi, India

Umeå Plant Science Center, Department of Plant Physiology, Umeå University, Umeå, Sweden

S. Shukla (✉)
Amity Institute of Environmental Toxicology, Safety and Management, Amity University, Noida, India

V. D. Rajput et al. (eds.), *Contemporary Suitability of Nanobionics in Agriculture*, Nanotechnology in the Life Sciences, https://doi.org/10.1007/978-3-031-82779-2_6

through smart delivery systems and gene transformation. The integration of nanomaterials with proteomic tools is also discussed, highlighting its potential for understanding plant-nanoparticle interactions and advancing genetic engineering for agricultural sustainability and food safety. By elucidating the multifaceted applications of nanotechnology in crop improvement, this review underscores the imperative for further research and adoption of nanobiotechnology to address the evolving demands of global agriculture.

Keywords Nanobionics · Nanomaterials · Nanobiotechnology · Proteomic · Crop improvement

6.1 Introduction

The worrisome rate of human population growth has led the United Nations to predict that by 2050, there will be approximately 10 billion people on the earth (Chouhan et al., 2021). To meet the food needs of this growing population, the currently accessible amount of arable land needs to be increased by more than 70%. Due to biotic and abiotic reasons, this land is rare and seriously threatened (Kumar et al., 2021a, b). There is now even less land available because most of this arable land has been converted to industrial, residential, and other uses, such as parks and gardens for leisure (Winding et al., 2020). To further ensure the maximum level of food security and safety, current agricultural systems that depend on the manual labor technique of food production as well as chemical pesticides and fertilizers are unreliable and cannot be maintained (Kumar et al., 2021a, b). The systematic decrease in global agricultural productivity in many regions is due to efforts in plant breeding and pesticide efficacy. It appears from this that a rise in agricultural productivity is not directly connected with the addition of more chemicals (Busby et al., 2017; Singh & Trivedi, 2017; Kumari et al., 2020). Because of this, the vast majority of governmental and non-governmental activities are made to use natural solutions to address these problems with food security and sustainable agriculture. food security and safety (Kumari et al., 2020). The frequency and intensity of specific abiotic stresses, such as drought, extreme heat, and flooding, as well as climate change and global warming, which result in a continuous rise in CO_2 levels in the atmosphere, also have a significant impact on the survival, growth, and yield of plants (Raza et al., 2023; Rivero et al., 2022). This frequently leads to multifactorial stress and abiotic environmental components, or a biotic stress combination phenomenon (Zandalinas & Mittler, 2022). Interventions using modern technologies are therefore necessary to enhance food security and safety by managing these biotic and abiotic stresses.

6.1.1 Overview of Nanobionics

Nanotechnology is setting the standard for technological developments when it comes to resolving important problems facing the food and agriculture industries. Crop adaptation to changing environmental conditions is one of the objectives of the UN's 2030 Agenda for Sustainable Development. As it optimizes resources through targeted farming, reduced yield loss, and the application of effective food storage and recycling techniques to reduce waste, this is an essential strategy for achieving food security. In an effort to boost crop productivity and nutritional quality, control and selectively release plant nutrients (Fellet et al., 2021; Grillo et al., 2021), protect crops from pests (Abdelaziz et al., 2022; Elmer et al., 2018; Lopez-Lima et al., 2021), and strengthen plants' resistance to extreme weather (Djanaguiraman et al., 2018; Jacobson et al., 2018). Many nanoparticle formulations have been found to improve the delivery of macro- and micronutrients to plant tissues and promote plant growth. These formulations include hydroxyapatite, nano-clay, silicon oxide (SiO_2), manganese (MnO), zinc oxide (ZnO), iron oxide (Fe_3O_4), cerium oxide (CeO_2), molybdenum (MoO_3), carbon nanotubes, and fullerenes. The solubility, release rate, and availability of the active components of soil nutrients have been reported to be enhanced by nanoformulations, according to reports (Kah et al., 2019; Khan & Duke, 2001). However, the application of nanoparticles in agriculture has been restricted due to a lack of knowledge on the absorption, mobilization, transport, and associated biological reactions in plants. Numerous factors, such as species diversity, age, effective dose, duration of exposure, and the composition, size, surface coating, shape, stability, and solubility of the particles, affect how plants react to nanoparticles. Thanks to recent developments in analytical tools such as electron microscopy, synchrotron-based imaging techniques, and single particle inductively coupled plasma mass spectrometry (sp-ICPMS) (Avellan et al., 2019; Castillo-Michel et al., 2017; Keller et al., 2018), it is now easier to trace and monitor the transport of nanoparticles in plant tissues following foliar or root application. Unlike their bulk counterparts, reveal that nanoparticles can cross biological membranes and enter plant roots or leaves. Still unknown, nevertheless, are the chemical mechanisms behind the uptake, absorption, and transport of nanoparticles in plants.

A thorough understanding of the various biological networks operating at various levels is required in order to fully use the benefits of nanotechnology for sustainable food production in agriculture. Due to the high sensitivity and accuracy of analytical and bioinformatics tools, recent studies on the interaction between nanoparticles and plants have shifted from standard single endpoint assays to discovery-oriented, high-throughput system biology techniques, known as "omics". An omics methodology is a method of screening genes, mRNA, proteins, metabolites, and other targeted or untargeted biomolecules in an organism (genomics, transcriptomics, proteomics, and metabolomics). Computational biology has been applied over the years to understand the molecular mechanisms of plants and to understand how genes, proteins, and metabolites behave in response to biotic or abiotic environmental stress. Proteomic approach has gained more interest in precision agriculture recently because of the urgent need to understand the mechanism underlying complex agronomic traits and plants' reactions to research including the use of nanoparticles. Proteins play a major role in plant signaling and stress response because they are intimately connected to cellular homeostasis. Because it sheds light on the functions and translation of certain plant genome regions, proteomics is a subject that complements transcriptome and metabolic research in plant biology. It is therefore helpful in comprehending the complex processes involved in many plants cellular functions. Plant genome sequencing, according to Agrawal et al. (2013), will fundamentally alter how we handle issues with human health, safety, and food insecurity in the twenty-first century. Oryza sativa and Arabidopsis thaliana are just two of the several plants whose genomes have previously been sequenced and annotated; many more are currently in development (Kumar et al., 2021a, b). Crop germplasm collections can be characterized in order to produce and sustainably supply food with the help of modern technology. For example, the use of genetic resources in crop development is demonstrated by the investigation of more than 20,000 genotypes of wild and domesticated barley using genotyping and bioinformatics methods (Langridge & Waugh, 2019). In order to fulfill the needs of the expanding global population, we therefore discuss the various ways that nanotechnology and proteomics techniques may be employed to produce food that is safer in this review. Crop scientists will be able to utilize the database created by this review to create new and enhanced crop types.

6.2 Importance of Crop Improvement in Agriculture

The potential of nanotechnology in agriculture has yet to be completely explored. Nanotechnology has numerous benefits in agriculture, including plant growth promoters (nanofertilizers), pesticides, insecticides, and antimicrobials, and food storage and packaging materials among others. Nanotechnology in agriculture aims to increase productivity by optimizing nutrient management, controlling fertilizer release, and reducing chemical use (Chen et al., 2013). Nanomaterials can impact plant features, including physiological and biochemical responses, affecting yield and product quality (Gardea-Torresdey et al., 2014). Plant scientists are increasingly interested in using nanotechnology in agriculture due to its benefits (Feizi et al., 2012; Haghighi & Teixeira da Silva, 2014). Batsmanova et al. (2013) and Scott and Chen et al. (2013) state that nano agrotechnology aims to enhance animal and agricultural productivity through the use of nanoparticles with unique properties. Nanotechnology benefits sustainable agriculture by reducing waste, improving plant protection, disease detection, and growth monitoring (Mishra et al., 2017).

Researchers are exploring the potential of nanobiotechnology for crop enhancement across several industries. Much study has examined the usefulness of carbon-based or metallic nanoparticles in absorption, integration, transport, persistence, and impacts on plant growth and development for various commercial purposes. Several research suggests that nanoparticle exposure promotes plant growth and development (Mishra et al., 2017).

6.2.1 Fundamentals of Nanobionics

Nanobionics is the foundation for the development of innovative biotechnological materials. The advent of nanobionics has enabled us to explore the architecture and mechanisms within live cells, as well as replicate and imitate biological molecules utilizing innovative nanomaterials (Kozhukharov & Machkova, 2013). Nanobionics is the field of research that focuses on the investigation of electrical interactions occurring at the nanoscale inside biological systems. Nanobionics refers to the development of engineering systems, specifically electronic ones, that are modeled after biological systems (Mansoori et al., 2017). The term "nanobionics"may be deconstructed into two constituent words: "nano" and "bionics." Nanotechnology is the scientific study and manipulation of matter at the nanoscale, which refers to dimensions ranging from 1 to 100 nanometers. Plant nanobionics is a novel field of study that focuses on the intersection of nanotechnology and plant biology. Plant nanobionics is a field within plant science that focuses on the utilization of nanoparticles and their interaction with plant systems to provide new and innovative functions. Plant bionics has great potential in several fields such as agriculture, biosensors, defense, and more. The fusion of nanotechnology and plant life has revolutionized the way we strive to improve our living environment. Plant nanobionics refers to the utilization of nanoparticles in plants to confer them with novel capabilities. According to Michael Strano, a professor of Chemical Engineering, nanobionics aims to give plants abilities that they do not naturally possess. Put simply, his aim is to bestow extraordinary abilities upon plants. Bionics is the scientific examination of the organization and operation of living systems, which serves as a basis for creating and developing materials and machinery. It shares the same meaning as biomimicry, biomimesis, and biognosis and is comparable to biologically inspired design. Biomimetic pertains to artificial processes, substances, strategies, or systems that mimic nature. Biomimetics refers to the practice of developing and constructing biomimetic equipment, which involves the application of principles from both art and science. This field is particularly significant to researchers in nanotechnology, robotics, artificial intelligence, the medical industry, and the military (Bar-Cohen, 2005; Neville, 2007). Nanotechnology can improve agricultural production through various applications such as developing nanoformulations of agrochemicals like pesticides and fertilizers, using plants as nano biosensors to protect crops from diseases, and employing nanodevices for genetic manipulation of plants (Khan & Rizvi, 2014). The utilization of nanotechnology in plant disease management is an innovative and promising strategy to effectively tackle plant diseases (Ghorbanpour & Fahimirad, 2017).

6.2.2 *Nanomaterials and their Properties*

Nano fertilizers are a new advancement in nanotechnology that involves the use of nano-based products to provide agricultural plants with nutrients in a regulated manner (Ingle, 2021). Nanoparticles possess distinctive attributes, including decreased particle size and expansive surface area, which can facilitate enhanced interaction and efficient absorption of nutrients for the purpose of agricultural fertilization. The use of nanotechnology in fertilizer formulation is believed to improve nutrient absorption, resulting in significant economic and environmental advantages (Kumar et al. 2021a, b). A recent study has shown that nanofertilizers have the ability to gradually and precisely release nutrients into the soil. This helps in delivering agrochemicals to particular areas and reduces the potential harm caused by nanomaterial toxicity. The references used are Heinisch et al. (2019). Au, SiO_2, ZnO, SnO_2, and TiO_2 nanoparticles have been shown to augment the growth and development of plants by facilitating more absorption of elements and nutrients. In addition, nanofertilizers have demonstrated the ability to enhance plant performance by promoting efficient nutrient absorption, increasing crop yield, boosting photosynthesis rate, and significantly expanding the surface area of leaves. To prevent eutrophication and water pollution, the discharge of nutrients in nano fertilizers is regulated (Ingle, 2021).

6.3 Interaction of Nanomaterials with Biological Systems

There is increasing apprehension over the issue of pest and disease infestation in agriculture (Mohamed & Abd-Elsalam, 2018). Pest and diseases, which are biotic factors, cause an annual crop loss of around 20–40%. The majority of strategies employed to manage plant pests and diseases include the utilization of harmful pesticides and antimicrobials, which have adverse effects on both human health and the environment. Another drawback is the problem of antimicrobial resistance in food crops, since many traditional medicines are no longer effective against these infections. This leads to significant expenditure with little or no positive outcome (Abd-Elsalam & Prasad, 2018; Mohamed & Abd-Elsalam, 2018). Additionally, the presence of antimicrobial and pesticide residue in crops is a drawback that arises from the utilization of traditional pesticides and antibiotics. These residues ultimately reach consumers in the food chain, namely humans and animals, leading to further health complications. In addition, the ecosystems, soil, water, and air are also disrupted due to the extensive utilization of these antibiotics and chemicals (Bansal, 2011). Given the multitude of difficulties arising from the utilization of traditional pesticides and antibiotics, it is imperative to seek alternate approaches that are more effective and safer.

6.4 Applications of Nanobionics in Crop Improvement

In modern times, plants have become significant not just for their nutritional worth or their role in transpiration but also for their potential use as biosensors. Plants possess a high level of environmental awareness and may anticipate the occurrence of a drought in advance. They possess the ability to detect alterations in the characteristics of soil and the potential of water. By delving into those chemical signaling pathways, there is a wealth of knowledge to acquire. Nitroaromatics are chemical compounds that are commonly included in explosives and landmines. These compounds may be identified using minuscule nanotubes that are placed into the leaves of plants such as spinach. Carbon nanotubes are cylindrical structures composed of carbon atoms that can be found naturally or synthesized in a laboratory setting. When transformed into fibers, they exhibit remarkable strength, surpassing even that of steel (He et al., 2013) (Fig. 6.1).

6.4.1 Enhancing Plant Growth and Development

They have remarkable optical characteristics, since they emit near-infrared light that is sensitive to subtle variations in the chemical surroundings. The nanotubes are absorbed into the mesophyll cell layer, where photosynthesis occurs (Giraldo et al., 2014). The nitroaromatic compounds were diluted with water and applied to the soil surrounding the spinach plant. The roots absorbed the water, which was then transported to the leaves in the form of droplets, a process that took around 10 minutes. The plants' large root networks make them an ideal method for monitoring groundwater. The scientists patiently waited for almost 10 minutes before directing laser light onto the leaves of the spinach plant. Upon laser light exposure, the implanted nanotubes emitted near-infrared fluorescent light. The little infrared camera, coupled with a Raspberry Pi computer, identified the near-infrared light. The signal can also be detected using a smartphone that does not have an infrared filter.

6.4.2 Improving Nutrient Uptake Efficiency

Nanoparticle accumulation in plants correlates with phytotoxicity in ecosystems (Giorgetti, 2019; Jogaiah et al., 2021). TiO_2 and SiO_2 are the most stable metallic nanoparticles during absorption, but ZnO and CuO may undergo transformation. Regardless of plant type, the root system delivers nanoparticles of ZnO_2+, Cu_2+,

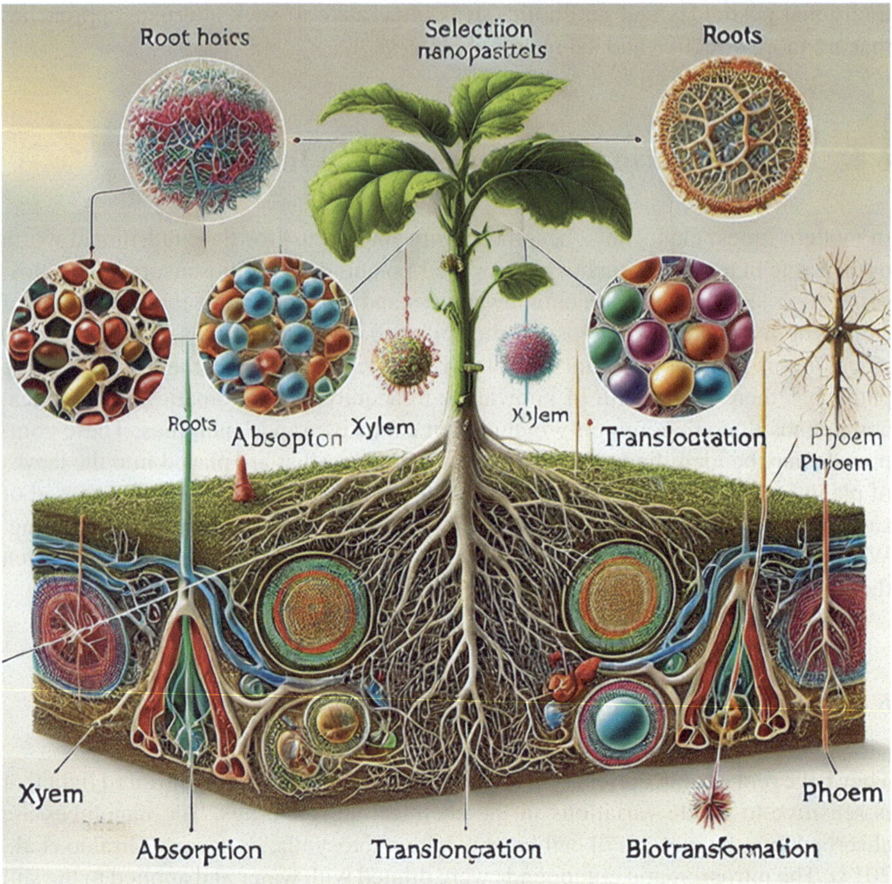

Fig. 6.1 The mechanisms behind the effects of nanoparticles

Al_3+, Ag2+, and Fe_3O_4 to the foliar region. Rico et al. (2011) found species-dependent nanomaterial absorption in plants, including CeO_2 and Ni $(OH)_2$ nanoparticles in stems and leaves, and Al, Ag, Cu, and ZnO nanoparticles in leaves.

Furthermore, magnetic nanoparticles exhibit distinct behavior in plant systems (Rico et al., 2011). The kind of growing media impacts magnetic nanoparticle absorption. Hydroponic medium absorbed more than sand, perhaps owing to magnetic nanoparticles' soil adherence. The size of the cell wall pore determines nanoparticle entry into plant cells. Effectiveness of translocation depends on nanoparticle size. Nanoparticles smaller than cell wall pores can easily penetrate and pass across the plasma membrane. Particles larger than cell wall pores may aggregate, preventing plant absorption of nanoparticles. The nanoparticle can be functionalized to increase or create new cell wall pores (Ingle, 2021; Rico et al., 2011). This allows big particles to enter the plant system more easily. Panpatte et al. (2016) and Mali et al. (2020) found that plant cells may absorb nanoparticles by binding to carrier proteins, aquaporins, ion channels, or endocytosis. Nanoparticles can enter plant cells via membrane transporters or root exudates (Guleria et al., 2022; Kumari et al., 2020).

Other studies show nanoparticles can enter plants via leaf stomata or trichomes (Burkhardt et al., 2012; Eichert et al., 2008; Fernandez & Eichert, 2009). Nanoparticles enter plants via biotransformation within or outside the cell. Solutes can enter the plant system by diffusion or aqueous pores (Lv et al., 2020). Nanoparticles infiltrate plant cells and disrupt metabolic processes by targeting several cytoplasmic organelles (Ingle, 2021). For example, TiO_2 nanoparticles were detected in wheat root parenchyma and vascular tissues following absorption (Larue et al., 2011). ZnO nanoparticles were shown to internalize and move up into *Lolium perenne* (ryegrass) cells (Lin & Xing, 2008). ZnO nanoparticles may migrate from ryegrass root cells to plant vascular tissues, according to their findings. Lopez-Moreno et al. (2010) showed ZnO nanoparticle absorption and accumulation in glycine max seedlings treated with 500–4000 ppm. The result indicates increased ZnO nanoparticle absorption at 500 ppm. High concentrations of nanoparticles can agglomerate, inhibiting uptake by seeds through cell pores (López-Moreno et al., 2010). Studies using X-ray absorption spectroscopy indicate that Zn_2. is found in the plant rather than ZnO nanoparticles, indicating that the root system is responsible for surface ionization of ZnO. Additionally, high-magnification transmission electron microscopy revealed ZnO nanoparticles in plant vascular cylinder and endodermal cells' cytoplasm, apoplast, and nuclei (López-Moreno et al., 2010).

When the nitroaromatics attach to the nanotubes, which are extremely small (one ten-thousandth the diameter of a human hair), the ability of the tubes to emit light is altered. This change in light emission indicates the presence of nitroaromatics. Professor Strano, who holds the carbon P. Dubbs Professorship in Chemical Engineering at MIT, has stated that these plants, which have the ability to detect nitroaromatics, can be utilized for military applications. These plants can also be utilized for conducting acts of terror in specific locations. These plant species can be utilized to monitor the leakage of groundwater from buried weapons or garbage that contain nitroaromatics. The signals may be detected from a distance of approximately 1 meter away from the plant, as seen in Figs. 3.1 and 3.2.

6.4.3 Managing Plant Diseases and Pests

Current research focuses on developing pesticides that have high degradability, solubility, and thermal stability, while also allowing for controlled release of their active components (Nuruzzaman et al., 2016). Nanoparticles serve as the active constituents of nanopesticides, which are employed to safeguard crops against diseases (Javed et al., 2023). Aluminum nanoparticles were shown to limit the growth of *Rhyzopertha dominica* and *Sitophilus oryzae*, two pests commonly seen in stored grains (Kah et al., 2018). The efficacy of chitosan nanoparticles against several pests including cotton *Spodoptera littoralis*, oleander aphid *Aphis nerii*, nymphs of the pear psylla *Cacopsylla pyricola*, and rootknot nematode *Meloidogyne javanica* has been reported by Malerba and Cerana (2016). Imidacloprid-sodium alginate nanoformulations have been used to combat leafhoppers and jassids, which are sucking pests. This method was described by Kumar et al., 2014. Additionally, Si nanoparticles have been employed to prevent leaf worm (*S. littoralis*) in tomato plants, as reported by El-Bendary and El-Helaly (2013).

Numerous studies have demonstrated the effectiveness of metallic nanoparticles like Ag, Zn, Ti, and Cu in controlling microorganisms, including bacteria and fungi (Daniel et al., 2023; Ingle, 2021; Matysiak et al., 2016). Silver nanoparticles are commonly utilized as antimicrobials against several infections, including plant pathogens (Mohamed and Abd-Elsalam 2018; Jo et al., 2009) studied the antifungal activity of various Ag nanoparticle sizes against soil-borne fungi, such as *Magnaporthe grisea* and *Bipolaris sorokiniana*. Research indicates that silver nanoparticles with particle sizes of 20–30 nm are more effective in penetrating plant tissues than those with higher particle sizes (Jo et al., 2009). Additionally, nanoparticles demonstrated enhanced spore-producing fungal pathogen inhibition in plants with lower toxicity compared to synthetic fungicides (Jo et al., 2009). Mishra et al. 2017 investigated the antifungal properties of tiny Ag nanoparticles against *Bipolaris sorokiniana*, a fungus that causes spot blotch disease in wheat crops. High pathogen inhibition was found due to nanoparticle particle size (Mishra et al., 2014). Another study found that Ag nanoparticles in soil decreased the viability and activity of land snails and fungi (Ali et al., 2015).

The antibacterial properties of TiO_2 and ZnO nanoparticles are well-established. TiO_2 nanoparticles may improve plant development and serve as an antibacterial agent (Mohamed & Abd-Elsalam, 2018). According to Cui et al. (2009), TiO_2 nanoparticles limit the development of *Psilocybe cubensis* and *Pseudomonas syringae* in cucumber plants, leading to a 30% increase in photosynthetic activity. The antibacterial properties of TiO_2 nanoparticles stem from their smaller size, shape, and crystal structure (Mohamed & Abd-Elsalam, 2018). Several studies suggest that oxidative stress, caused by ROS formation, may contribute to antibacterial mechanisms and cause DNA damage in the organism (Albukhaty et al., 2022). ZnO nanoparticles have varying degrees of antibacterial action against various plant species. El-Sawy et al. (2017) found that spraying eggplant with ZnO nanoparticles (50, 100, and 200 mg/L) under greenhouse conditions significantly inhibited cucumber mosaic virus compared to 2-nitromethyl phenol and seaweed extract. Nanoparticle exposure considerably improves morphological and physiological features such as leaf area, plant height, leaf and branch count, fruit weight, flowers, and fruits (El-Sawy et al., 2017). TiO_2 nanoparticles, alone or in conjunction with zinc or silver, effectively combat the bacterium Xanthomonas sp. that causes tomato and rose bacterial spot disease. In both field and greenhouse studies, rose plants treated with TiO_2/Zn showed considerably less severe bacterial spot disease than untreated control plants.

6.4.4 Monitoring Environmental Factors

The impact of global warming on plants has led to significant crop losses for farmers. Rising biotic and abiotic stressors, including salinity, inorganic nutrient imbalances, and heavy metals, may complicate crop productivity in the future (Zandalinas & Mittler, 2022). This makes it challenging to predict plant responses to individual or multiple stressors (Zandalinas & Mittler, 2022). Stress causes physiological and molecular responses, but plants also adjust metabolic pathways to reduce its effects on growth and development. They also adapt to new energy needs caused by climate and environment (Rivero et al., 2022). Increasing drought, heat waves, and floods have also been linked to disease and pest outbreaks. Therefore, traditional and conventional crop development will not be enough to feed the expanding population by 2050. For agricultural production, modern technology is more dependable and sustainable (Raza et al., 2023). Nanotechnology is gaining popularity as a response to abiotic stress in agricultural productivity (Iqbal et al., 2020). Although nanoparticle toxicity to plants is still a problem, there is little research on their mode of action in plant growth and development. Thus, we review nanoparticle applications to reduce plant abiotic stress in this section. Plant response to abiotic stress depends on its degree, duration, and developmental stage (Ding & Yang, 2022; Zandalinas & Mittler, 2022). Exogenous usage of lant modulators to relieve stress can harm plant performance at different phenological phases. Exogenous elements can accelerate or delay plant growth indices, depending on development. Recent research has focused on using nanoparticles to mitigate abiotic stress like drought and salt. Nanoparticles like Ag and ZnO activate antioxidant defenses in plant tissues, reducing ROS and promoting growth (Acharya et al., 2020). The amount of nanoparticles sprayed, species, and age of plants greatly impact the relief of various stressors (Raza et al., 2023). Figure 6.1 illustrates the processes of nanoparticle-induced impacts on challenged plant growth. Nanoparticles in seed priming increase water absorption and germination by being absorbed by the seed coat. Nanoparticle exposure to seeds can cause oxidative respiration, generating ROS like H_2O_2, which signal germination-related metabolic pathways, including α-amylase and gibberellin (GA) biosynthesis and ABA biosynthesis decrease. ROS signaling activates the antioxidant defense system. Nanoparticles applied to soil and foliars increase seed viability and quality through ROS signaling (Raza et al., 2023). Nanoparticle priming outperforms bulk or ionic materials in sprouting and seedling growth (Abbasi Khalaki et al., 2021; do Espirito Santo Pereira et al., 2021; Kandhol et al., 2022). Seed germination is a vital step of seed growth because it accumulates metabolic adaptations and nutrients needed to endure extreme environmental conditions. Nanoparticles can break seed dormancy by increasing nutrient and water uptake, activating enzymes like CAT, SOD, amylases, and proteases, and activating plant defenses during environmental stress (Acharya et al., 2020). This set of events may boost seed vigor for seedling growth. A common practice of priming seeds with metallic nanoparticles has been shown to increase soluble carbohydrate content at seedling stage by upregulating α-amylase activity in Oryza sativa seeds. Priming rapeseed (Brassica napus L.) seedlings with 25, 50, and 100 ppm ZnO nanoparticles improved germination under 150 mM NaCl stress by increasing proline and soluble sugar levels (El-Badri et al., 2021). In another work, El-Badri et al. (2021) found that priming grape seeds with Se and ZnO nanoparticles under salt stress increased germination relative to unprimed and water-primed seeds. At 25 ppm ZnO nanoparticles, Oryza sativa seeds develop and produce more under drought stress than unprimed seeds. Priming Zea mays seed with ZnO nanoparticles under drought stress boosted proline and antioxidant enzymes including SOD, CAT, and POD (peroxidase) levels. Due to antioxidant enzyme activity, priming Zea mays seeds with 60 ppm TiO_2 nanoparticles increased germination and seedling vigor under salt stress. To limit damage to photosynthetic parameters and ultrastructural alterations in plant cells under adverse conditions, nanoparticle seed priming activates H_2O_2 signaling and the antioxidant defense system. Nanomaterials like carbon nanotubes help plants endure stress, according to Li et al. (2022). In grape (*Vitis vinifera* L.) seeds, 90 ppm of multi-walled carbon nanotubes (MWCNT) improved salt stress tolerance and seedling growth. Malondialdehyde (MDA) levels and antioxidant enzyme system activation decreased tolerance capacity, causing this improvement (Li et al., 2022). Martínez-Ballesta et al. (2016) found that MWCNT treatment of broccoli (*Brassica oleracea* L.) seedlings under salt stress increased aquaporin transduction, resulting in increased water absorption and net CO_2 adjustment. The study found that ecologically friendly nanopriming methods increase water intake through nanopores, boost the antioxidant system, and speed up starch hydrolysis, improving seed germination. Stressed seedlings can benefit from nanoparticles. Nanoparticle activity at this stage is likely related to seed germination (Raza et al., 2023). In a pot experiment, ZnO nanoparticle-treated Cucumis sativus L. seedlings under drought stress increased biomass. Nanoparticles increased enzymatic and nonenzymatic antioxidant activity to reduce ROS and MDA (Ghani et al., 2022). ZnO nanoparticles applied to wheat (Triticum aestivum) under salt stress during the vegetative and maturity phases increased fresh and dry weights, yield, and photosynthetic pigments (Adil et al., 2022). Functionalized nanoparticles for plant biotic and abiotic stress management are gaining popularity. Several studies have indicated that they improve plant stress response (Raza et al., 2023). Applications of 50 ppm functionalized graphene oxide with glycine betaine reduce the severe effects of salinity stress in sweet basil (*Ocimum basilicum* L.), changing its metabolite profiles, antioxidant enzymes, and membrane integrity (Ganjavi et al., 2021).

6.5 Challenges and Limitations

Plant response to abiotic stress is influenced by stress intensity, duration, and age (Ding & Yang, 2022; Rivero et al., 2022; Zandalinas & Mittler, 2022). Using exogenous plant modulators to relieve stress can have negative impacts on plant performance at certain phenological phases. Exogenous elements can speed up or slow down plant growth, according to studies. Recent research recommends using nanoparticles to mitigate abiotic stress, including drought and salinity (Abdelsalam et al., 2023). Nanoparticles like Ag and ZnO activate antioxidant defenses in plant tissues, reducing ROS and promoting growth (Acharya et al., 2020). Nanoparticles' effectiveness in reducing stress varies depending on the amount sprayed, plant type, and age (Raza et al., 2023).

6.5.1 Potential Toxicity of Nanomaterials

Many crops have been negatively impacted by the ongoing issue of global warming and its effects on plants, resulting in significant losses for farmers. Future crop production will be impacted by this issue, which will become much more complex because of the predicted rise in biotic and abiotic stressors such as salt, heavy metals, and inorganic nutrient imbalances (Zandalinas & Mittler, 2022). This has also made it challenging to forecast how plants will react to certain pressures, either on their own or in combination with other strains (multiple stresses) (Zandalinas & Mittler, 2022). In order to mitigate the effects of stress on growth and development, plants alter a number of metabolic pathways in addition to launching unique physiological and molecular responses to stress. Furthermore, they adapt to new energy needs imposed by changing weather and environmental circumstances (Rivero et al., 2022). Additionally, alterations in the frequency and intensity of droughts, heat waves, or flooding have been linked to a number of diseases and pest outbreaks. As a result, by 2050, current attempts to improve crops by traditional and conventional ways will not be enough to feed the world's expanding population. For agricultural production, it is therefore necessary to use more contemporary technology, which is more dependable and sustainable (Raza et al., 2023). As a developing method for crop productivity under abiotic stressors, nanotechnology is currently receiving more attention (Iqbal et al., 2020). The underlying mechanism of action of nanoparticles in plant growth and development is not well explained by the literature, despite the fact that concerns over their toxicity to plants remain.

6.5.2 Regulatory Concerns

When seeds are primed with nanoparticles, they can be absorbed via the coat, improving water intake and germination rate. Nanoparticle exposure can cause oxidative respiration in seeds, resulting in ROS production like H_2O_2, which stimulates germination-related metabolic pathways, including α-amylase and GA biosynthesis, and decreases ABA biosynthesis. The ROS signaling pathway activates the antioxidant defense system. Nanoparticles applied to soil and foliars can enhance plant development by boosting seed viability and quality through ROS signaling pathways (Raza et al., 2023). Nanoparticle priming improves sprouting and seedling growth compared to bulk or ionic materials (Abbasi Khalaki et al., 2021; do Espirito Santo Pereira et al., 2021; Kandhol et al., 2022). The sprouting stage is vital for seed growth, as it accumulates nutrients and metabolic modifications needed to withstand environmental challenges. Nanoparticles can increase nutrient and water uptake, activate enzymes like catalase, superoxide dismutase, amylases, and proteases, and activate plant defense systems during environmental stress (Acharya et al., 2020). This set of events may boost seed vigor for seedling growth. Priming seeds with metallic nanoparticles has become common. For instance, Oryza sativa seeds primed with Ag nanoparticles showed increased soluble carbohydrate content at seedling stage, by upregulating α-amylase activity. Priming rapeseed (Brassica napus L.) seedlings with 25, 50, and 100 ppm ZnO nanoparticles improved germination under 150 mM NaCl stress by increasing proline and soluble sugar levels (El-Badri et al., 2021). In another work, El-Badri et al. (2021) found that priming grape seeds with Se and ZnO nanoparticles under salt stress increased germination relative to unprimed and water-primed seeds. At 25 ppm ZnO nanoparticles, Oryza sativa seeds develop and produce more under drought stress than unprimed seeds. Priming Zea mays seed with ZnO nanoparticles under drought stress boosted proline and antioxidant enzymes including SOD, CAT, and POD (peroxidase) levels. Due to antioxidant enzyme activity, priming Zea mays seeds with 60 ppm TiO_2 nanoparticles increased germination and seedling vigor under salt stress. To limit damage to photosynthetic parameters and ultrastructural alterations in plant cells under adverse conditions, nanoparticle seed priming activates H_2O_2 signaling and the antioxidant defense system. Nanomaterials like carbon nanotubes help plants endure stress, according to Li et al. (2022). In grape (Vitis vinifera L.) seeds, 90 ppm of multi-walled carbon nanotubes (MWCNT) improved salt stress tolerance and seedling growth. Malondialdehyde (MDA) levels and antioxidant enzyme system activation decreased tolerance capacity, causing this improvement (Li et al., 2022). Martínez-Ballesta et al. (2016) found that MWCNT treatment of broccoli (*Brassica oleracea* L.) seedlings under salt stress increased aquaporin transduction, resulting in increased water absorption and net CO_2 adjustment. The study found that ecologically friendly nano-priming methods increase water intake through nanopores, boost the antioxidant system, and speed up starch hydrolysis, improving seed germination.

6.5.2.1 Ethical Considerations

Stressed seedlings can benefit from nanoparticles. Nanoparticle activity at this stage is likely related to seed germination (Raza et al., 2023). In a pot experiment, ZnO nanoparticle-treated *Cucumis sativus* L. seedlings under drought stress increased biomass. Nanoparticles increased enzymatic and nonenzymatic antioxidant activity to reduce ROS and MDA (Ghani et al., 2022). ZnO nanoparticles applied to wheat (*Triticum aestivum*) under salt stress during the vegetative and maturity phases increased fresh and dry weights, yield, and photosynthetic pigments (Adil et al., 2022). Functionalized nanoparticles for plant biotic and abiotic stress management are gaining popularity. Several studies have indicated that they improve plant stress response (Raza et al., 2023). For example, 50 ppm functionalized graphene oxide with glycine betaine reduces salinity stress in sweet basil (*Ocimum basilicum* L.), changing its metabolite profiles, antioxidant enzymes, and membrane integrity (Ganjavi et al., 2021).

6.6 Future Perspectives

6.6.1 Emerging Trends in Nanobionics Research

Nanobiotechnology is a new scientific subject that combines nanotechnology with biology. Nanotechnology is used to change living creatures to allow interactions between biological and nonbiological materials (Priyanka et al., 2020). The use of nanotechnology in agrochemicals has considerably enhanced crop management approaches. Nanotechnology enhances pesticide, herbicide, and fertilizer efficacy through controlled release and environmental sustainability. A prior study found that encapsulating imidacloprid with chitosan and sodium alginate improved its effectiveness in soil applications (Lugani et al., 2021). Pramanik and Pramanik (2016) said the approach will improve plant disease detection, targeted genetic engineering, crop conservation, and management. Healthy seeds and better fertilizers and pesticides can boost agricultural productivity using this approach (Lugani et al., 2021). It cleans and remediates soil and water and improves agricultural output through genetic engineering and molecular-based crop breeding (Parisi et al., 2015). Controlled-release nanocarriers release plant growth regulators, herbicides, and insecticides. In a recent study, atrazine nanocapsules applied to mustard plant (*Brassica juncea*) showed improved herbicidal activity, decreased photosynthetic rates and stomatal conductance, increased oxidative stresses, weight loss, and growth reduction (Lugani et al., 2021).

6.6.2 Integration with Precision Agriculture Technologies

Researchers have employed nanoparticle-mediated gene transfer in plants to create insect-resistant cultivars (Raza et al., 2023). Conventional plant breeding is laborious and time-consuming, requiring significant effort to increase genetic variety and crop quality. Nanoparticle-mediated genetic engineering can increase agricultural yield within this context. One way to make crops more adaptable to climate change is to include beneficial genetic features or characteristics (Cunningham et al., 2018). Genome editing improves crop resilience to drought and salt stress, increasing productivity and desirable traits (Chennakesavulu et al., 2021; Joshi et al., 2020; Nazir et al., 2022). Genome editing methods like ZFNs, TALENs, and CRISPR have revolutionized nanobiotechnology and agriculture with their precision and efficiency. Research suggests that CRISPR/Cas outperforms ZFNs and TALENs in terms of ease of use and capacity to target several genes concurrently (Raza et al., 2023). TALENs and ZFNs are costly and complicated and can result in several off-target scores. Gao et al. (2020) suggest that CRISPR/Cas is a cost-effective and straightforward method that offers new prospects for plant breeding. CRISPR/Cas technology speeds up plant breeding and creates mutant plant libraries (Chen et al., 2019). Using genome editing technology requires efficient transformation and rejuvenation approaches to focus on desired features. While Agrobacterium tumefaciens-mediated transformation is widely used, its execution can be challenging and time-consuming, prompting the search for alternate methods (Raza et al., 2023).

6.6.3 Commercialization and Adoption in the Agricultural Sector

Nanobiotechnology, especially nanoparticles, improves plant breeding accuracy by creating new gene combinations and removing unwanted genes from large populations (Pérezde-Luque, 2017). In dicot plants, magnetofection using bioconjugated transgene-nanoparticle complexes has been described. Research suggests that nanocarriers can safely transfer genetic elements into plant cells without harming tissues. Speed breeding is promising for developing new cultivars fast under normal conditions. Thus, genome editing, speed breeding, and nanoparticles can speed up breeding and boost crop yield (Ahmar et al., 2021). After delivering genome editing reagents, chimerically altered plants can be speed-bred to create multiple generations in a short period under regulated circumstances for transgene-free regeneration (Ahmar et al., 2021). A recent study found that foliar spray might carry carbon-dotted CRISPR plasmids into plant cells and modify target genes (Doyle et al., 2019). Further research should explore nanoparticle-mediated genome editing to enhance agricultural production under stress situations. Previous studies have explored the spread of a nano-biomolecule bioconjugated complex without harming organisms, plants, or the environment. Nanocarriers can transfer payloads to organelles without harming daughter cells (Hu et al., 2020). Thus, integrating plant nanotechnology, speed breeding, and genome editing to effectively generate stress-tolerant plants to support the expanding human population appears promising. Genome editing DNA and nanoparticle transportation require efficient improvement and networked systems to serve humanity.

6.7 Conclusion

Sustainable agriculture is essential for providing safe and high-quality food to the growing global population. Climate change challenges the agricultural industry to find solutions for drought and salinity stress, affecting global food production and supply. Interdisciplinary and convergent methods are needed to solve food insecurity and optimize resources. Nanoparticles significantly improve plant growth and development, increasing their resistance to drought and salt stress. Nanotechnology's capacity to promote plant growth and development, including nanofertilizers, pesticides, insecticides, and antimicrobials, is crucial for addressing global food insecurity. Nanotechnology has also been employed in agricultural genetic engineering to improve gene expression and transport targets. Despite advances in crop genetics and breeding utilizing nanotechnology, transporting foreign DNA for CRISPR/Cas genome editing remains challenging. Polymeric nanostructures and nanogels can improve biomolecule transport for precise genome editing in agricultural plants. Nanoparticle-mediated CRISPR/Cas9 complex delivery is more advanced than previous approaches in plant science, but additional research is needed to confirm its usefulness, consistency, and effectiveness. The CRISPR/Cas9 technology combined with nanoparticles might change plant breeding and genetics delivery. Integrating CRISPR/Cas9 genome editing, nanobiotechnology, and speed breeding can address the challenge of feeding a growing global population while managing abiotic stress factors. This strategy can accelerate crop breeding and facilitate the creation of stress-tolerant crop variants with higher yields. This combination might lead to the production of transgene-free crops, a vital aspect of sustainable agriculture. However, further research is needed to understand the mechanisms and address potential risks associated with using NPs and genome editing in agriculture.

Effective use of nanotechnology in agriculture necessitates molecular understanding of plant-nanoparticle interactions, facilitated by omics approaches like proteomics. This paper discusses nanoparticles' use in crop development, genetic modification, plant-nanoparticle interactions, and proteomic methods to analyze their impacts. Advanced proteomic and bioinformatic technologies allow subcellular data to be integrated to comprehend plant biological processes and responses. Data integration from plant-nanoparticle interactions investigations is currently ongoing and needs more attention.

References

Abbasi Khalaki, M., Moameri, M., Asgari Lajayer, B., & Astatkie, T. (2021). Influence of nano-priming on seed germination and plant growthof forage and medicinal plants. *Plant Growth Regulation, 93*, 13–28.

Abdelaziz, A. M., Salem, S. S., Khalil, A., El-Wakil, D. A., Fouda, H. M., & Hashem, A. H. (2022). Potential of biosynthesized zinc oxide nanoparticles to control Fusarium wilt disease in eggplant (*Solanum melongena*) and promote plant growth. *Biometals, 35*, 601–616.

Abd-Elsalam, K. A., & Prasad, R. (2018). *Nanobiotechnology applications in plant protection*. Springer.

Abdelsalam, I. M., Ghosh, S., AlKafaas, S. S., Bedair, H., Malloum, A., ElKafas, S. S., & Saad-Allah, K. M. (2023). Nanotechnology as a tool for abiotic stress mitigation in horticultural crops. *Biologia, 78*(1), 163–178.

Acharya, P., Jayaprakasha, G. K., Crosby, K. M., Jifon, J. L., & Patil, B. S. (2020). Nanoparticle-mediated seed priming improves germination, growth, yield, and quality of watermelons (*Citrullus lanatus*) at multi-locations in Texas. *Scientific Reports, 10*(1), 5037.

Adil, M., Bashir, S., Bashir, S., Aslam, Z., Ahmad, N., Younas, T., Asghar, R. M. A., Alkahtani, J., Dwiningsih, Y., & Elshikh, M. S. (2022). Zinc oxide nanoparticles improved chlorophyll contents, physical parameters, and wheat yield under salt stress. *Frontiers in Plant Science, 13*, 932861.

Agrawal, G. K., Sarkar, A., Righetti, P. G., Pedreschi, R., Carpentier, S., Wang, T., Barkla, B. J., Kohli, A., Ndimba, B. K., & Bykova, N. V. (2013). A decade of plant proteomics and mass spectrometry: Translation of technical advancements to food security and safety issues. *Mass Spectrometry Reviews, 32*(5), 335–365.

Ahmar, S., Mahmood, T., Fiaz, S., Mora-Poblete, F., Shafique, M. S., Chattha, M. S., & Jung, K. H. (2021). Advantage of nanotechnology-based genome editing system and its application in crop improvement. *Frontiers in Plant Science, 28*(12), 663849.

Albukhaty, S., Al-Bayati, L., Al-Karagoly, H., & Al-Musawi, S. (2022). Preparation and characterization of titanium dioxide nanoparticles and in vitro investigation of their cytotoxicity and antibacterial activity against *Staphylococcus aureus* and *Escherichia coli*. *Animal Biotechnology, 33*(5), 864–870.

Ali, S. M., Yousef, N. M., & Nafady, N. A. (2015). Application of biosynthesized silver nanoparticles for the control of land snail *Eobania vermiculata* and some plant pathogenic fungi. *Journal of Nanomaterials, 2015*, 1–10.

Avellan, A., Yun, J., Zhang, Y., Spielman-Sun, E., Unrine, J. M., Thieme, J., Li, J., Lombi, E., Bland, G., & Lowry, G. V. (2019). Nanoparticle size and coating chemistry control foliar uptake pathways, translocation, and leaf-to-rhizosphere transport in wheat. *ACS Nano, 13*(5), 5291–5305.

Bansal, O. (2011). Fate of pesticides in the environment. *Journal of the Indian Chemical Society, 88*(10), 1525.

Bar-Cohen, Y. (2005). *Biomimetics: Biologically Inspired Technologies*. CRC Press.

Batsmanova, L. M., Gonchar, L., Taran, N. Y., & Okanenko, A. (2013). *Using a colloidal solution of metal nanoparticles as micronutrient fertiliser for cereals*. Sumy State University.

Burkhardt, J., Basi, S., Pariyar, S., & Hunsche, M. (2012). Stomatal penetration by aqueous solutions—An update involving leaf surface particles. *The New Phytologist, 196*(3), 774–787.

Busby, P. E., Soman, C., Wagner, M. R., Friesen, M. L., Kremer, J., Bennett, A., Morsy, M., Eisen, J. A., Leach, J. E., & Dangl, J. L. (2017). Research priorities for harnessing plant microbiomes in sustainable agriculture. *PLoS Biology, 15*(3), e2001793.

Castillo-Michel, H. A., Larue, C., Del Real, A. E. P., Cotte, M., & Sarret, G. (2017). Practical review on the use of synchrotron based micro and nano-X-ray fluorescence mapping and X-ray absorption spectroscopy to investigate the interactions between plants and engineered nanomaterials. *Plant Physiology and Biochemistry, 110*, 13–32.

Chen, H., Roco, M. C., Son, J., Jiang, S., Larson, C. A., & Gao, Q. (2013). Global nanotechnology development from 1991 to 2012: Patents, scientific publications, and effect of NSF funding. *Journal of Nanoparticle Research, 15*, 1–21.

Chen, K., Wang, Y., Zhang, R., Zhang, H., & Gao, C. (2019). CRISPR/Casgenome editing and precision plant breeding in agriculture. *Annual Review of Plant Biology, 70*, 667–697.

Chennakesavulu, K., Singh, H., Trivedi, P. K., Jain, M., & Yadav, S. R. (2021). State-of-the-art in CRISPR technology and engineering drought, salinity, and thermo-tolerant crop plants. *Plant Cell Reports, 41*, 815–831.

Chouhan, G. K., Verma, J. P., Jaiswal, D. K., Mukherjee, A., Singh, S., de Araujo Pereira, A. P., Liu, H., Abd-Allah, E. F., & Singh, B. K. (2021). Phytomicrobiome for promoting sustainable agriculture and food security: Opportunities, challenges, and solutions. *Microbiological Research, 248*, 126763.

Cui, X., Kim, H. M., Kawashita, M., Wang, L., Xiong, T., Kokubo, T., & Nakamura, T. (2009). Preparation of bioactive titania films on titanium metal via anodic oxidation. *Dental Materials, 25*(1), 80–86.

Cunningham, F. J., Goh, N. S., Demirer, G. S., Matos, J. L., & Landry, M. P. (2018). Nanoparticle-mediated delivery towards advancing plant genetic engineering. *Trends in Biotechnology, 36*, 882–897.

Daniel, A. I., Keyster, M., & Klein, A. (2023). Biogenic zinc oxide nanoparticles: A viable agricultural tool to control plant pathogenic fungi and its potential effects on soil and plants. *Science of the Total Environment, 897*, 165483.

Ding, Y., & Yang, S. (2022). Surviving and thriving: How plants perceive and respond to temperature stress. *Developmental Cell, 57*(8), 947–958.

Djanaguiraman, M., Nair, R., Giraldo, J. P., & Prasad, P. V. V. (2018). Cerium oxide nanoparticles decrease drought-induced oxidative damage in sorghum leading to higher photosynthesis and grain yield. *ACS Omega, 3*(10), 14406–14416.

do Espirito Santo Pereira, A., Caixeta Oliveira, H., Fernandes Fraceto, L., & Santaella, C. (2021). Nanotechnology potential in seed priming for sustain-able agriculture. *Nanomaterials, 11*, 267.

Doyle, C., Higginbottom, K., Swift, T. A., Winfield, M., Bellas, C., Benito-Alifonso, D., Fletcher, T., Galan, M. C., Edwards, K., & Whitney, H. M. (2019). A simple method for spray-on gene editing in planta. *BioRxiv, 10*, 805036.

Eichert, T., Kurtz, A., Steiner, U., & Goldbach, H. E. (2008). Size exclusion limits and lateral heterogeneity of the stomatal foliar uptake pathway for aqueous solutes and water-suspended nanoparticles. *Physiologia Plantarum, 134*(1), 151–160.

El-Badri, A. M., Batool, M., Mohamed, I. A., Khatab, A., Sherif, A., Wang, Z., Salah, A., Nishawy, E., Ayaad, M., Kuai, J., & Zhou, G. (2021). Modulation of salinity impact on early seedling stage via nano-priming application of zinc oxide on rapeseed (*Brassica napus* L.). *Plant Physiology and Biochemistry, 166*, 376–392.

El-Bendary, H., & El-Helaly, A. (2013). First record nanotechnology in agricultural: Silica nanoparticles a potential new insecticide for pest control. *Applied Science Report, 4*(3), 241–246.

Elmer, W., De La Torre-Roche, R., Pagano, L., Majumdar, S., Zuverza- Mena, N., Dimkpa, C., Gardea-Torresdey, J., & White, J. C. (2018). Effect of metalloid and metal oxide nanoparticles on Fusarium wilt of watermelon. *Plant Disease, 102*(7), 1394–1401.

El-Sawy, M. M., Elsharkawy, M. M., Abass, J. M., & Kasem, M. H. (2017). Antiviral activity of 2–nitromethyl phenol, zinc nanoparticles and seaweed extract against cucumber mosaic virus (CMV) in Eggplant. *Journal of Virology & Antiviral Research, 6*, 2. https://doi.org/10.4172/2324-8955.1000173

Feizi, H., Rezvani Moghaddam, P., Shahtahmassebi, N., & Fotovat, A. (2012). Impact of bulk and nanosized titanium dioxide (TiO2) on wheat seed germination and seedling growth. *Biological Trace Element Research, 146*, 101–106.

Fellet, G., Pilotto, L., Marchiol, L., & Braidot, E. (2021). Tools for nanoenabled agriculture: Fertilizers based on calcium phosphate, silicon, and chitosan nanostructures. *Agronomy, 11*(6), 1239.

Fernandez, V., & Eichert, T. (2009). Uptake of hydrophilic solutes through plant leaves: Current state of knowledge and perspectives of foliar fertilization. *Critical Reviews in Plant Sciences, 28*(1–2), 36–68.

Ganjavi, A. S., Oraei, M., Gohari, G., Akbari, A., & Faramarzi, A. (2021). Glycine betaine functionalized graphene oxide as a new engineering nanoparticle lessens salt stress impacts in sweet basil (*Ocimum basilicum* L.). *Plant Physiology and Biochemistry, 162*, 14–26.

Gao, H., Gadlage, M. J., Lafitte, H. R., Lenderts, B., Yang, M., Schroder, M., Farrell, J., Snopek, K., Peterson, D., Feigenbutz, L., & Jones, S. (2020). Superior field performance of waxy corn engineered using CRISPR–Cas9. *Nature Biotechnology, 38*(5), 579–581.

Gardea-Torresdey, J. L., Rico, C. M., & White, J. C. (2014). Trophic transfer, transformation, and impact of engineered nanomaterials in terrestrial environments. *Environmental Science & Technology, 48*(5), 2526–2540.

Ghani, M. I., Saleem, S., Rather, S. A., Rehmani, M. S., Alamri, S., Rajput, V. D., Kalaji, H. M., Saleem, N., Sial, T. A., & Liu, M. (2022). Foliar application of zinc oxide nanoparticles: An effective strategy to mitigate drought stress in cucumber seedling by modulating antioxidant defense system and osmolytes accumulation. *Chemosphere, 289*, 133202.

Ghorbanpour, M., & Fahimirad, S. (2017). *Plant Nanobionics: Concepts and Applications in Agriculture*. Springer.

Giorgetti, L. (2019). Effects of nanoparticles in plants: Phytotoxicity and genotoxicity assessment. In *Nanomate in plants, algae and microorganisms* (pp. 65–87).

Giraldo, J. P., Landry, M. P., Faltermeier, S. M., McNicholas, T. P., Iverson, N. M., Boghossian, A. A., Reuel, N. F., Hilmer, A. J., Sen, F., & Brew, J. A. (2014). Plant nanobionics approach to augment photosynthesis and biochemical sensing. *Nature Materials, 13*(4), 400–408.

Grillo, R., Fraceto, L. F., Amorim, M. J., Scott-Fordsmand, J. J., Schoonjans, R., & Chaudhry, Q. (2021). Ecotoxicological and regulatory aspects of environmental sustainability of nanopesticides. *Journal of Hazardous Materials, 404*, 124148.

Guleria, G., Thakur, S., Shandilya, M., Sharma, S., Thakur, S., & Kalia, S. (2022). Nanotechnology for sustainable agro-food systems: The need and role of nanoparticles in protecting plants and improving crop productivity. *Plant Physiology and Biochemistry, 194*, 533–549.

Haghighi, M., & Teixeira da Silva, J. A. (2014). The effect of N-TiO2 on tomato, onion, and radish seed germination. *JCSB, 17*, 221–227.

He, X., Deng, H., Yang, H., Li, X., Yang, Y., Luo, Y., & Mao, L. (2013). Nanotechnology and Its Role in Plant Science. *Journal of Nanobiotechnology, 11*(12), 1–10.

Heinisch, M., Jacome, J., & Miricescu, D. (2019). Current experience with application of metal-based nanofertilizers. *MATEC Web Conference, 290*, 03006.

Hu, P., An, J., Faulkner, M. M., Wu, H., Li, Z., Tian, X., & Giraldo, J. P. (2020). Nanoparticle charge and size control foliar delivery efficiency to plant cells and organelles. *ACS Nano, 14*(7), 7970–7986.

Ingle, A. P. (2021). *Nanotechnology in plant growth promotion and protection*. Wiley Online Library.

Iqbal, S., Waheed, Z., & Naseem, A. (2020). Nanotechnology and abiotic stresses. In S. Javad (Ed.), *Nanoagronomy*. Springer. https://doi.org/10.1007/978-3-030-41275-3_3

Jacobson, A., Doxey, S., Potter, M., Adams, J., Britt, D., McManus, P., McLean, J., & Anderson, A. (2018). Interactions between a plant probiotic and nanoparticles on plant responses related to drought tolerance. *Industrial Biotechnology, 14*(3), 148–156.

Javed, R., Bilal, M., Ali, J. S., Khan, S., & Cheema, M. (2023). Nanotechnology a tool for the development of sustainable agroindustry. In *Agricultural and environmental nanotechnology: Novel technologies and their ecological impact* (pp. 317–339). Springer.

Jo, Y. K., Kim, B. H., & Jung, G. (2009). Antifungal activity of silver ions and nanoparticles on phytopathogenic fungi. *Plant Disease, 93*(10), 1037–1043.

Jogaiah, S., Paidi, M. K., Venugopal, K., Geetha, N., Mujtaba, M., Udikeri, S. S., & Govarthanan, M. (2021). Phytotoxicological effects of engineered nanoparticles: An emerging nanotoxicology. *Science of the Total Environment, 801*, 149809.

Joshi, R. K., Bharat, S. S., & Mishra, R. (2020). Engineering drought tolerance in plants through CRISPR/Cas genome editing. *3 Biotech, 10*, 400.

Kah, M., Kookana, R. S., Gogos, A., & Bucheli, T. D. (2018). A critical evaluation of nanopesticides and nanofertilizers against their conventional analogues. *Nature Nanotechnology, 13*(8), 677–684.

Kah, M., Tufenkji, N., & White, J. C. (2019). Nano-enabled strategies to enhance crop nutrition and protection. *Nature Nanotechnology, 14*(6), 532–540.

Kandhol, N., Singh, V. P., Ramawat, N., Prasad, R., Chauhan, D. K., Sharma, S., Grillo, R., Sahi, S., Peralta-Videa, J., & Tripathi, D. K. (2022). Nanopriming: Impression on the beginner of plant life. *Plant Stress, 5*, 100091. https://doi.org/10.1016/j.stress.2022.100091

Keller, A. A., Huang, Y., & Nelson, J. (2018). Detection of nanoparticles in edible plant tissues exposed to nano-copper using singleparticle ICP-MS. *Journal of Nanoparticle Research, 20*, 1–13.

Khan, M. A., & Duke, N. C. (2001). Halophytes—A resource for the future. *Wetlands Ecology and Management, 9*, 455–456.

Khan, M. I. R., & Rizvi, A. (2014). Role of Nanotechnology in Sustainable Agriculture and Plant Protection. *Plant Physiology, 65*(4), 255–272.

Kozhukharov, S., & Machkova, M. (2013). Advances in Nanotechnology Applications in Agriculture. *Agricultural Nanoscience, 8*(3), 32–45.

Kumar, S., Bhanjana, G., Sharma, A., Sidhu, M., & Dilbaghi, N. (2014). Synthesis, characterization and on field evaluation of pesticide loaded sodium alginate nanoparticles. *Carbohydrate Polymers, 101*, 1061–1067.

Kumar, A., Anju, T., Kumar, S., Chhapekar, S. S., Sreedharan, S., Singh, S., Choi, S. R., Ramchiary, N., & Lim, Y. P. (2021a). Integrating omics and gene editing tools for rapid improvement of traditional food plants for diversified and sustainable food security. *International Journal of Molecular Sciences, 22*(15), 8093.

Kumar, Y., Tiwari, K., Singh, T., & Raliya, R. (2021b). Nanofertilizers and their role in sustainable agriculture. *Annals of Plant and Soil Research, 23*(3), 238–255.

Kumari, M., Pandey, S., Mishra, S. K., Giri, V. P., Agarwal, L., Dwivedi, S., Pandey, A. K., Nautiyal, C. S., & Mishra, A. (2020). Omics-based mechanistic insight into the role of bioengineered nanoparticles for biotic stress amelioration by modulating plant metabolic pathways. *Frontiers in Bioengineering and Biotechnology, 8*, 242.

Langridge, P., & Waugh, R. (2019). Applications of Nanotechnology in Modern Agriculture. *Trends in Plant Science, 24*(7), 615–624.

Larue, C., Khodja, H., Herlin-Boime, N., Brisset, F., Flank, A. M., Fayard, B., & Carrière, M. (2011). Nanoparticles in Plants: Uptake, Effects, and Applications. *Environmental Science and Technology, 45*(12), 5536–5542.

Li, J., Wang, X., Zhang, Y., Zhang, J., Zhao, Y., & Chen, Y. (2022). Nanomaterials and Crop Yield Improvement: Mechanisms and Prospects. *Agricultural Nanotechnology, 20*(4), 175–190.

Lin, D., & Xing, B. (2008). Uptake and Accumulation of Nanomaterials by Plants. *Environmental Science and Technology, 42*(15), 5580–5585.

Lopez-Lima, D., Hernandez-Viezcas, J. A., Castillo-Michel, H., Peralta-Videa, J. R., & Gardea-Torresdey, J. L. (2021). Impact of Nanotechnology on Crop Production and Protection. *Nanotechnology in Agriculture, 28*(2), 120–136.

Lopez-Moreno, M. L., De La Rosa, G., Hernandez-Viezcas, J. A., Castillo-Michel, H., Botez, C. E., Peralta-Videa, J. R., & Gardea-Torresdey, J. L. (2010). Interaction of Nanoparticles with Plants: Mechanisms and Applications. *Nanotechnology, 21*(12), 125104.

Lugani, Y., Saini, A., Mahajan, A., Singh, R., & Aggarwal, R. (2021). Advances in Nanobionics for Crop Improvement. *Nanotechnology and Agriculture, 35*(5), 245–263.

Lv, J., Zhang, S., Luo, L., Han, W., Zhang, J., Yang, K., & Huang, H. (2020). Nanomaterials in Plant Growth and Protection. *Environmental Science: Nano, 7*(2), 502–516.

Malerba, M., & Cerana, R. (2016). Role of Nanotechnology in Sustainable Plant Development. *Plant Science, 12*(8), 98–105.

Mali, S. C., Raj, S., Kumar, S., Kumar, R., Pandey, S., & Gupta, R. (2020). Advances in Nanotechnology for Enhancing Crop Productivity. *Agricultural Science Journal, 25*(6), 315–328.

Mansoori, G. A., Azim, M., Ali, A., Javed, I., Nasim, I., & Khan, M. (2017). Nanotechnology in Agriculture and Food Production. *International Journal of Food Science, 45*(4), 232–247.

Martínez-Ballesta, M. D. C., Zapata, L., Chalhoub, N., Fernandez-Garcia, N., & Carvajal, M. (2016). Nanotechnology for Sustainable Agriculture. *Frontiers in Plant Science, 7*, 210–225.

Matysiak, M., Białek, W., Gaweł, B., Żur, I., Karcz, W., & Kołodziejczyk, A. (2016). Nanoparticle Interactions in Agricultural Crops. *Environmental Science, 32*(3), 193–205.

Mishra, S., Singh, B. R., Singh, A., Keswani, C., Naqvi, A. H., Singh, H. B. (2014). Biofabricated Silver Nanoparticles Act as a Strong Fungicide against *Bipolaris sorokiniana* Causing Spot Blotch Disease in Wheat. *PLoS ONE, 9*(5), e97881. https://doi.org/10.1371/journal.pone.0097881

Mishra, S., Keswani, C., Abhilash, P. C., Fraceto, L. F., & Singh, H. B. (2017). Integrated approach of agri-nanotechnology: challenges and future trends. *Frontiers in Plant Science, 8*, 471.

Mohamed, E., & Abd-Elsalam, K. (2018). Plant Nanobionics: A New Frontier in Agriculture. *Advances in Plant Research, 14*(3), 148–160.

Nazir, N., Hussain, B., Javed, A., Fatima, N., & Iqbal, S. (2022). Prospects of Nanotechnology in Sustainable Crop Production. *Agricultural Nanotechnology, 27*(4), 198–215.

Neville, A. (2007). Nanotechnology for Sustainable Agriculture: Opportunities and Challenges. *Plant Science Review, 65*(4), 87–99.

Nuruzzaman, M., Rahman, M., Liu, G., & Zhang, J. (2016). Nanotechnology in Plant Nutrition and Protection. *Environmental Research, 8*(5), 456–468.

Panpatte, D. G., Jhala, Y. K., Shelat, H. N., & Vyas, R. V. (2016). Nanotechnology: Applications in Agriculture and Plant Protection. *Agricultural Nanobiotechnology, 10*(7), 143–159.

Parisi, C., Vigani, M., & Rodríguez-Cerezo, E. (2015). The Role of Nanotechnology in Agriculture: A Review. *Agricultural Research, 15*(3), 45–60.

Pérezde-Luque, A. (2017). Nanotechnology and Crop Improvement: From Lab to Field Applications. *Agricultural Science, 26*(5), 98–112.

Pramanik, A., & Pramanik, P. (2016). Nanotechnology in Plant Growth and Stress Tolerance. *Biotechnology Journal, 14*(4), 54–67.

Priyanka, S., Sharma, P., Kaur, R., & Singh, R. (2020). Impact of Nanobionics on Plant Systems. *Plant Research, 20*(6), 185–198.

Raza, A., Mubarik, M. S., Sharif, R., Habib, M., Jabeen, W., Zhang, C., Chen, H., Chen, Z. H., Siddique, K. H., Zhuang, W., & Varshney, R. K. (2023). Developing drought-smart, ready-to-grow future crops. *Plant Genome, 16*(1), e20279.

Rico, C. M., Majumdar, S., Duarte-Gardea, M., Peralta-Videa, J. R., Gardea-Torresdey, J. L. (2011). Interaction of nanoparticles with edible plants and their possible implications in the food chain. *Journal of Agricultural and Food Chemistry, 59*(8), 3485–3498. https://doi.org/10.1021/jf104517j

Rivero, R. M., Mittler, R., Blumwald, E., & Zandalinas, S. I. (2022). Developing climate-resilient crops: Improving plant tolerance to stress combination. *The Plant Journal, 109*(2), 373–389.

Singh, B. K., & Trivedi, P. (2017). Microbiome and the future for food and nutrient security. *Microbial Biotechnology, 10*(1), 50.

Winding, A., Singh, B. K., Bach, E., Brown, G., Zhang, J., Cooper, M., Dion, P., Mele, P., Eisenhauer, N., Pena-Neira, S., & Lindo, Z. (2020). State of knowledge of soil biodiversity: Status, challenges, and potentialities.

Zandalinas, S. I., & Mittler, R. (2022). Plant responses to multifactorial stress combination. *The New Phytologist, 234*(4), 1161–1167.

Chapter 7
Nanomaterials and Nanotechnology for the Delivery of Agrochemicals

Demet Altindal and Nüket Altindal

Abstract The use of nanomaterials and nanotechnology in the delivery of agricultural chemicals has potential benefits and challenges. Traditional agricultural practices, along with excessive use of chemical fertilizers and pesticides, negatively impact soil, water, and air quality. Nanotechnology, as a sustainable solution, is an important tool to minimize environmental harm and increase crop yield. Different types of nanoparticles (NPs), such as carbon-based, inorganic-based, organic-based, and composite-based materials, have unique properties like large surface area and reactivity. NPs can enhance the effectiveness of pesticides and fertilizers while having the potential to reduce environmental damage. However, there is insufficient knowledge regarding the effects of NPs on ecosystems and the risks of bioaccumulation. Nanotechnology-enhanced phytoremediation ("nano-phytoremediation") is a promising, eco-friendly method for pollutant removal. Further field studies are needed to assess the ecological impacts of nanotechnology and its sustainable applications in agriculture.

Keywords Nanoparticles (NPs) · Phytoremediation · Nano-phytoremediation · Environmental sustainability · Agrochemicals

D. Altindal
Organic Farming Programme, Department of Plant and Animal Production, Fethiye Ali Sıtkı Mefharet Koçman Vocational School, Muğla Sıtkı Koçman University, Muğla, Türkiye

N. Altindal (✉)
Medical and Aromatic Plants Programme, Department of Plant and Animal Production, Sivaslı Vocational School, Uşak University, Uşak, Türkiye

V. D. Rajput et al. (eds.), *Contemporary Suitability of Nanobionics in Agriculture*, Nanotechnology in the Life Sciences, https://doi.org/10.1007/978-3-031-82779-2_7

7.1 Introduction

In traditional agricultural activities carried out throughout history for plant and animal production, environmental problems such as excessive consumption of plant nutrients in the soil, decrease, and deterioration of soil fertility, accumulation of diseases and pests in the soil, and erosion occur. Moreover, unconscious and intensive use of chemical fertilizers in the agricultural system has a negative impact on soil structure and reaction, soil fauna, and the accumulation of toxic substances in the soil. Besides, chemical fertilizers and pesticides pose a threat to water, air, and plant quality. As a result of the development of technology, population growth in the world, and inefficient use of resources, the environment and humanity are harmfully affected.

Agricultural production is the oldest production activity in the history of mankind. With the new policies of the world and the introduction of advanced irrigation, fertilizers, pesticides, agricultural tools, and technological power in agriculture, new approaches and techniques such as nanotechnology have been needed in agricultural production to minimize the damage to the environment and ensure the sustainability of agriculture. The idea of nanotechnology was conceived by Dr. Richard Feynman in 1959 with his theory based on how to store more information in nanoscale devices to be produced (Avati, 2023; Formoso et al., 2023; Pitkethly, 2008). Nanomaterials have extraordinary specific functions at nanoscale dimensions, which has led to a major advancement in nanotechnology studies. In agriculture, the application of nanotechnology especially targets reducing the excessive use of chemical-containing plant protection products and increasing nutrient uptake and yield through the effective use of fertilization. Due to their properties such as pharmacological, drug delivery, optical properties, pollution removal, and antimicrobial activity properties, nanoparticles (NPs) with sizes ranging from 0.1 nm to 100 nm produced by nanotechnology method are employed in multiple fields such as textile, biotechnology, pharmacy, medical, defense industry, materials, manufacturing sector, and agriculture to solve environmental problems. In agriculture, nanoparticles (NPs) can be utilized in plant growth, herb, and pest control, and plant and animal breeding.

The environmental impact of nanoparticles used in many fields is not fully appreciated. Studies have reported that although NPs have detrimental effects on the environment, they are an innovative approach that can be economically utilized in agriculture. Nanoparticles can be used as bactericides/fungicides, nano-based fertilizers, plant disease diagnostics, and agrochemicals due to their physicochemical properties such as being nanoscale and reactive and having large surface area. Nanotechnological applications are growing with each passing day, and this has led to an increase in the concentration of nanoparticles in nature. Despite their prevalence in the ecosystem, the reactions of plants, animals, and microorganisms to NPs are still not well understood. The increasing use of nanoparticles in many different fields is thought to cause negative effects on the ecosystem. Hence, research has been focused on reducing the negative effects of nanoparticles due to their interactions with the environment. NPs that are dense in the ecosystem interact directly with various substances or mixtures of substances (pollutants, components) that are undesirable in the environment, affecting their diffusion and transport into the environment. The elimination of pollutants in the environment is critical for the prevention of environmental pollution and the protection of ecosystems. Plants, which are the main components of terrestrial ecosystems such as forests, mountains, caves, and valleys, and aquatic ecosystems such as rivers, seas, and marshes, are thought to play an important role in the transport of nanoparticles because they absorb and accumulate pollutants. Various technologies such as plant breeding, immobilization, water management strategies, and microbial inoculation practices are employed to determine the negative impact of polluting components such as soil, water, and air on the ecosystem and to find solutions for these problems. However, because of the high cost of these technologies, phytoremediation technology has been proposed as a new alternative/complementary technology. The phytoremediation system utilizes phytoremediator plants as purifiers to remove soil pollutants, transform toxins into less harmful substances, and clean up groundwater pollution. Green and inexpensive "nano-phytoremediation"technology is also implemented for the remediation of the polluted environment with the help of nanoparticles in the phytoremediation method.

Phytoremediation is a bioremediation technique that leverages the natural capacity of plants to remove environmental pollutants. Nano-phytoremediation explores the interaction between plants and nanomaterials, focusing on how these materials can be utilized for the remediation of contaminated environments.

The application of nanotechnology and nanomaterials in agriculture offers opportunities not only for enhancing productivity but also for promoting environmental sustainability. While nanoparticles can improve the efficiency of pesticides and fertilizers, their misuse or excessive accumulation can lead to environmental pollution. Nano-phytoremediation provides a potential solution to this problem. Through this process, plants can absorb or stabilize nanoparticles, effectively removing them from environmental systems.

7.2 Nanoparticles (NPs)

The development of nanotechnology, which emerged as the sixth technological revolution following the Green Revolution (1960) and the Biotechnology Revolution of the 1990s, in which various technologies such as the use of pesticides and herbicides, seed breeding, mechanization, irrigation, and use of chemical fertilizers were applied to obtain more crops in agriculture, nanoparticles (NPs), which are nano-sized materials such as nanocrystals and nanotubes, have based on nanotechnology and constituted the basis of nanoscale materials. Nanomaterials are technological materials with the structure of particles, fibers, tubes, thin sheets, etc., with potential applications in biology due to the size of cells ranging from approximately 1 nanometer to 100 nanometers (nm). Nanoparticles with high volume-to-surface ratio and solubility, specific properties, low toxicity, and quantum-scale effects are utilized in the materials and nanoelectrical, manufacturing sector, computer technologies, medical and health sector, aerospace research, environment and energy, defense sector, biotechnology, medicine, agriculture, and food sectors. Nanoparticles are available in different types such as polymeric nanoparticles, metal nanoparticles, and magnetic nanoparticles. NPs are categorized into four main classes: composite-based NPs, carbon-based NPs, inorganic-based NPs and organic-based NPs (Fig. 7.1 and Table 7.1).

Nanoparticles are classified into three classes according to their source: natural NPs, incidental NPs, and engineered NPs (Table 7.2). Specially designed (engineered) NPs are widely used NPs that are industrially and scientifically produced by changing their atomic arrangement and specific properties according to the purpose. Industrial (engineered) nanoparticles are specifically designed and manufactured by human beings to serve specific purposes.

Examples of carbon-based nanoparticles, include fullerenes and graphene; inorganic-based NPs include metal oxides, gold, silver, and iron oxide; organic-based NPs include micelles, liposomes, and dendrimers; composite-based NPs include chitosan and polylactic acid (Table 7.1). Among carbon-based nanoparticles, fullerene can have a spherical, tubular, or elliptical structure, as well as the cylindrical version of carbon nanotubes. Fullerene has low density, high boiling point, electrochemical stability, and high electrical and thermal conductivity (Li

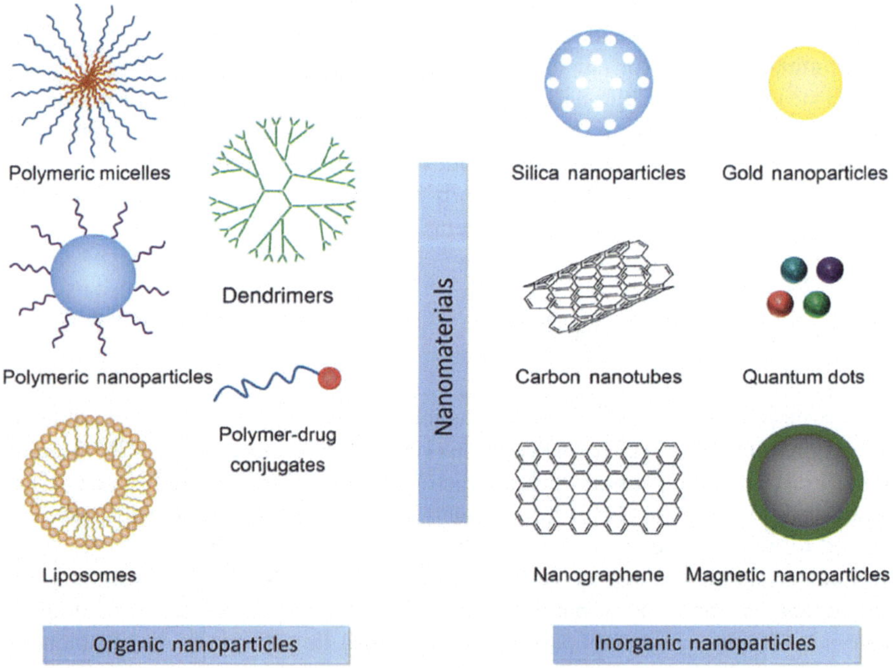

Fig. 7.1 Nanoparticles types (Zhou et al., 2017)

Table 7.1 Classification of nanoparticles

Nanoparticles (NPs)—nanostructured materials	
Carbon-based nanomaterials	Fullerenes (C60) Carbon nanotubes (CNT) Carbon nanofibers Graphene (Gr)
Inorganic-based nanomaterials	Gold Silver Titanium dioxide Zinc oxide Iron oxide Silicon Quantum dots
Organic-based nanomaterials	Dendrimers Micelles Liposomes Polymer nanoparticles
Composite-based nanomaterials	Chitosan Polylactic acid Hydroxy ethyl methacrylate

Table 7.2 Classes and examples of nanoparticles according to sources

Nanoparticle sources	Nanoparticle examples
Natural nanoparticles	They are nanoparticles found spontaneously in nature in organisms such as soil, water, plants, and animals and are formed by both geological and biological processes that are naturally produced in the ecosystem (Katiyar et al., 2021).
Engineered nanoparticles (concise description)	This class of nanoparticles refers to NPs that are specifically designed and manufactured to serve the purposes for which they are intended to be consciously used by humans (Gubert et al., 2023; Quazi et al., 2023).
Incidental nanoparticles (anthropogenic)	They are nanostructures formed through many industrial productions. Some of these incidentally produced nanoparticles contaminate ecosystems (Bhat et al., 2023; Kaur, 2021).

et al., 2019; Kalyani & Dhoble, 2023). Because of its hydrophobic nature, it is used in surface modifications, organic photovoltaic materials, antioxidant systems, biopharmaceutical materials and energy systems, cosmetics, medical applications, water treatment systems, and electronic applications (Baskar et al., 2022; Bhakta & Barthunia, 2020; Inshakova et al., 2020). Some fullerene structures exhibit antiviral properties and are successfully utilized against Human Immunodeficiency Virus (HIV) replication (Ahmadi et al., 2023).

Metal NPs are widely used in many fields such as biomedical applications including disease diagnostic analysis and radiotherapy, biotechnology studies, as well as biological applications to increase microbial resistance to antibiotics because of their high surface area and antibacterial properties. The most commonly manufactured metal NPs are composed of pure metals such as titanium, silver, gold, zinc, cerium, and iron or compounds of metals with oxides, fluorides, chlorides, hydroxides, sulfides and phosphates. Natural NPs found in all ecosystems such as soil, air, water, and volcanic dust are the result of both geological and biological events. Even if these NPs are toxic to the ecosystem, most organisms have evolved adaptations to them. Nanomaterials that are accidentally, coincidentally, or unexpectedly released into nature as a result of the activities of the pharmaceutical, defense, food, construction, textile, cosmetics, automotive, etc. sectors, accompanied by machines driven by the development of technology, cause environmental pollution. Gold nanoparticles, also one of the metal nanoparticles, are nanomaterials frequently used in chemistry, physics, biomedicine, and biotechnology. Gold nanoparticles, one of the most stable metal nanostructures, are effectively used in determining the properties of other nanoparticles. Gold nanoparticles ranging in size between 1 and 100 nm may be in the form of spherical, octahedral, decahedral, icosahedral multiple twined, amorphous, tetrahedral, nanotriangles, nano prisms, hexagonal thrombocytes, wire, and sheet. Gold nanoparticles are widely used in areas such as enhancing the effect of drugs, treatment of genetic diseases and cancer, tumor therapy, bio-imaging, diagnosis, and identification of diseases thanks to their electrical and optical properties, shape, and particle sizes. Iron oxide compounds found in nature come in different forms such as magnetite (Fe_3O_4), hematite (α-Fe_2O_3), and maghemite (γ-Fe_2O_3), and have low toxicity. These compounds are used in technological and biomedical applications, biosensors, data storage systems, and as catalysts in various reaction mechanisms such as NH_3 production and dehydration, and as pigments in paints, ceramics, and porcelain materials. The magnetic properties and low toxicity properties of these nanoparticles, especially magnetite, are used in biomedical applications, magnetic resonance imaging (MRI), cancer treatment, and biosensors (Alangari et al., 2022; Jagadeesh et al., 2023; Montiel Schneider et al., 2022; Pullar, 2023). Silver nanoparticles are used in chemical and biological sensors (Hassan et al., 2019), biopharmaceuticals, biosignals, electronic circuitry, and water purification systems (Marimuthu et al., 2020) thanks to their optical, thermal, and electrical properties. Also, its antimicrobial activity raises interest in the food industry (Ali et al., 2023; Vorobyova, 2023). Silver nanoparticles, which are commercially used in food packaging technologies, small household appliances, fabric coatings, and disinfectant products, have an important place in our daily lives. Titanium dioxide (TiO_2), which is obtained from mineral deposits in nature, is used in plastics, paints, foods, food additives, and cosmetics, as well as in medical applications due to its photocatalysis properties and as an antimicrobial against antibiotic-resistant bacteria (Wu et al., 2023). It also has the potential to be used in solar energy systems due to its low toxicity and semiconductivity (Xie et al., 2023).

Magnesium oxide (MgO) NPs have recently attracted attention as an environmentally friendly, economical, and industrially valuable material. This hygroscopic solid mineral has a high surface area (Hornak, 2021) and a hollow structure. It can be used to remove organic pollutants and heavy metals from the environment due to its reactive nature, selectivity, and high surface area. Magnesium oxide NPs are characterized by structural strength, high thermal conductivity, low electrical conductivity, and optical transmittance (Ameur et al., 2023). In industry, magnesium oxide NPs are used in the production of semiconductor materials and electrochemical sensors (Sagadevan et al., 2022).

Nanotechnology, which is called the fastest-spreading technical revolution at the global level, has a significant potential for use in agricultural fields beside the great benefits it provides to many fields. Since nanotechnology will come into our lives from seed planting to production, food processing, preservation, and packaging, until reaching the consumer, it is estimated that nano-foods will be produced by 2040 (Hasmin et al., 2021). Therefore, agricultural applications hold an advanced priority in nanotechnology and can be used to detect and eliminate the detrimental effect of overused pesticide residues in the fight against various agricultural pests, thus reducing environmental pollution and reducing application costs. Nanoparticles can be utilized as biomarkers (Fe_3O_4/AgBr, Au/Ag NPs, ZnO NPs, and clay-chitosan NP) to detect various pathogens in the fight against plant diseases (Alghuthaymi et al., 2021). Nanoparticles are effective in efficient water use in plant cultivation practices of sustainable agriculture due to their high ability to retain water and absorb and diffuse water and nutrients. They are also important in improving soil properties to retain water and in the production of nano fertilizers (Khalid et al., 2022). Nanotechnology is also used in solving problems especially in agriculture by identifying pollutants. Pesticides used in the fight against plant diseases and pests, as well as artificial fertilizers used in plant cultivation, cause the deterioration of water and soil structure. The agricultural system faces problems such as reduced production, food shortages, and climate change. Therefore, the use of innovative technologies such as biopesticides and nano-based materials such as nano pesticides, nano compounds, and nano sensors increases the control of plant diseases and pests, nutrient uptake, and the development of plants (Altındal & Altındal, 2020). Unfortunately, the use of industrial nanoparticles in many fields in different ways creates environmental and health concerns by spreading into wastewater and air. Unfortunately, the rapid increase in the industrial use of nanoparticles has led to uncontrolled releases into the environment, creating problems about which we know little about the pollutant properties of nanoparticles.

7.3 Phytoremediation

In the world, the increase in industrial activities, deforestation, climate change, excessive use of chemicals (pesticides and fertilizers) in agriculture, industry, mining, military activities, oil or diesel spills, nanoparticles (NPs), improper disposal of wastes including technological wastes, man-made chemicals, and unnatural changes cause soil pollution or land pollution. The main soil pollutants are metals such as lead, cadmium, selenium, zinc, cobalt, copper, nickel, chromium and mercury, nanoparticles (NPs), radioactive compounds such as cesium, uranium, or strontium, and other inorganic compounds such as arsenic, ammonia, nitrate, sodium or phosphate. The main contaminants of agricultural soils are (i) pesticides; (ii) organic fertilizers (manure and sewage waste); (iii) mineral fertilizers; (iv) wastewater for irrigation; (v) drip irrigation pipes and plastic materials used in agriculture, mulching, and greenhouses; (vi) urban waste. Furthermore, pesticides used in agriculture to control plant diseases and pests mix with groundwater and pollute water and soil. Trace elements and polycyclic aromatic hydrocarbons produced by forest fires also pollute the environment. Besides pesticides and nanoparticles (NPs) that spread rapidly in the soil and environment, heavy metals such as chromium, mercury, lead, copper, zinc, nickel, arsenic and cadmium are among the most important factors causing soil pollution. Heavy metals accumulated in the soil negatively impact plant yield and quality as well as microbial organisms in the soil. Different technologies such as plant breeding, immobilization, water management strategies, and microbial inoculation applications are utilized to ascertain the harmful effects of environmental pollutants such as air, water and soil on the ecosystem and to find solutions for them. Nevertheless, due to the high cost of these technologies, new alternative/complementary technologies such as phytoremediation have been introduced. The phytoremediation system is a green technology that utilizes phytoremediator plants as purifiers to remove other soil contaminants including nanoparticles (NPs), convert toxins into less harmful substances and remove groundwater contamination. Phytoremediation is a natural method of eliminating pollutants in the environment by utilizing the absorption properties and metabolic systems of plants. The most ideal plants as purifiers used in phytoremediation are those that are accumulators of heavy metals and nutrients and have a large root system that allows them to reach large soil and water surfaces. Plants grown in terrestrial ecosystems are exposed to pollutants mostly through their roots, whereas aquatic plants are exposed through all vegetative parts. Heavy metals are absorbed and accumulated in all plants, mostly by roots, and are retained in the tissues of the plants and removed from the environment. The purification system using phytoremediator plants employs various mechanisms depending on the plant species and environmental conditions, including degradation (rhizodegradation, phytodegradation), accumulation (phytoextraction, rhizofiltration), dispersion (phytovolatilization), and immobilization (phytostabilization and hydraulic control) to break down, remove or immobilize toxic substances in the soil (Kafle et al., 2022) (Fig. 7.2).

Phytoremediation is influenced by several parameters such as contaminant, type of plant, duration of exposure to the contaminant, environmental parameters, and concentration of contaminants.

7.4 Phytoremediation Mechanism

7.4.1 Phytoextraction (Phytoaccumulation, Herbal Absorption)

After absorption of toxic substances in the soil, especially heavy metals, by plant roots, the absorbed toxic substances deposit in the cell membrane and wall, tissues, stem, leaves, and shoots. Hyperaccumulator plants that accumulate heavy metals in

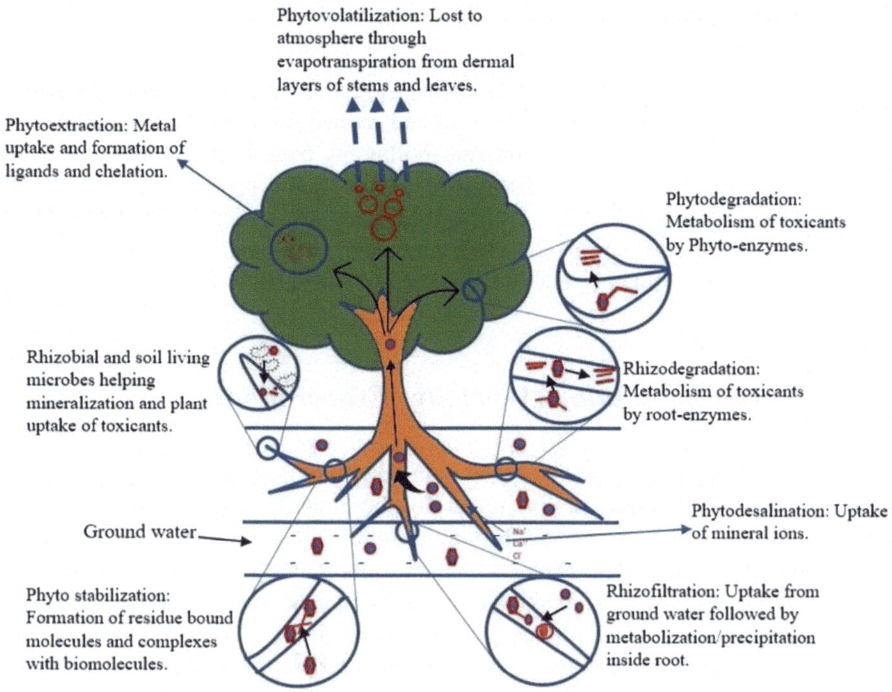

Fig. 7.2 Phytoremediation mechanism (Kafle et al., 2022)

their roots and stems absorb heavy metal ions and form new metal-ligand complexes and metal-phytochelatin complex with chemicals resulting in various water-soluble complexes such as organic acids entering the root cells (Kafle et al., 2022). After harvesting, the hyperaccumulator plants involved in phytoaccumulation are disposed of together with the toxic substances they contain by appropriate methods such as drying, incineration, and composting. (Hamutoğlu et al., 2012).

7.4.2 Phytodegradation (Phytotransformation, Plant Degradation)

In phytodegradation, plants metabolize or mineralize the organic and hydrocarbon compounds they absorb via their roots by using sunlight, through chemical reactions and break down toxic substances such as waste, chlorinated solvents, or herbicides with the help of some enzymes (Newman & Reynolds, 2004); However, in this method, toxic organic pollutants that are absorbed and stored in this manner are degraded either by the release of enzymes produced by the plant from the roots or directly through metabolic activities. (Donde et al., 2023).

7.4.3 Phytostabilization (Stabilization with Roots)

The phytostabilization method is implemented to reduce the mobilization of heavy metals in the rhizosphere, prevent erosion and reduce the solubility of pollutants, avoid leaching into groundwater, prevent pollutants from being absorbed by the roots of the plant, being accumulated in the roots, being precipitated in the root zone, and to prevent heavy metals from entering the food chain. By phytostabilization, copper (Cu), chromium (Cr), and lead (Pb) levels and metal leaching are reduced and low amounts of radionuclides are stabilized (Singh et al., 2022).

7.4.4 Phytovolatilization (Vegetative Evaporation)

Phytovolatilization is the uptake of pollutants by plant roots, their conversion into gaseous form, and their release into the atmosphere, while heavy metal (metalloid) toxins are eliminated from the soil through evapotranspiration without the need for their elimination by crop harvesting. Plants absorb organic pollutant metals such as mercury, selenium, arsenic (Sakakibara et al., 2010), and water and release them to the atmosphere through evapotranspiration from the leaves. The water taken up by the plant undergoes changes while being transported in the xylem (wood tubes), which are tissue elements, and some of the toxic substances are transported to the leaves and evaporated and taken away from the environment (Kvesitadze et al., 2015). Various volatile organic compounds can also evaporate into the atmosphere from roots, stems, and leaves (Limmer & Burken, 2016).

7.4.5 Rhizodegradation (Phytostimulation, Degradation by Roots)

The mechanism of rhizodegradation is the decrease of the toxicity of organic compounds in the rhizosphere by the action of plant roots or soil microorganisms. By this method, contaminated soils can be remediated, and for this purpose, bacterial inoculation with strains capable of degrading the harmful substance can be used. To increase microbial activity (biodegradation) in the rhizosphere of plants, a process called biostimulation is used to increase the population of root bacteria, thereby increasing oxygen in the rhizosphere of the plant and releasing sugars, alcohols, and organic acids into the rhizosphere, thus increasing the biodegradation of contaminants in the soil. Furthermore, mycorrhizal fungi, which settle on the root of the plant and live symbiotically with the plant by releasing hyphae in the root, are effective in the chemical metabolism of toxic organic substances and in improving soil structure (Fernández-Luqueño et al., 2017).

7.4.6 *Rhizofiltration (Filtration with Roots)*

Rhizofiltration is a phytoremediation method in which contaminants are absorbed and filtered by plant roots from contaminated groundwater to remove organic and inorganic contaminants. In highly oxygenated waters, the phytoremediator does not dissolve heavy metals in plant roots but sinks them to the bottom, transforms them into a new compound, and concentrates them.

7.4.7 *Hydraulic Control*

Hydraulic control is a method of treating groundwater contaminated with heavy metals, pesticides, crude oil, polyaromatic hydrocarbons, and residues by using plants to prevent and control the accumulation and transport of contaminants (Banarjee, 2018). Plants inhibit the leaching of contaminants into clean areas by reducing groundwater mobilization. The deep-rooted poplar tree acts as a natural pump, drawing 350 gallons of water a day from the soil (Bisht et al., 2020; Sharma et al., 2018; Yadav et al., 2018). Trees and plants that can absorb excess water are desirable for hydraulic control. Toxic substances are prevented from leaking into the environment by transpiration in plants, and the mixing of toxic substances into groundwater and drinking water is diminished. In addition to controlling erosion, plants decrease the infiltration of pollutants and contaminated water into the soil (Koda et al., 2022; Nagendran et al., 2006).

7.5 Phytoremediator Field Crops

In the choosing of some plants to be used as phytoremediators in the phytoremediation process; plants that are suitable for the treatment and cleaning of water and soil pollutants and adapted to climatic conditions are preferred. In soil contaminated with toxic compounds, plants have resistance to some heavy metals, good growth, heavy metal storage, and high biomass production. These plants grown in the field are herbaceous or shrubby, annual or woody perennial plants that differ from each other in terms of roots, leaves, stems, flowers, and fruits, and are categorized into four main groups including cereals, edible legumes, industrial crops, and forage crops; especially barley, sorghum (*Sorghum* spp.), cotton (*Gossypium* spp.), wheat (*Triticum* spp.) are widely used as phytoremediators in saline-sodic lands. However, alfalfa, rye, sorghum, sunflower, beans, corn, paddy, cotton, hemp, medicinal aromatic plants, and sugar beet are widely used as phytoremediators for the phytoremediation of contaminated soils (Altındal & Altındal, 2023).

*7.5.1 Phytoremediator Hemp (***Cannabis sativa***)*

The feasibility of using hemp (*Cannabis sativa* L.) for the treatment of soils contaminated with organic pollutants (e.g., benzo(*a*) pyrene, naphthalene and chrysene and aqueous solutions) or heavy metals (e.g., selenium and cobalt) has also been demonstrated (Stonehouse et al., 2020), indicating that the use of hemp for the remediation of soils contaminated with organic compounds is sustainable, environmentally benign and cost-effective. It is also used as a purifier in soils contaminated with nuclear radiation.

*7.5.2 Phytoremediator Sunflower (***Helianthus annus** *L.)*

The phytoremediator sunflower, which is one of the field crops with high biomass production, is reported to be effective in the removal of heavy metals (phytoextraction), has a high tolerance to heavy metals, and is environmentally friendly in cleaning-up toxic substances and therefore can be used in phytoremediation studies (Nguyen et al., 2021; Rizwan et al., 2016). Most plants can absorb at least one type of radiation. Sunflower is one of the rare plants that can eliminate all radioactive substances from the atmosphere and soil. The phytoremediator sunflower can be used to eliminate toxins including lead, arsenic, and uranium from the soil. After the Chernobyl nuclear disaster (1986), the sunflower was successfully utilized to remove cesium and strontium (radioactive elements) accumulated in soil and water. Recent research also shows that sunflower is suitable for use as a phytoremediator for the environment (Aman et al., 2022).

*7.5.3 Fitoremediatör Lavanta (***Lavandula officinalis** *L.)*

Lavender is a purifying plant that can be grown in toxic soil and is tolerant to heavy metals and has been successfully used as a hyperaccumulator against lead, cadmium, and zinc. The lavender plant (*Lavandula officinalis* L.) has been reported to reduce cadmium uptake and prevent its translocation in the shoot by allowing cadmium to accumulate in the root, and it can be grown as a phytoremediator by biochar application of lavender plant to remediate cadmium contaminated soils (Hashemi et al., 2017). It has been reported that lavender (*Lavandula angustifolia* L.) inoculated with arbuscular mycorrhiza (AMF) acts as a phytoremediator in the lead and nickel-contaminated soils and increases plant growth, so heavy metal accumulation occurs intensively in vegetative parts (Pirsarandib et al., 2022). Medicinal aromatic plants contribute to the elimination of inorganic and organic pollutants by reducing the mobilization of salts, metals, organic compounds, and radionuclides accumulated in soil and water through physiological mechanisms (Greff et al., 2023; Shmaefsky & Husen, 2023).

*7.5.4 Phytoremediator Paddy (**Oryza sativa L.**)*

Paddy can be used as a phytoremediator to reduce the amount of cadmium in the fertile soil and thus can be used as a phytoremediator in the phytoremediation process (Takahashi et al., 2016). Plants inoculated with root bacteria (*Rhizobacteria*) establishing symbiotic relationships contribute to the phytoremediation of soil including heavy metals (Liu et al., 2022). The water-holding capacity of the soil and the presence of excess arsenic in paddy fields are serious problems in cultivation. Paddy plant absorbs more arsenic than other plants during the development. In the accumulation of arsenic metal, it is important that water and soil phytoremediation should be sustainable and cost-effective. NP (nanoparticle) application in paddy cultivation increases the phytoremediation of arsenic (As) and reduces As accumulation in the plant (Srivastava et al., 2021).

*7.5.5 Phytoremediator Wheat (**Triticum aestivum**)*

Heavy metals that cause permanent damage to the ecosystem are a serious threat. Fungi (endophytes) inhabiting the intracellular spaces of plant tissues increase their uptake by their host plants or cause a decrease by binding to the intracellular/extracellular parts of the plant. Tolerance to lead toxicity was determined in wheat infected with *Trametes hirsuta*, an endophytic fungus isolated from soil with high lead content. Inoculated plant shoots were found to accumulate 50% more Pb. In metal-accumulated soil, host plants increase metal uptake and contribute to the phytoextract (Malik et al., 2020). In a study investigating the interaction between wheat and Microplastic/Nanoplastic (MP/NP); it was stated that MP/NP absorption varied based on plant species, transported in tissues and MP/NP pollution affected phytoremediation, and polystyrene nano plastics (PSNP) increased root length, carbon and nitrogen content and plant biomass in wheat, thus wheat was indicated to be used as a phytoremediator plant in the treatment of MP/NPs (Lian et al., 2020). In a study investigating the growth parameters, oxidoreductase activity against stress and copper accumulation in wheat inoculated with three varieties of rhizobacteria (*Azospirillum*); all rhizobacteria had a positive effect on copper toxicity accumulated in roots and shoots of wheat and a significant effect on oxidoreductase belirlenmiştir (Muratova et al., 2022).

7.5.6 *Phytoremediator Clover (*Medigo sativa*)*

The use of alfalfa as a phytoremediator for the phytoremediation of toxic element-contaminated soils is sustainable and appropriate because of its high productivity, toxic tolerance, and high absorption and accumulation rate of toxic pollutants in the roots. Furthermore, the use of alfalfa inoculated with soil bacteria for phytoremediation of toxic substances is a positive and environmentally friendly method (Chen et al., 2022). In a study investigating the phytoremediation of soil artificially contaminated with Cd and Pd, it was stated that alfalfa should be included in the rotation cropping system as an accumulator in soils with high heavy metal accumulation (Morsy et al., 2022).

7.5.7 *Phytoremediator Cotton (*Gossypium hirsitum*)*

Cotton is a plant with high plant biomass, dense roots, and high tolerance to heavy metals. Heavy metal residues in cotton, especially in cotton fiber, are minimal. Therefore, it can be used as a phytoremediator in soil remediation in ecological and socioeconomic terms (Li et al., 2020). The cotton plant has high Pb tolerance in Pb-contaminated soil remediation and the population of actinomycete bacteria increases in its root. Again, the cotton plant is effective for Pb elimination and phytostabilization of Pb-contaminated soils (Ramana et al., 2021).

7.5.8 *Phytoremediator Sugar Beet (*Beta vulgaris *L.)*

In a study on the growth, tolerance, and soil remediation of sugar beet (*Beta vulgaris* L.) in heavy metal-contaminated soils with nickel and cadmium, it was found that Ni inhibited plant growth, Ni and Cd accumulated in sugar beet tissues, and it was tolerant to Cd toxicity (Papazoglou & Fernando, 2017). Sugar beet (*Beta vulgaris* L.) is intensively cultivated in China as it is resistant to soil salinity and effective in reducing salinity and can be used in phytoremediation (Yan et al., 2021). Removing pollutants with magnetic nanocomposite obtained by using sugar beet pulp and Fe_3O_4 together, a low-cost method, can be used effectively in the remediation of contaminated soils (Sadat et al., 2022). The amount of boron (B) needed for the development of sugar beet is high and this plant can be used as a phytoremediator in boron-contaminated soils due to its tolerance to boron stress and high capacity for translocation of boron metal to the plant through diffusion (Huo et al., 2022).

7.6 Phytoremediation of Nanoparticles (NPs)

While there is a large body of research examining the phytoremediation of pollutant nanometals or the interaction between plants and metals, their absorption and accumulation by plants, information on the interaction between nanoparticles and organisms is scarce. While nano-based materials positively affect germination and shoot seedling development in plants, they may have positive or negative effects on rhizobacteria and legume-rhizobia symbiosis. Again, these materials have a negative effect on arbuscular mycorrhiza, and compounds secreted from the plant root affect the amount of nanomaterial (Oliveira et al., 2023). With the development of technology, the production and use of nanoscopic devices or nano-based materials are faced with the problem of nano contamination. Nano-pollutants include carbon monoxide, chlorofluorocarbons (CFCs), heavy metals (cadmium, arsenic, chromium, mercury, lead, and zinc), nitrogen oxides, hydrocarbons, organic compounds (dioxins and volatile organic compounds), particulates and sulfur dioxide. Phytoremediation, an environmentally friendly and inexpensive method, is used to eliminate NPs, which can be synthesized by chemical, physical, or biological methods. Ornamental plants can be used in the phytoremediation of contaminated soils due to their ability to store metal (loid) in their shoots and roots. In a study on the remediation of soils contaminated with arsenic (As), mercury (Hg) and lead (Pb) the phytoaccumulation efficiency of cosmos, vinca, globose, and balsamina plant species was investigated. In the study, it was stated that the cosmos is the most powerful phytoremediation plant and can be used in phytoremediation in an effective sustainable way in a growing period of 2 months (Majumdar et al., 2022).

Although the removal of nano contaminants (phytoremediation) varies according to the phytoremediator plant species, this mechanism is realized by utilizing the uptake and accumulation capacity of NPs by the roots. Since titanium dioxide (TiO_2) NPs, which are commonly used in sunscreen materials and are washed into sewage, are highly toxic, nano contaminants should be reduced to prevent ecosystem pollution in irrigated landscapes. In a study using the halophyte plant *Suaeda glauca*, a special wetland habitat, as a phytoremediator to purify a marsh from TiO_2 NPs and ZnO NPs nano contaminants, *Suaeda glauca* was compared with the phytoremediator rape (*Brassica campestris*). In the study, the uptake of both NPs by the salt marsh plant was higher than by rapeseed. It was also reported that dry matter weight and phytoremediator activity were higher than rapeseed (Song et al., 2022). Industrially and scientifically widely used, specially designed NPs, which are produced by changing their atomic sequences and specific properties according to the purpose, cause environmental pollution by unexpectedly mixing into water. To determine the success of the aquatic plant *Typha latifolia* and *Carex rostrata* species in phytoremediation, metallic NPs were added to the plant growth medium for 15 weeks. In the study, the accumulation of Ag, TiO_2, and ZnO industrial NPs in *Carex rostrata* shoots and the accumulation of engineered NPs in *Typha latifolia* roots were found in high amounts. It was concluded that the difference between the species was due to root morphology and that further research should be conducted on the chemical behavior and toxicity of NPs taken up by aquatic plants (Ebrahimbabaie et al., 2023).

7.7 Nano-Phytoremediation

In sustainable agriculture, which is an innovative agricultural system, nanotechnological applications have important contributions to the utilization of biomolecules such as nutrient content and antioxidants. In sustainable agriculture, nanoparticles are promising in the production of pesticides, chemical drugs used in the fight against pests, diseases, and weeds, plant and animal breeding, diagnosis and prevention of plant diseases, heavy metal detection in soil, and water and soil management. Moreover, controlling the plant nutrients needed by plants and the pH, moisture, and water levels of the soil from a long distance contributes to reducing agricultural inputs and preventing environmental pollution. The increasing use of nanomaterials in many fields significantly increases the interaction of organisms such as plants, animals, and bacteria with the environment. However, nanoparticles can be used to enhance the plant's effect in soil and water purification and contribute to environmental rehabilitation by removing pollutants such as salts, metals, minerals, and inorganic substances, pesticides (toxaphene, dieldrin, endrin, aldrin, chlordane, DDT, heptachlor, hexachlorobenzene, and mirex), industrial chemicals (polychlorinated biphenyls (PCBs), hexachlorobenzene (HCB), mireks, etc.) from the environment with nano-phytoremediation technology. Nano-phytoremediation, which is still the latest technology, is thought to be a solution to environmental pollution. Although there is information on the treatment of most pollutants, there are few studies on the use of nanoparticles in the nano-phytoremediation method and the determination of the treatment efficiency of phytoremediator plants. Zand et al. (2020) applied PGPR *Pseudomonas fluorescens* and TiO_2 NPs together to determine the effectiveness of the phytoremediator *T. repens* in cadmium (Cd) contaminated soils and reported that cadmium accumulation in *T. repens* was high, TiO_2 nanoparticles increased germination and plant growth, and co-application of plant growth-promoting rhizobacteria (PGPR) and TiO_2 NPs could increase photosynthesis in plants. In the study, it was stated that NPs should be applied together with biological methods and TiO_2 NPs will contribute to the success of nano-phytoremediation.

Lead (Pb), a toxic heavy metal and lethal to living organisms, contaminates soil and water as a result of natural and anthropogenic activities. In a study conducted to determine plant growth, antioxidant ability, and lead (Pb) accumulation of zinc oxide (ZnO) nanoparticles in *Persicaria hydropiper* L., a perennial plant for effective phytoremediation of Pb-contaminated soil and water, it was reported that Pb inhibited seedling growth, but ZnO NPs promoted plant growth and reduced oxidative stress and Pb-induced stress. Again, lead accumulation in stems, roots, and leaves increased with ZnO NP application. The positive correlation between proline and secondary metabolites in Pb accumulation in plants was determined, and it was indicated that ZnO NPs can be used in the treatment of heavy metal-contaminated waters through nano-phytoremediation using plants (Hussain et al., 2021). The ability of *D. stramonium* to eliminate Cd from contaminated soils was investigated by incorporating chitosan (CS) and chitosan nanoparticles (CSNPs) into phytoremediation, so their effects on Cd-induced oxidative stress were explored. To determine the tolerance of the phytoremediator *D. stramonium* to Cd, different doses of 0, 75, 150, and 225 mg kg^{-1} cadmium were added to the soil. Cd accumulation in roots and shoots increased parallel to the increase in Cd doses. Again, excess CS and CSNPs in the soil increased Cd accumulation in the shoots and roots of the plant. In addition, peroxidase and polyphenol oxidase activity increased in phytoremediator leaves in Cd-contaminated soil. It was found that CSNPs were more effective than CS in enhancing Cd phytoremediation (Shirkhani et al., 2021). Heavy metals (HMs) are not biodegradable and cannot be destroyed in nature, so they remain in soil and water for a long time, causing pollution. Again, nutrient and mineral deficiency in plants negatively affects various physiological and biological events. Studies on the effectiveness of phytoremediation by incorporating nanoparticles or plant growth-promoting rhizobacteria (PGPR) in combination with phytoremediation for the elimination of HMs have revealed that NP application can increase the phytoremediation of HMs such as lead, cadmium, zinc, nickel, arsenic, and chromium, while PGPR inoculation increases the plant biomass rate in metal-contaminated soils (Gulzar & Mazumder, 2022). Rajput and Minkina (2023) indicate that nanoparticles absorb and break down most toxic substances and render them harmless due to their surface properties. Therefore, they prevent the accumulation and dispersion of pollutants in the environment. The use of NPs in nano-phytoremediation neutralizes pollutants by various biological pathways, including adsorption, redox reaction, precipitation, and co-precipitation, and can also assist hyperaccumulator plant and soil microorganisms. Therefore, nano-phytoremediation investigations, which have been limited to the laboratory, should be carried out more extensively in the field to determine the effects of the environment. Nanomaterials (NMs) make a significant contribution to the remediation of acidic waters by reducing the amount of micro-sized pollutants that occur in human and aquatic life and reduce the quality value of the water source with toxic undesirable effects and by decreasing the amount of micro-sized pollutants that pass into the environment with human activity. Nano-sized zero-valent iron (nZVI) is the most widely used nanoparticle in environmental pollution remediation. It neutralizes dissolved heavy metals in mineral water containing dissolved mineral salts, elements, and gas. Carbon Nanotubes (CNTs) bind to pollutants such as heavy metals and react chemically, thus, they can be used as nanomaterials in the removal of various pollutants due to their strong absorption ability, morphology, reactivity, and large specific surface areas. Again, metal oxide nanomaterials rapidly remove pollutants from the environment thanks to their physicochemical properties as well as their surface area as soil pH decreases (Sharma et al., 2022). In a study investigating the effects of nano-sized zero-valent iron (nZVI) on the mobilization of As, which causes toxic effects, the As and Hg accumulation properties of phytoremediator alfalfa (*Medicago sativa*) cultivated in soil supplemented with organic fertilizer and/or nZVI were determined. In the study, although the treatment of nano-sized zero-valent iron (nZVI) to the soil reduced oxidative stress, it decreased the activity of As in the soil, and Hg accumulated more in the phytoremediator alfalfa. In addition, the accumulation of non-protein thiol compounds (NPT), which have a role in defense against oxidative stress, increased in alfalfa roots in nZVI soil. When organic fertilizer and nZV1 were applied to the soil together, As accumulation decreased, and when nZVI was implemented alone, the biomass of alfalfa increased due to the decrease in As and Hg accumulation in the soil. In the study, it was reported that plant growth augmented when organic fertilizer and nZVI were applied together (Baragaño et al., 2022). Since successful chemical and physical methods to clean soils from plastic pollutants are inadequate, environmentally friendly low-cost techniques based on phytoremediation and nano-phytoremediation applications have gained importance. In a study investigating the effectiveness of nano-phytoremediation by adding phytoremediator vetch, *Rhizobium leguminosarum* strain, and nanomagnetite to the soil in the treatment of soils contaminated with petroleum-derived low-density polyethylene (LDPE, thermoplastic), the height and dry matter weight of the phytoremediator plant and the number of root nodules increased 2.2 fold. In the study, it was stated that vetch and rhizobia can be used successfully as phytoremediators in thermoplastic-contaminated soil because they are symbiotic with each other (Prisacari et al., 2022). Eight different metal oxide nanoparticles (ZnO, MgO, CuO, CeO_2, MnO_2, TiO_2, MnO_2, Fe_3O_4, and Fe_20_3 NPs) were sprayed in combination on wheat leaves grown on land contaminated with Cd by wastewater and over-irrigation. Thus, in the study in which the contribution of metal oxide NPs to nano-phytoremediation was determined, it was reported that metal oxide NPs affected other growth parameters except for plant height and thousand-grain weight. According to the obtained results, ZnO NPs applied together with MgO and Fe_3O_4 NPs increased spike dry weight, number of grains per spike, grain yield, and productivity. ZnO NPs applied in a mixture with MgO, CeO_2, or TiO_2 NPs increased Cd uptake in stems, leaves, and shoots, while ZnO NPs applied with MnO_2, Fe_2O_3, Fe_3, O_4, or CuO NPs decreased Cd accumulation in grain. It was stated that the mechanisms of the molecules involved in transcription, translation, and regulation and changes in the root part of the phytoremediator plant by adding NPs in combination in Cd-contaminated soils should be investigated (Lian et al., 2022).

7.8 Conclusion

Humans make new small things by cutting, dividing, and shrinking big things. In this context, nanotechnology is the use of nano-sized atoms and molecules to make something bigger, that is, to obtain new things by designing from small to large. Nanoparticles (NPs) produced by the technology of assembling, creating, and modifying things at the nanoscale (nanotechnology) are used in almost every field in physics, chemistry, medicine, engineering, mechanics, and agriculture. In the nanotechnology technique that allows the substance to go directly to the site of action, nanoparticles are synthesized by acidification and oxidation with ferric nitrate following the addition of alkali to aqueous solutions of iron oxide powders Fe(III) and Fe(II) salts.

Nanoparticles hold the potential to completely change our lives. These technologies are used in agricultural applications such as water treatment for environmental protection, desalination of water, removal of industrial pollutants, air pollution detection, solar and fuel energy, plant disease and pathogen diagnosis, nano pesticides against plant diseases and pests, and nano fertilizers. Using nanotechnology in agricultural production, nano-phytoremediation is a promising tool for reducing soil pollution with heavy elements as well as environmental pollution from materials such as chemical fertilizers and pesticides. More than 90% of agricultural chemicals are indirectly released into the environment due to uncontrolled application methods. Nanoparticles (NPs), which can contain not only one type of element or atom but also many different types of elements and many different types of molecules, have advantages in terms of electrical, magnetic, morphological, thermal, optical, large surface area, reactivity with substances, and chemical and physical properties. In this context, effective, low-cost, and environmentally friendly nano-phytoremediation technology plays a major role in reducing environmental pollution and thus risks to human health by enabling the use of nanoparticles to increase the success of phytoremediation technology.

References

Ahmadi, S., Warkiani, M. E., Rabiee, M., Iravani, S., & Rabiee, N. (2023). Carbon-based nanomaterials against SARS-CoV-2: Therapeutic and diagnostic applications. *OpenNano, 10*, 100121.

Alangari, A., Alqahtani, M. S., Mateen, A., Kalam, M. A., Alshememry, A., Ali, R., et al. (2022). Iron oxide nanoparticles: Preparation, characterization, and assessment of antimicrobial and anticancer activity. *Adsorption Science & Technology, 2022*, 1–9.

Alghuthaymi, M. A., Kalia, A., Bhardwaj, K., Bhardwaj, P., Abd-Elsalam, K. A., Valis, M., & Kuca, K. (2021). Nanohybrid antifungals for control of plant diseases: Current status and future perspectives. *Journal of Fungi, 7*(1), 48.

Ali, M., Ullah, M., Khan, M. M., Rasheed, A., Khan, M., Aswad, S., & Zaman, M. Q. (2023). Role of silver nanoparticles (AgNPs) in food surface packaging: Effects of AgNPs on packaging. *Biological Sciences-PJSIR, 66*(1), 65–73.

Altındal, D., & Altındal, N. (2023). Phytoremediation and phytoremediator field crops. International theories, research, and reviews in the field of agriculture, forestry, and fisheries. In E. Küçüker & K. Özrenk (Eds.), *Serüven* (pp. 101–106).

Altındal, N., & Altındal, D. (2020). Agriculture and nanoparticles. In *Biogenic nano-particles and their use in agro-ecosystems* (pp. 59–66).

Aman, A. S., Kumar, A., Kumar, P., Mishra, P. K., & Mohan, M. (2022). Implementation of an effective and successful management tactics against diamondback moth (Plutella xylostella L.) on cabbage. In *International year of millets 2023* (p. 47).

Ameur, I., Khantoul, A. R., Boudine, B., Brien, V., Horwat, D., Sebais, M., & Halimi, O. (2023). Effects of erbium incorporation on structural, surface morphology, and degradation of methylene blue dye of magnesium oxide nanoparticles. *Journal of Inorganic and Organometallic Polymers and Materials, 33*(1), 30–46.

Avati, A. (2023, February). A survey on nanotechnology classification and utilization. In *AIP conference proceedings* (Vol. 2427, No. 1, p. 020068). AIP Publishing LLC.

Banarjee, P. (2018). Phytoremediation: Using natural strength for curing nature. *Acta Scientific Agriculture, 2*(2), 44–153.

Baragaño, D., Forján, R., Álvarez, N., Gallego, J. R., & González, A. (2022). Zero valent iron nanoparticles and organic fertilizer assisted phytoremediation in a mining soil: Arsenic and mercury accumulation and effects on the antioxidative system of Medicago sativa L. *Journal of Hazardous Materials, 433*, 128748.

Baskar, A. V., Benzigar, M. R., Talapaneni, S. N., Singh, G., Karakoti, A. S., Yi, J., et al. (2022). Self-assembled fullerene nanostructures: Synthesis and applications. *Advanced Functional Materials, 32*(6), 2106924.

Bhakta, P., & Barthunia, B. (2020). Fullerene and its applications: A review. *Journal of Indian Academy of Oral Medicine and Radiology, 32*(2), 159–163.

Bhat, M. A., Gedik, K., & Gaga, E. O. (2023). Environmental impacts of nanoparticles: Pros, cons, and future prospects. In *Synthesis of bionanomaterials for biomedical applications* (pp. 493–528). Elsevier.

Bisht, R., Chanyal, S., & Srivastava, R. K. (2020). A systematic review on phytoremediation technology: Removal of pollutants from waste water and soil. *International Journal of Research in Engineering, Science and Management, 3*, 54–59.

Chen, L., Beiyuan, J., Hu, W., Zhang, Z., Duan, C., Cui, Q., et al. (2022). Phytoremediation of potentially toxic elements (PTEs) contaminated soils using alfalfa (Medicago sativa L.): A comprehensive review. *Chemosphere, 133577.*

Donde, O. O., Otieno, A. O., & Muia, A. W. (2023). Wetlands for remediation in Africa: Threats and opportunities. In *Wetlands for remediation in the tropics: Wet ecosystems for nature-based solutions* (pp. 175–195).

Ebrahimbabaie, P., Smith, A., Zahran, E. M., & Pichtel, J. (2023). Phytoremediation of engineered nanoparticles using Typha latifolia and Carex rostrata. *Applied and Environmental Soil Science*, 2023.

Fernández-Luqueño, F., López-Valdez, F., Sarabia-Castillo, C. R., García-Mayagoitia, S., & Pérez-Ríos, S. R. (2017). Bioremediation of polycyclic aromatic hydrocarbons-polluted soils at laboratory and field scale: A review of the literature on plants and microorganisms. In *Enhancing cleanup of environmental pollutants: Volume 1: Biological approaches* (pp. 43–64).

Formoso, I., Shegokar, R., Zielinska, A., Eder, P., Silva, A. M., & Souto, E. B. (2023). Progress of nanotechnology in the development of medicines. In *Nanotechnology and regenerative medicine* (pp. 1–21). Academic Press.

Greff, B., Sáhó, A., Lakatos, E., & Varga, L. (2023). Biocontrol activity of aromatic and medicinal plants and their bioactive components against soil-borne pathogens. *Plants, 12*(4), 706.

Gubert, P., Gubert, G., Oliveira, R. C. D., Fernandes, I. C. O., Bezerra, I. C., Ramos, B. D., et al. (2023). Caenorhabditis elegans as a prediction platform for nanotechnology-based strategies: Insights on analytical challenges. *Toxics, 11*(3), 239.

Gulzar, A. B. M., & Mazumder, P. B. (2022). Helping plants to deal with heavy metal stress: The role of nanotechnology and plant growth promoting rhizobacteria in the process of phytoremediation. *Environmental Science and Pollution Research, 29*(27), 40319–40341.

Hamutoğlu, R., Dinçsoy, A. B., Cansaran-Duman, D., & Aras, S. (2012). Biyosorpsiyon, adsorpsiyon ve fitoremediasyon yöntemleri ve uygulamaları. *Türk Hijyen ve Deneysel Biyoloji Dergisi, 69*(4), 235–253.

Hashemi, S. B., Momayezi, M., & Taleei, D. (2017). Biochar effect on cadmium accumulation and phytoremediation factors by lavender (Lavandula stoechas L.). *Open Journal of Ecology, 7*(07), 447.

Hasmin, N. A., Zainol, Z. A., Ismail, R., & Matmin, J. (2021). Disclosure of nanomaterials under nanotechnology product inventory, voluntary certification, and voluntary labelling. *Pertanika Journal of Social Sciences & Humanities, 29*(1), 157–174.

Hassan, B. S., Islam, G. M. N., & Haque, A. N. M. A. (2019). Applications of nanotechnology in textiles: A review. *Advance Research in Textile Engineering, 4*(2), 1038.

Hornak, J. (2021). Synthesis, properties, and selected technical applications of magnesium oxide nanoparticles: A review. *International Journal of Molecular Sciences, 22*(23), 12752.

Huo, J., Song, B., Riaz, M., Song, X., Li, J., Liu, H., et al. (2022). High boron stress leads to sugar beet (Beta vulgaris L.) toxicity by disrupting photosystem II. *Ecotoxicology and Environmental Safety, 248*, 114295.

Hussain, F., Hadi, F., & Rongliang, Q. (2021). Effects of zinc oxide nanoparticles on antioxidants, chlorophyll contents, and proline in Persicaria hydropiper L. and its potential for Pb phytoremediation. *Environmental Science and Pollution Research, 28*, 34697–34713.

Inshakova, E., Inshakova, A., & Goncharov, A. (2020, November). Engineered nanomaterials for energy sector: Market trends, modern applications and future prospects. In *IOP conference series: Materials science and engineering* (Vol. 971, No. 3, p. 032031). IOP Publishing.

Jagadeesh, P., Sanjay, M. R., & Siengchin, S. (2023). Advanced characterisation techniques for nanostructured materials in biomedical applications. *Advanced Industrial and Engineering Polymer Research.*

Kafle, A., Timilsina, A., Gautam, A., Adhikari, K., Bhattarai, A., & Aryal, N. (2022). Phytoremediation: Mechanisms, plant selection and enhancement by natural and synthetic agents. *Environmental Advances*, 100203.

Kalyani, N. T., & Dhoble, S. J. (2023). Introduction to nano materials. In *Quantum dots* (pp. 3–40). Woodhead Publishing.

Katiyar, N. K., Goel, G., Hawi, S., & Goel, S. (2021). Nature-inspired materials: Emerging trends and prospects. *NPG Asia Materials, 13*(1), 56.

Kaur, T. (2021). Fate and transport of engineered nanoparticles as an emerging agricultural contaminant. In *Sustainable agriculture reviews 50: Emerging contaminants in agriculture* (pp. 283–308).

Khalid, M. F., Iqbal Khan, R., Jawaid, M. Z., Shafqat, W., Hussain, S., Ahmed, T., et al. (2022). Nanoparticles: The plant saviour under abiotic stresses. *Nanomaterials, 12*(21), 3915.

Koda, E., Winkler, J., Wowkonowicz, P., Černý, M., Kiersnowska, A., Pasternak, G., & Vaverková, M. D. (2022). Vegetation changes as indicators of landfill leachate seepage locations: Case study. *Ecological Engineering, 174*, 106448.

Kvesitadze, G., Khatisashvili, G., Sadunishvili, T., & Kvesitadze, E. (2015). Plants for remediation: Uptake, translocation and transformation of organic pollutants. In *Plants, pollutants and remediation* (pp. 241–308).

Li, C., Zheng, C., Zhou, K., Han, W., Tian, C., Ye, S., et al. (2020). Toleration and accumulation of cotton to heavy metal-potential use for phytoremediation. *Soil and Sediment Contamination: An International Journal, 29*(5), 516–531.

Li, Z., Wang, L., Li, Y., Feng, Y., & Feng, W. (2019). Carbon-based functional nanomaterials: Preparation, properties and applications. *Composites Science and Technology, 179*, 10–40.

Lian, J., Cheng, L., Zhai, X., Wu, R., Liu, W., Pan, J., et al. (2022). Foliar spray of combined metal-oxide nanoparticles alters the accumulation, translocation and health risk of Cd in wheat (Triticum aestivum L.). *Journal of Hazardous Materials, 440*, 129857.

Lian, J., Wu, J., Xiong, H., Zeb, A., Yang, T., Su, X., et al. (2020). Impact of polystyrene nanoplastics (PSNPs) on seed germination and seedling growth of wheat (Triticum aestivum L.). *Journal of Hazardous Materials, 385*, 121620.

Limmer, M., & Burken, J. (2016). Phytovolatilization of organic contaminants. *Environmental Science & Technology, 50*(13), 6632–6643.

Liu, A., Wang, W., Zheng, X., Chen, X., Fu, W., Wang, G., et al. (2022). Improvement of the Cd and Zn phytoremediation efficiency of rice (Oryza sativa) through the inoculation of a metal-resistant PGPR strain. *Chemosphere, 302*, 134900.

Majumdar, A., Upadhyay, M. K., Ojha, M., Afsal, F., Giri, B., Srivastava, S., & Bose, S. (2022). Enhanced phytoremediation of Metal (loid) s via spiked ZVI nanoparticles: An urban clean-up strategy with ornamental plants. *Chemosphere, 288*, 132588.

Malik, A., Butt, T. A., Naqvi, S. T. A., Yousaf, S., Qureshi, M. K., Zafar, M. I., et al. (2020). Lead tolerant endophyte Trametes hirsuta improved the growth and lead accumulation in the vegetative parts of Triticum aestivum L. *Heliyon, 6*(7), e04188.

Marimuthu, S., Antonisamy, A. J., Malayandi, S., Rajendran, K., Tsai, P. C., Pugazhendhi, A., & Ponnusamy, V. K. (2020). Silver nanoparticles in dye effluent treatment: A review on synthesis, treatment methods, mechanisms, photocatalytic degradation, toxic effects and mitigation of toxicity. *Journal of Photochemistry and Photobiology B: Biology, 205*, 111823.

Montiel Schneider, M. G., Martín, M. J., Otarola, J., Vakarelska, E., Simeonov, V., Lassalle, V., & Nedyalkova, M. (2022). Biomedical applications of iron oxide nanoparticles: Current insights progress and perspectives. *Pharmaceutics, 14*(1), 204.

Morsy, M., Nossier, M., Elsebaay, A. E., & Abd-Elrahman, S. (2022). Phytoremediation of Pb and Cd by Alfalfa (Medicago sativa L.): An applied study in the presence of lettuce plants (Lactuca sativa L.). *Arab Universities Journal of Agricultural Sciences, 30*(1), 163–174.

Muratova, A. Y., Lyubun, E. V., Golubev, S. N., & Turkovskaya, O. V. (2022). Effect of copper ions on the associations of Azospirillum bacteria with wheat seedlings (Triticum aestivum L.). *Vavilov Journal of Genetics and Breeding, 26*(5), 477.

Nagendran, R., Selvam, A., Joseph, K., & Chiemchaisri, C. (2006). Phytoremediation and rehabilitation of municipal solid waste landfills and dumpsites: A brief review. *Waste Management, 26*(12), 1357–1369.

Newman, L. A., & Reynolds, C. M. (2004). Phytodegradation of organic compounds. *Current Opinion in Biotechnology, 15*(3), 225–230.

Nguyen, D. T. C., Nguyen, T. T., Le, H. T., Nguyen, T. T. T., Bach, L. G., Nguyen, T. D., et al. (2021). The sunflower plant family for bioenergy, environmental remediation, nanotechnology, medicine, food and agriculture: A review. *Environmental Chemistry Letters, 19*, 3701–3726.

Oliveira, H. C., Seabra, A. B., Kondak, S., Adedokun, O. P., & Kolbert, Z. (2023). Multilevel approach of plant-nanomaterial relationship: From cells to living ecosystems. *Journal of Experimental Botany*, erad107.

Papazoglou, E. G., & Fernando, A. L. (2017). Preliminary studies on the growth, tolerance and phytoremediation ability of sugarbeet (Beta vulgaris L.) grown on heavy metal contaminated soil. *Industrial Crops and Products, 107*, 463–471.

Pirsarandib, Y., Hassanpouraghdam, M. B., Rasouli, F., Aazami, M. A., Puglisi, I., & Baglieri, A. (2022). Phytoremediation of soil contaminated with heavy metals via arbuscular mycorrhiza (funneliformis mosseae) inoculation ameliorates the growth responses and essential oil content in lavender (lavandula angustifolia L.). *Agronomy, 12*(5), 1221.

Pitkethly, M. (2008). Nanotechnology: Past, present, and future. *Nano Today, 3*(3–4), 6.

Prisacari, S., Todiraş, V., & Corcimaru, S. (2022). The influence of nanomagnetite on the processes of growth, development, and formation of the legume-rhizobia complex in vetch plants under soil conditions of plastics pollution. In *Microbial biotechnology* (p. 41).

Pullar, R. C. (2023). Applications of magnetic oxide nanoparticles in hyperthermia. *Magnetic Nanoparticles for Biomedical Applications, 143*, 76–101.

Quazi, S., Rodrigo, S., Fatima, Z., & Prakash, A. (2023). Nanoparticles: Types, applications, and commercial products. In *Implications of nanoecotoxicology on environmental sustainability* (pp. 15–41). IGI Global.

Rajput, V. D., & Minkina, T. (2023). Editorial for special issue "Nano-bioremediation approaches for degraded soils and sustainable crop production". *Nanomaterials, 13*(7), 1138.

Ramana, S., Tripathi, A. K., Bharati, K., Singh, A. B., Kumar, A., Sahu, A., et al. (2021). Tolerance of cotton to elevated levels of Pb and its potential for phytoremediation. *Environmental Science and Pollution Research, 28*, 32299–32309.

Rizwan, M., Ali, S., Rizvi, H., Rinklebe, J., Tsang, D. C., Meers, E., et al. (2016). Phytomanagement of heavy metals in contaminated soils using sunflower: A review. *Critical Reviews in Environmental Science and Technology, 46*(18), 1498–1528.

Sadat, S. M. O., Kucukcongar, S., & Turkyilmaz, M. (2022). Nickel adsorption from waters onto Fe3O4/sugar beet pulp nanocomposite. *International Journal of Phytoremediation*, 1–14.

Sagadevan, S., Lett, J. A., Fatimah, I., Selvi, K. T., Sivasankaran, R. P., Weldegebrieal, G. K., & Oh, W. C. (2022). Photocatalytic and electrochemical activity of magnesium oxide nanoballs synthesized via a hydrothermal route. *PRO, 10*(10), 2098.

Sakakibara, M., Watanabe, A., Inoue, M., Sano, S., & Kaise, T. (2010, January). Phytoextraction and phytovolatilization of arsenic from as-contaminated soils by Pteris vittata. In *Proceedings of the annual international conference on soils, sediments, water and energy* (Vol. 12, No. 1, p. 26).

Sharma, A., Kaswan, V., & Kaushik, A. (2018). Bioremedition and phyto–remidiation. *Journal of Pharmacognosy and Phytochemistry, 7*(1), 723–726.

Sharma, N., Singh, G., Sharma, M., Mandzhieva, S., Minkina, T., & Rajput, V. D. (2022). Sustainable use of nano-assisted remediation for mitigation of heavy metals and mine spills. *Water, 14*(23), 3972.

Shirkhani, Z., Chehregani Rad, A., & Mohsenzadeh, F. (2021). Improving Cd-phytoremediation ability of Datura stramonium L. by Chitosan and Chitosan nanoparticles. *Biologia, 76*(8), 2161–2171.

Shmaefsky, B. R., & Husen, A. (2023). Phytoremediation potential of medicinal plants. In *Medicinal plants: Their response to abiotic stress* (pp. 443–469). Springer.

Singh, B. M., Singh, D., & Dhal, N. K. (2022). Enhanced phytoremediation strategy for sustainable management of heavy metals and radionuclides. *Case Studies in Chemical and Environmental Engineering, 5*, 100176.

Song, U., Kim, B. W., Rim, H., & Bang, J. H. (2022). Phytoremediation of nanoparticle-contaminated soil using the halophyte plant species Suaeda glauca. *Environmental Technology & Innovation, 28*, 102626.

Srivastava, S., Shukla, A., Rajput, V. D., Kumar, K., Minkina, T., Mandzhieva, S., et al. (2021). Arsenic remediation through sustainable phytoremediation approaches. *Minerals, 11*(9), 936.

Stonehouse, G. C., McCarron, B. J., Guignardi, Z. S., El Mehdawi, A. F., Lima, L. W., Fakra, S. C., & Pilon-Smits, E. A. (2020). Selenium metabolism in hemp (Cannabis sativa L.)-potential for phytoremediation and biofortification. *Environmental Science & Technology, 54*(7), 4221–4230.

Takahashi, R., Ito, M., Katou, K., Sato, K., Nakagawa, S., Tezuka, K., et al. (2016). Breeding and characterization of the rice (Oryza sativa L.) line "Akita 110" for cadmium phytoremediation. *Soil Science and Plant Nutrition, 62*(4), 373–378.

Vorobyova, V. I. (2023). Plant extract based on deep eutectic solvent-mediated biosynthesis of silver nanoparticles: Cytotoxicity and antibacterial effects. *Bioinorganic Chemistry and Applications*, 2023.

Wu, X. H., Liew, Y. K., Lim, W. M., Mai, C. W., & Then, Y. Y. (2023). Blood compatible and noncytotoxic superhydrophobic graphene/titanium dioxide coating with antibacterial and antibiofilm properties. *Journal of Applied Polymer Science, 140*(11), e53629.

Xie, L., Wu, P., Lei, Q., Xu, C., Huang, W., Chen, X., et al. (2023). Constructing Z-scheme 0D/2D TiO2 nanoparticles/Bi2O3 nanosheet heterojunctions with enhanced visible light induced photocatalytic antibiotics degradation and hydrogen evolution. *Catalysts, 13*(3), 583.

Yadav, K. K., Gupta, N., Kumar, A., Reece, L. M., Singh, N., Rezania, S., & Khan, S. A. (2018). Mechanistic understanding and holistic approach of phytoremediation: A review on application and future prospects. *Ecological Engineering, 120*, 274–298.

Yan, F., Zhang, F., Fan, J., Hou, X., Bai, W., Liu, X., et al. (2021). Optimization of irrigation and nitrogen fertilization increases ash salt accumulation and ions absorption of drip-fertigated sugar beet in saline-alkali soils. *Field Crops Research, 271*, 108247.

Zand, A. D., Mikaeili Tabrizi, A., & Vaezi Heir, A. (2020). Application of titanium dioxide nanoparticles to promote phytoremediation of Cd-polluted soil: Contribution of PGPR inoculation. *Bioremediation Journal, 24*(2–3), 171–189.

Zhou, Q., Zhang, L., & Wu, H. (2017). Nanomaterials for cancer therapies. *Nanotechnology Reviews, 6*(5), 473–496.

Chapter 8
Concern and Outputs of Nano-enabled Agricultural Strategies on Food Quality

Avinash Singh, Esha Yadav, Hara Prasad, and Giridhari Pal

Abstract Improving the effectiveness of agrochemical distribution to crop plants is becoming more achievable with the use of nano-enabled agricultural goods like nanopesticides and nanofertilizers. According to the findings of the experiments, nano-agrochemicals are highly promising for boosting crop yields while drastically decreasing the environmental toll of conventional agrochemicals. On the other hand, new evidence suggests that nano-enabled goods can improve crop quality while simultaneously improving productivity. As a result, new, more nutritious alternatives have emerged in the agriculture sector through the use of nanotechnology and nano-fertilizers (NFs). To improve nutrient utilization efficiency, these novel NFs use the extraordinary characteristics of nanoparticles, which can be anywhere from 1 to 100 nm in size, like zeolites and nanoclays. There are numerous benefits of using NFs instead of their traditional alternatives. These include the following: improved targeted action, reduced eco-toxicity, controlled release mechanisms, consistent and effective performance, and safe and easy administration and disposal. Nanofertilizers (NFs) reduce the risk of environmental damage by preventing nutrients from being wasted and allowing plants to absorb them quickly and completely. Some reports indicate that there is variation in the amount of metallic necessary elements, proteins, carbohydrates, and starch. Exposure of rice to CeO_2-engineered nanoparticles (ENPs) has been found to increase *Verbi gratia*, albumin, globulin, and prolamin, while exposure to CuO, ZnO, and CeO2 ENPs has been found to increase Ca, Mg, and P in various crops. Alternatively, it has been observed that kidney beans exposed to ZnO-manufactured nanoparticles had a decrease in Ni,

A. Singh (✉) · E. Yadav
Department of Biotechnology, Axis Institute of Higher Education, Kanpur, Uttar Pradesh, India

H. Prasad
Department of Pharmacology, University College of Medical Sciences, University of Delhi, Delhi, India

G. Pal
Department of Pharmacology, Vallabh Patel Chest Institute University of Delhi, Delhi, India

V. D. Rajput et al. (eds.), *Contemporary Suitability of Nanobionics in Agriculture*, Nanotechnology in the Life Sciences, https://doi.org/10.1007/978-3-031-82779-2_8

while cucumbers treated with CeO_2 had an increase. Nevertheless, there has been a lack of information regarding the precise health impacts of consuming agricultural products that have been exposed to nano-agrochemicals.

Keywords Sustainable agriculture · Nanoparticle · Nanofertilizers · Crop quality · Food safety

8.1 Introduction

A greater emphasis on agricultural production and developments in technology has helped to enhance food security. The fact that 795 million people are currently affected by hunger, while two billion people suffer from inadequate nutrition, highlights the severity of the global food crisis. According to the United Nations Department of Economic and Social Affairs (2017), the majority of the expected increases in the world population—from 7.7 billion in 2017 to 9.8 billion in 2050 and 11.2 billion in 2100—will take place in emerging nations. By mid-century, global food production would have to surge 50% to meet the need of a rapidly expanding population (Xin et al., 2020). The subsequent rise in land, water, and energy inputs is directly proportional to the growth in food output. Nevertheless, according to the Global Soil Partnership, 75 billion tons of soil are lost annually from croplands around the world due to factors such as salinization, wind and water erosion, and nutrient depletion (Borrelli et al., 2017). Due to over-pumping and poisoning of aquifers, two-thirds of the world's population might be under stress conditions by 2025, and 1.8 billion people would live in countries or regions with water scarcity. Changes in precipitation patterns and extreme weather events are two other ways in which modern agriculture is expected to be profoundly impacted by climate change in the decades to come. As an example, the net revenue in northwest Vietnam, which is mainly dependent on agriculture, was anticipated to fall by 17.7% in 2050 and by 21.28% in 2100 as a result of climate change.

Consequently, feeding the world's growing population while reducing our impact on the environment and maintaining or improving farmers' incomes is an enormous task (Huong et al., 2019).

Nanotechnology is the application of technology to manipulate nanomaterials for various purposes. It is vital to the food and agriculture sectors, improving food quality and safety, improving crop productivity, and promoting human health through creative and inventive methods (Neme et al., 2021). Due to their distinct physical, chemical, and biological characteristics, which include a high surface-to-volume ratio and modified solubility and toxicity compared to larger particles, engineered nanometer-sized particles have become increasingly important in various fields such as medicine, agro-food sectors, sewage water treatments, and other industries (Parisi et al., 2015; El-Temsah & Joner, 2012). Silver (Ag), gold (Au), zinc oxide (ZnO), titanium dioxide (TiO2), and carbon nanoparticles are produced in significantly larger quantities compared to other nanomaterials due to their potential antimicrobial properties. These materials are utilized in various applications such as air filters, food storage containers, deodorants, bandages, toothpastes, paints, and other household appliances (Parisi et al., 2015; El-Temsah & Joner, 2012). In addition, the strong antibacterial properties of nanosized copper oxides (nCuO) have led to their extensive use in commercial nano-biocide products (Nair et al., 2010). Nanomaterials are minuscule particles with sizes ranging from 1 to 100 nm. They are insoluble or bio-persistent and are created by several methods. These materials have many applications in disciplines such as health, electronics, agriculture, and the food industry (Duncan, 2011). Nanotechnologies in food science utilize nanoparticles of various sizes to potentially produce and process foods that are healthier, safer, and of superior quality. Increased food availability and improved goods with significant benefits in agriculture, water, the environment, medication development, and health are two ways in which nanotechnology has shown great promise as a solution to food security challenges (Azeredo et al., 2009). Several food-related issues, including nutrient distribution, protein bioseparation, solubilization, and nutraceutical nanoencapsulation, might be greatly helped by a handful of the new areas of nanotechnology. Nanotechnology in food refers to the use of nanocarrier methods and technologies to enhance bioactive substances, alter their biological availability, and create a barrier against various chemical or environmental changes (Naoto et al., 2009).

Many benefits have been demonstrated using nanotechnology. The complicated legislative processes, limited regulations resulting from a lack of research reports on the toxicity and risk associated with nanomaterials, and the complexity of nanomaterials themselves are some of the reasons why it has not gained wider acceptance (Yadav, 2017). Because of this, studies into nanomaterials' effects on ecosystems, toxicity, bioaccumulation, disintegration, and absorption in humans are urgently needed. Many areas of the food sector can benefit from nanomaterials, including processing, manufacturing, shipping, and storage (Nile et al., 2020).

The progress of nano-based technology research in the field of commercial food application has been rapid; however, the development of nanostructures has been somewhat slower. As research in food nanotechnology progresses, there is a growing public concern regarding the safety of nanotechnology-based products designed for human consumption and use (Rizvi et al., 2022). Before nanofood items may be commercialized, it is necessary to conduct a comprehensive assessment of the possible threats they pose to human health. In order to ensure the safety of our food, it is imperative to carry out and advance research on human and environmental health. Although a complete understanding of the fate and potential toxicity of nanoparticles is currently lacking, it is crucial to support the expansion and use of nanotechnology in food science and industry. With advances in nanobiotechnology, devices and materials derived from it are becoming increasingly compact and sensitive. In addition, nanoparticles have been successfully used to preserve food by protecting it against moisture, lipids, gases, unpleasant flavors, and odors. They play a crucial role in transporting bioactive substances to targeted tissues. Although advancements in nanotechnology are occurring regularly, there remain several challenges and prospects for enhancing current technology (Amini et al., 2014). Additionally, concerns regarding the potential consequences of nanotechnology must be acknowledged and resolved in order to alleviate client apprehensions. Despite the introduction of nanotechnology, the difficulties involved in constructing a robust and environmentally friendly food industry remain. The control of the linked health, safety, and environmental implications should be prioritized. Prioritizing transparency on safety risks and environmental impact is crucial in the development of nanotechnology in food systems. Therefore, it is imperative to conduct required testing of nanofoods before their release to the market. Ensuring proper public education is essential for the sustained success of introducing and advancing nanotechnology in the food system (Yadav et al., 2023).

8.2 Nano-enabled Agricultural Strategies on Food Quality

The advent of nanotechnology has significantly transformed the fields of food preservation and production shown in Fig. 8.1. The application of nanotechnology in the food industry has several benefits and drawbacks. The complexity of this sector arises from the inherent alteration of nanoparticles' chemical and physical properties upon entering the human body. It is crucial to exercise great care when adding nanoparticles to food products. Furthermore, it is imperative to exclusively utilize nanoparticles that are both non-toxic and biodegradable. In addition, due to their diminutive size, they can easily infiltrate cells, resulting in DNA harm and potentially triggering an immune reaction. Developing environmentally friendly nanomaterials is crucial, and regulatory organizations worldwide are implementing safe legislations and standards to address the associated risks (Singh et al., 2017,Grobe et al., 2012).

Nanotechnology has the capacity to greatly influence various areas such as packaging, food processing, and agriculture. Nanocomposites used in food packaging materials can offer benefits such as enzyme immobilization, oxygen-scavenging capabilities, and antibacterial properties. They improve the quality of the packing material and protect food from environmental dangers. Nanotechnology is being used in the food industry to provide advanced diagnostic devices and selective sensors (Biswas et al., 2022). These technologies help ensure the quality of food throughout the manufacturing process and assist in monitoring food production. In addition, edible coatings and films are used to maintain the quality of food and extend its storage duration. Regional and international institutions are working together to establish clear and specific criteria for the utilization of nanotechnology in food, including its packaging and additives. In addition, a comprehensive assessment of the environmental, water, land, and public health implications will be necessary to evaluate the effects of forthcoming advancements in nanotechnology in the fields of food, agriculture, and business. Unique properties and possible uses in many areas have made nanotechnology very interesting. Nanotechnology is a new way to increase agricultural production while also improving efficiency, safety for

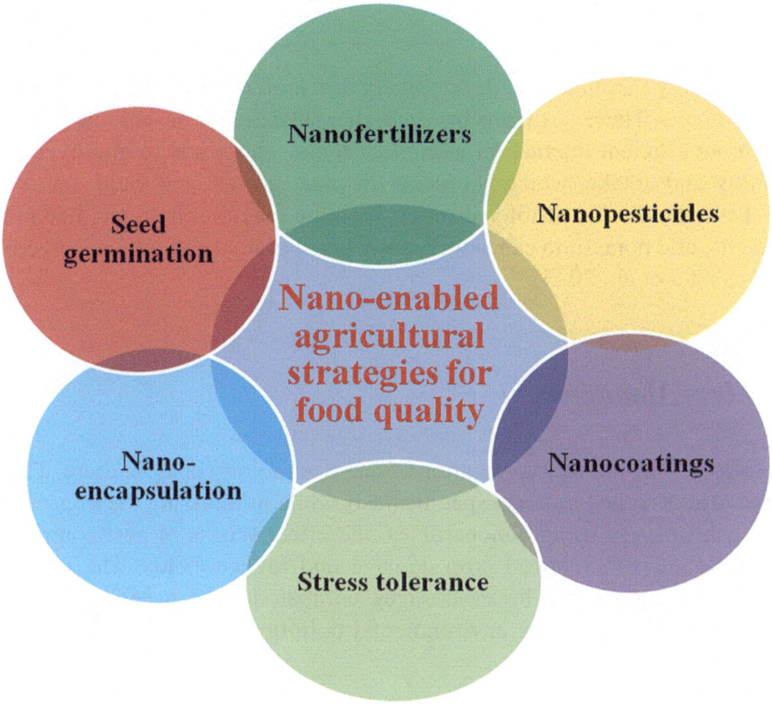

Fig. 8.1 Nano-enabled agricultural strategies for food quality

the environment, safety for living things, and safety for money. Furthermore, modern farming relies on methods that are better for the environment and can be used instead of conventional pesticides and fertilizers. With the help of nanotechnology, the problems that come with regular farming can be fixed. Since nanotechnology has the potential to make the world economy stronger, it is important to look into how nanoparticles can be used in farming. In this review, we talk in detail about how nanoparticles are made by microbes, where and how they work in living cells and plants, how nanofertilizers are made and how they improve nutrients, how they help plants deal with abiotic stresses and diseases, and how nanoparticles affect the metabolic processes of both plants and microbes (Mohammad et al., 2022; Gupta et al., 2023).

8.3 Nano-based Strategies Improve Crop Yields

Nano-based strategies significantly enhance crop yields through various mechanisms that improve nutrient delivery, pest management, and overall plant health. Below are the key ways in which these strategies contribute to agricultural productivity.

8.3.1 Nutrient Delivery

Nanotechnology facilitates the development of nanofertilizers that enhance nutrient uptake by plants. These fertilizers have a higher surface area-to-volume ratio, allowing for more efficient interaction with plant roots. This leads to improved nutrient availability and uptake, which is crucial for plant growth and yield. For instance, studies have shown that nanofertilizers containing essential nutrients like nitrogen, phosphorus, and potassium can significantly boost plant development and crop production (Shelar et al., 2023).

8.3.2 Pest Management

Nanopesticides are another application of nanotechnology in agriculture. These formulations are designed to target specific pests while minimizing the impact on non-target organisms. By using nanoparticles, the effectiveness of pest control can be increased, leading to reduced crop damage and higher yields. The precision of nanopesticides allows for lower dosages compared to conventional pesticides, which also helps in reducing environmental pollution (Ijaz et al., 2023; Gangwar et al., 2023).

8.3.3 Stress Resilience

Nanotechnology also plays a crucial role in enhancing plants' resilience to biotic and abiotic stresses. For example, engineered nanomaterials can improve plants' resistance to diseases and environmental stresses such as drought and salinity. This resilience is vital for maintaining crop yields, especially in the face of climate change and increasing pest pressures (Gangwar et al., 2023).

8.3.4 Genome Editing and Smart Sensors

Advancements in nano-enabled genome editing techniques allow for more precise modifications to plant genomes, which can lead to improved traits such as disease resistance and enhanced growth rates. Additionally, nanosensors can monitor plant health and nutrient levels in real time, enabling farmers to make informed decisions about resource application, further optimizing crop yields (Shang et al., 2019).

In summary, nano-based strategies improve crop yields by enhancing nutrient delivery, improving pest management, increasing stress resilience, and enabling advanced agricultural practices through genome editing and monitoring technologies. These innovations are essential for developing sustainable agricultural systems capable of meeting the demands of a growing global population (Shelar et al., 2023).

8.4 Nano-enabled Agricultural Strategies Impact the Nutritional Content of Crops

The nutritional value of crops is significantly altered by agricultural practices that are allowed by nanotechnology, and the results of these modifications can be both positive and possibly problematic. Based on the most recent research, the following is a summary of the effects:

8.4.1 Positive Impacts on Nutritional Content

8.4.1.1 Increased Nutrient Availability

Nanofertilizers enhance the efficiency of nutrient uptake in plants. This can lead to higher concentrations of essential nutrients such as calcium (Ca), magnesium (Mg), and phosphorus (P) in crops. For example, studies have shown that engineered nanoparticles can significantly increase the levels of these nutrients in various crops, improving their overall nutritional profile (Gomez et al., 2021).

8.4.1.2 Alterations in Protein Content

The application of certain engineered nanomaterials, such as cerium oxide (CeO_2), has been linked to increased levels of specific proteins in crops. For instance, research indicates that the protein content in rice can be enhanced through exposure to CeO_2 nanoparticles, which may contribute to improved food quality and nutritional value (Gomez et al., 2021).

8.4.1.3 Improved Bioavailability of Nutrients

Nanotechnology can enhance the bioavailability of nutrients in food. By improving the delivery and absorption of micronutrients, nano-agrochemicals can help fortify crops with essential vitamins and minerals, potentially addressing nutritional deficiencies in populations that rely heavily on staple crops (Gomez et al., 2021).

8.4.2 Concerns Regarding Nutritional Content

8.4.2.1 Variability in Nutrient Composition

While some nutrients may increase, there is evidence that the application of nano-based agrochemicals can lead to variability in the composition of crops. For example, reductions in certain trace elements, such as molybdenum (Mo) and nickel (Ni), have been observed in crops like cucumbers and kidney beans exposed to specific nanoparticles. This variability raises concerns about the overall balance of nutrients in the food supply (Gomez et al., 2021).

8.4.2.2 Potential Toxicity and Health Implications

The long-term effects of consuming crops treated with nano-based agrochemicals are not fully understood. There is a lack of comprehensive studies on how these materials may affect human health through dietary intake. The possibility of altered nutrient profiles and the introduction of engineered materials into the food chain necessitate further research to assess safety and health implications (Gomez et al., 2021).

8.4.2.3 Public Acceptance and Ethical Concerns

The introduction of nanotechnology in agriculture has raised ethical concerns similar to those surrounding genetically modified organisms (GMOs). Public perception of potential health risks associated with nano-enabled food products can influence acceptance and market dynamics, which is critical for the successful implementation of these technologies (Gomez et al., 2021).

In summary, nano-enabled agricultural strategies can significantly enhance the nutritional content of crops through improved nutrient delivery and increased bioavailability. However, these benefits come with concerns regarding variability in nutrient composition and potential health implications. Ongoing research is essential to fully understand these impacts and to develop appropriate regulatory frameworks to ensure the safe use of nanotechnology in agriculture.

8.5 Potential Health Benefits of Crops Enhanced by Nano-enabled Strategies

There are a number of possible health benefits that could be derived from crops that have been altered by agricultural practices that are enabled by nanotechnology. These benefits principally include higher nutritional content and enhanced resilience against environmental challenges. The following is a list of the most important health benefits attributable to these advancements.

A. *Improved Nutritional Content*

Enhanced Nutrient Uptake Nano-enabled strategies, particularly the use of nanofertilizers, can significantly increase the availability and uptake of essential nutrients in crops. For instance, nanoparticles can enhance the levels of vital minerals such as calcium, magnesium, and phosphorus in various crops, contributing to a more nutritious food supply (Jakhar et al., 2022).

Increased Protein Levels Research has shown that exposure to certain engineered nanoparticles, like cerium oxide (CeO_2), can lead to higher concentrations of proteins in crops such as rice. This increase in protein content can improve the overall dietary quality of these foods (Daniel et al., 2024).

B. *Enhanced Crop Resilience*

Stress Tolerance Nano-agrochemicals can improve plants' resilience to abiotic stresses such as drought, salinity, and extreme temperatures. By enhancing the plants' antioxidant systems and nutrient uptake, these strategies can lead to healthier crops that are more capable of withstanding environmental challenges, ultimately resulting in more reliable food sources (Ijaz et al., 2023).

Disease Resistance Nanoparticles can also enhance plant immunity against pathogens. By improving the physical barriers of crops and modulating plant defense mechanisms, nano-enabled strategies can reduce the incidence of crop diseases, leading to healthier plants and safer food products (Tortella et al., 2023).

8.6 Sustainable Food Production

Reduced Chemical Residues The efficient delivery and targeted action of nano-agrochemicals can minimize the overall use of pesticides and fertilizers, leading to lower chemical residues in food products. This reduction can contribute to safer food for consumers and less environmental contamination (Ali et al., 2023).

Mitigation of Nutrient Deficiencies By addressing specific nutrient deficiencies in crops through targeted nano-delivery systems, these strategies can help combat malnutrition in populations reliant on staple crops, thereby improving public health outcomes. The integration of nano-enabled agricultural strategies holds significant promise for enhancing the nutritional quality and health benefits of crops. By improving nutrient uptake, increasing resilience to environmental stresses, and reducing chemical residues, these technologies can contribute to a more sustainable and health-conscious food production system. However, ongoing research is necessary to fully understand the long-term effects of these innovations on human health and the environment (Sandhu et al., 2023).

8.7 Concerns of Nano-enabled Agricultural Strategies on Food Quality

Nano-enabled agricultural techniques encompass the utilization of nanotechnology in various aspects of agriculture, including crop production, pest control, and food processing. These technologies are specifically created to enhance the efficiency and efficacy of agricultural techniques. However, they also give rise to worries regarding potential effects on the quality and safety of food.

8.7.1 Changes in Nutritional Composition

Nano-agrochemicals, including nanofertilizers and nanopesticides, can significantly alter the nutritional profile of crops. Research indicates that these engineered nanomaterials can affect the levels of proteins, sugars, and essential minerals in plants.

8.7.1.1 Protein Content

Exposure to cerium oxide nanoparticles has been shown to increase the levels of albumin, globulin, and prolamin in rice, which are important for human nutrition (Gomez et al., 2021).

8.7.1.2 Mineral Composition

Nanoparticles like CeO_2, CuO, and ZnO have been reported to enhance the concentrations of calcium, magnesium, and phosphorus in various crops, potentially improving their nutritional value (Gomez et al., 2021; Ji et al., 2022).

8.7.2 Potential Health Implications

Despite the potential benefits in crop yield and quality, there are concerns about the long-term health implications of consuming food produced with nano-agrochemicals. Current studies have not sufficiently addressed the specific effects on human health from the consumption of these products, raising questions about their safety (Singh & Gurjar, 2022).

8.7.3 Environmental and Ethical Considerations

The application of nanotechnology in agriculture raises ethical concerns regarding environmental impact and food safety. The introduction of nanoparticles into ecosystems could lead to unforeseen consequences, including bioaccumulation and disruption of soil microbial communities. Moreover, public acceptance of nano-enabled food products is crucial, similar to the concerns surrounding genetically modified organisms (GMOs) (Kah et al., 2019).

8.7.4 Regulatory Challenges

The rapid advancement of nanotechnology in agriculture has outpaced the development of comprehensive regulatory frameworks. Current risk assessment methods may not adequately address the unique properties and risks associated with nanoparticles, necessitating the establishment of specific guidelines to ensure food safety and environmental protection (Kah et al., 2019). While nano-enabled agricultural strategies hold promise for enhancing crop production and nutritional quality, they also present significant concerns regarding food safety, health implications, and environmental impacts. Ongoing research and regulatory oversight are essential to address these challenges and ensure the responsible use of nanotechnology in agriculture (Wahab et al., 2024).

8.8 Comparison of Nano-based Agrochemicals with Conventional Agrochemicals in Terms of Effectiveness and Environmental Impact

It has been demonstrated that nano-based agrochemicals have a number of advantages over conventional agrochemicals in terms of their effectiveness and their influence on the environment. On the basis of the findings of current investigations, the following comparison may be made.

8.8.1 Increased Efficacy

Nano-based agrochemicals, such as nanopesticides, have been found to be more effective against target organisms. They can enhance the efficacy of active ingredients (AIs) by up to 31.5%, with field trials showing an 18.9% improvement in pest control compared to conventional pesticides. This increased effectiveness is attributed to their ability to encapsulate and deliver AIs in a controlled manner, improving adhesion and reducing premature loss before reaching the target pests (Abdollahdokht et al., 2022; An et al., 2022).

8.8.2 Targeted Delivery

The nanoscale formulation allows for more precise targeting of pests and diseases, which can lead to better crop protection and, consequently, higher yields. This targeted approach minimizes the quantity of chemicals needed, which is a significant advantage over traditional methods that often require larger volumes of active ingredients (Rajput et al., 2022).

8.8.3 Reduced Toxicity

Studies indicate that nano-based agrochemicals generally exhibit lower toxicity to non-target organisms compared to their conventional counterparts. For example, the toxicity of nanopesticides to non-target aquatic species is reported to be 43.1% lower, suggesting a reduced risk of collateral environmental damage (Wang et al., 2022).

8.8.4 Lower Leaching Potential

Nano-based agrochemicals also demonstrate a 22.1% reduction in leaching potential in soils, which means they are less likely to contaminate water bodies compared to conventional agrochemicals. This characteristic is crucial for protecting aquatic ecosystems from chemical runoff (Wang et al., 2022).

8.8.5 Sustainable Practices

The use of nano-based agrochemicals aligns with sustainable agricultural practices by potentially reducing the overall chemical load in the environment. However, the long-term impacts of these materials on soil health and microbial communities are still being studied, indicating that while they offer immediate benefits, further research is necessary to fully understand their ecological implications (Wang et al., 2022).

In summary, nano-based agrochemicals provide enhanced effectiveness through improved pest control and targeted delivery, while also presenting lower environmental risks compared to conventional agrochemicals. However, ongoing research is essential to address the knowledge gaps regarding their long-term ecological impacts and to ensure that their use promotes sustainable agricultural practices.

8.9 Recent Advancements of Bionanocomposites for the Food Quality

The primary purpose of food packaging materials is to ensure both the preservation of food quality and the assurance of food safety. Petroleum-derived plastic materials have been extensively utilized in food packaging owing to their exceptional features, high quality, ease of processing, and affordability. Plastic materials have diverse use in packaging, but their primary drawback is their potential to produce significant environmental issues. These items are non-biodegradable, and their disposal pollutes the soil, water, and air. Hence, there is a requirement for an alternative, and, recently, novel packaging materials have been formulated and are available in the market. Emerging packaging materials, such as bio-based polymers, possess distinct and novel characteristics. Biopolymers are derived from sustainable sources and are classified as biodegradable. In addition, by using active molecules, certain characteristics can be enhanced, such as antibacterial and antioxidant qualities. Bionanocomposites are a type of composite material made up of a polymeric matrix derived from biological sources, which is strengthened by fillers that have dimensions ranging from 1 to 100 nm on the nanometer scale. In addition, various bioactive molecules can be added to provide active or intelligent characteristics that can either maintain or detect/indicate changes in food products, respectively. Novel materials are being used more and more in food packaging to decrease food losses and enhance safety shown in Fig. 8.2. These materials have numerous advantages

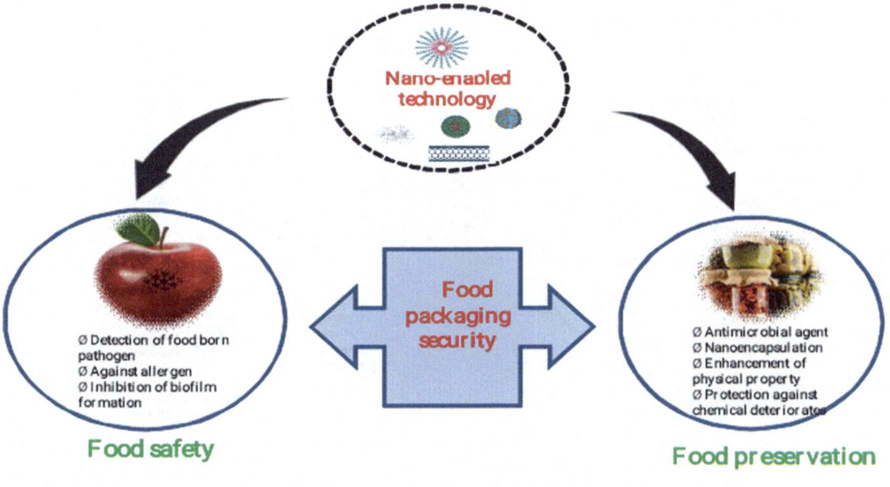

Fig. 8.2 Recent advancements of bionanocomposites for the food quality

compared to traditional biopolymers, including better water vapor and gas barrier properties, as well as thermal stability, mechanical strength, UV barrier, biodegradability, and antimicrobial properties.

8.10 Nanomaterials Activate Defense Responses in Crops Against Pathogens

A plant's physiological response to an attack by a pathogen is much like a defense mechanism. To start immune reactions, phytohormones like salicylic acid, methyl jasmonate, and jasmonic acid are thought to play a role. Molecules like reactive oxygen species send and receive messages. Another way this works is that liposomes, which are round structures made of two layers of phospholipids, can carry these chemicals directly to plant cells. This action starts the process of turning on genes that are part of the plant's defense system. Poly(lactic-co-glycolic acid) (PLGA) nanoparticles are one example. These are made up of many smaller particles that can carry phytohormones and signaling molecules to plants. Also, chitosan nanoparticles have been looked into to see if they could be used as carriers for signaling molecules and phytohormones. Figure 8.3 shows that these NPs can protect the molecules being carried from breaking down and guide them to specific plant cells, making the defense system stronger.

On top of that, host plants can stop infections by activating their innate immune system in reaction to pathogen-associated molecular patterns (PAMPs). During

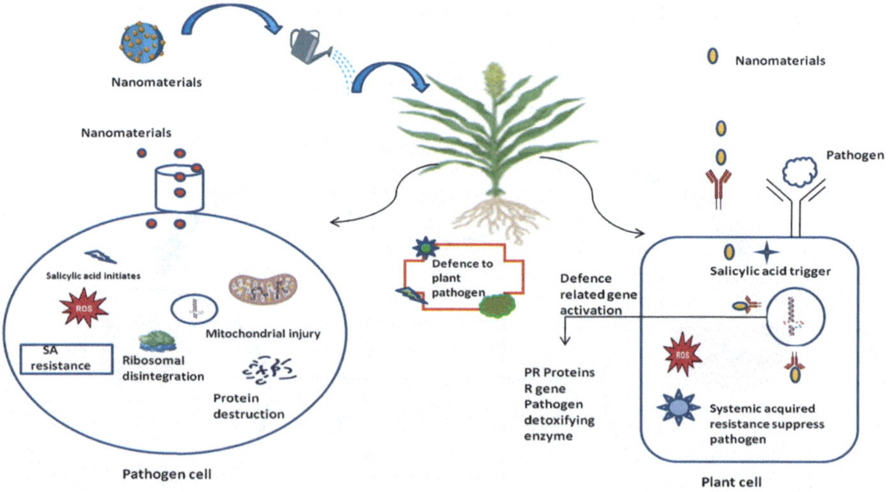

Fig. 8.3 Nanomaterials activate defense responses in crops against pathogens. The diagram provides a clear illustration of the role of nanomaterials in activating immune responses and suppressing plant pathogens

plant growth and development, hormones control many of the processes that affect the plant. Scientists have found that the production of peptides, auxins, cytokinin, gibberellins, brassino steroids, ethylene, salicylic acid, and abscisic acid (ABA) changes in plants that are affected. Some polymeric nanoparticles can also make plants less likely to get different kinds of plant diseases. For example, chitosan nanoparticles stop the rice blast fungus *Pyricularia grisea* from growing. Researchers treated pearl millet seeds with chitosan nanoparticles and found that changes in gene expression for enzymes such as peroxidase, phenylalanine ammonia lyase, catalase, superoxide, and polyphenol oxidase were linked to better protection to downy mildew. When copper and silicon nanoparticles are put on tomatoes, they make more enzymatic and non-enzymatic plant immune chemicals. This makes the plants more resistant to *Clavibacter michiganensis* and, in the end, makes better crops. In the same way, some research looked at how magnesium oxide nanoparticles (MgO NPs) affected the *Ralstonia solanacearum* bacteria's ability to infect tomato plants. Tomato plants that were exposed to MgO nanoparticles in the soil and on the stems' upper parts had higher levels of salicylic acid-inducible PR1, jasmonate LoxA, and systemic resistance-related GluA.

8.11 Conclusion and Future Perspective

As a result of nanotechnology's contributions to high-quality food production, processing, packaging, and long-term preservation, the food industry has reaped huge benefits. The food sector has expanded as a result of better food quality, as measured by taste and texture. Nanomaterials and nanosensors aid customers by informing them of the food's condition and nutritional value, while also improving food safety through the detection of pathogens. Because most dietary bioactives used to treat diseases are hydrophobic and hence have poor bioavailability and stability, delivery systems based on nanotechnology have allowed for the targeted distribution of these compounds with much improved bioavailability. Assuring public trust and approval of commercially available nanofoods present formidable obstacles for governments and businesses alike. Recent news has been rife with reports of nanocolloidal particles' active use in several food business domains, including processing, packaging, nutrition, food safety, and quality. Designing safer, healthier, higher-quality, and more sustainably produced meals requires an understanding of the characteristics and behavior of colloidal particles. Although nanoparticles are produced globally, only a handful of nations have established universal regulations for the use of nanotechnology in food. It is challenging to draw any conclusions about the effectiveness of nanosystems due to lack of scientific investigation. Using nanoparticles as an ingredient in food is more damaging than using them in food packaging. The production and use of nanomaterials pose an ongoing risk of DNA damage, cell membrane disruption, and cell death making their way into the food chain via the soil, water, and air. Nanofoods' impacts on human and animal health have been the subject of few in vivo research. For nanofoods to be more widely accepted by consumers, there needs to be clear labeling and rules around their marketing.

So, these nanotechnologies, when used properly, have the potential to greatly enhance food processing and product quality, which in turn benefits human health and well-being.

References

Abdollahdokht, D., Gao, Y., Faramarz, S., Poustforoosh, A., Abbasi, M., Asadikaram, G., & Nematollahi, M. H. (2022). Conventional agrochemicals towards nano-biopesticides: An overview on recent advances. *Chemical and Biological Technologies in Agriculture, 9*(1), 13.

Ali, S., Ahmad, N., Dar, M. A., Manan, S., Rani, A., Alghanem, S. M., Khan, K. A., Sethupathy, S., Elboughdiri, N., Mostafa, Y. S., & Alamri, S. A. (2023). Nano-agrochemicals as substitutes for pesticides: Prospects and risks. *Plants, 13*(1), 109.

Amini, S. M., Gilaki, M., & Karchani, M. (2014). Safety of nanotechnology in food industries. *Electronic Physician, 6*(4), 962.

An, C., Sun, C., Li, N., Huang, B., Jiang, J., Shen, Y., Wang, C., Zhao, X., Cui, B., Wang, C., & Li, X. (2022). Nanomaterials and nanotechnology for the delivery of agrochemicals: Strategies towards sustainable agriculture. *Journal of Nanobiotechnology, 20*(1), 11.

Azeredo, H. M., Mattoso, L. H., Wood, D., Williams, T. G., Avena-Bustillos, R. J., & McHugh, T. H. (2009). Nanocomposite edible films from mango puree reinforced with cellulose nanofibers. *Journal of Food Science, 74*(5), N31–N35.

Biswas, R., Alam, M., Sarkar, A., Haque, M. I., Hasan, M. M., & Hoque, M. (2022). Application of nanotechnology in food: Processing, preservation, packaging and safety assessment. *Heliyon, 8*(11).

Borrelli, P., Robinson, D. A., Fleischer, L. R., Lugato, E., Ballabio, C., Alewell, C., Meusburger, K., Modugno, S., Schütt, B., Ferro, V., & Bagarello, V. (2017). An assessment of the global impact of 21st century land use change on soil erosion. *Nature Communications, 8*(1), 1–3.

Daniel, A. I., Hüsselmann, L., Shittu, O. K., Gokul, A., Keyster, M., & Klein, A. (2024). Application of nanotechnology and proteomic tools in crop development towards sustainable agriculture. *Journal of Crop Science and Biotechnology*, 1–21.

Duncan, T. V. (2011). The communication challenges presented by nanofoods. *Nature Nanotechnology, 6*(11), 683–688.

El-Temsah, Y. S., & Joner, E. J. (2012). Impact of Fe and Ag nanoparticles on seed germination and differences in bioavailability during exposure in aqueous suspension and soil. *Environmental Toxicology, 27*(1), 42–49.

Gangwar, J., Kadanthottu Sebastian, J., Puthukulangara Jaison, J., & Kurian, J. T. (2023). Nanotechnological interventions in crop production—A review. *Physiology and Molecular Biology of Plants, 29*(1), 93–107.

Gomez, A., Narayan, M., Zhao, L., Jia, X., Bernal, R. A., Lopez-Moreno, M. L., & Peralta-Videa, J. R. (2021). Effects of nano-enabled agricultural strategies on food quality: Current knowledge and future research needs. *Journal of Hazardous Materials, 401*, 123385.

Grobe, A., Rissanen, E., & M. (2012). Nanotechnologies in agriculture and food-An overview of different fields of application, risk assessment and public perception. *Recent Patents on Food, Nutrition Agriculture, 4*(3), 176–186.

Gupta, R. K., Abd El Gawad, F., Ali, E. A., Karunanithi, S., Yugiani, P., & Srivastav, P. P. (2023). Nanotechnology: Current applications and future scope in food packaging systems. *Measurement: Food*, 100131.

Huong, N. T., Bo, Y. S., & Fahad, S. (2019). Economic impact of climate change on agriculture using Ricardian approach: A case of Northwest Vietnam. *Journal of the Saudi Society of Agricultural Sciences, 18*(4), 449–457.

Ijaz, M., Khan, F., Ahmed, T., Noman, M., Zulfiqar, F., Rizwan, M., Chen, J., Siddique, K. H., & Li, B. (2023). Nanobiotechnology to advance stress resilience in plants: Current opportunities and challenges. *Materials Today Bio*, 100759.

Jakhar, A. M., Aziz, I., Kaleri, A. R., Hasnain, M., Haider, G., Ma, J., & Abideen, Z. (2022). Nano-fertilizers: A sustainable technology for improving crop nutrition and food security. *NanoImpact, 27*, 100411.

Ji, H., Guo, Z., Wang, G., Wang, X., & Liu, H. (2022). Effect of ZnO and CuO nanoparticles on the growth, nutrient absorption, and potential health risk of the seasonal vegetable Medicago polymorpha L. *PeerJ, 10*, e14038.

Kah, M., Tufenkji, N., & White, J. C. (2019). Nano-enabled strategies to enhance crop nutrition and protection. *Nature Nanotechnology, 14*(6), 532–540.

Mohammad, Z. H., Ahmad, F., Ibrahim, S. A., & Zaidi, S. (2022). Application of nanotechnology in different aspects of the food industry. *Discover Food, 2*(1), 12.

Nair, R., Varghese, S. H., Nair, B. G., Maekawa, T., Yoshida, Y., & Kumar, D. S. (2010). Nanoparticulate material delivery to plants. *Plant Science, 179*(3), 154–163.

Naoto, S., Hiroshi, O., & Mitsutoshi, N. (2009). Micro and nanotechnology for food processing. (Food safety series) resource: Engineering and technology for a sustainable. *World American Society of Agricultural and Engineers, 16*, 19.

Neme, K., Nafady, A., Uddin, S., & Tola, Y. B. (2021). Application of nanotechnology in agriculture, postharvest loss reduction and food processing: Food security implication and challenges. *Heliyon, 7*(12).

Nile, S. H., Baskar, V., Selvaraj, D., Nile, A., Xiao, J., & Kai, G. (2020). Nanotechnologies in food science: Applications, recent trends, and future perspectives. *Nano-micro Letters, 12*, 1–34.

Parisi, C., Vigani, M., & Rodríguez-Cerezo, E. (2015). Agricultural nanotechnologies: What are the current possibilities? *Nano Today, 10*(2), 124–127.

Rajput, V. D., Minkina, T., Upadhyay, S. K., Kumari, A., Ranjan, A., Mandzhieva, S., Sushkova, S., Singh, R. K., & Verma, K. K. (2022). Nanotechnology in the restoration of polluted soil. *Nanomaterials, 12*(5), 769.

Rizvi, S. S., Moraru, C. I., Bouwmeester, H., Kampers, F. W., & Cheng, Y. (2022). Nanotechnology and food safety. In *Ensuring global food safety* (pp. 325–340). Academic Press.

Sandhu, R., Chaudhary, N., Shams, R., Singh, K., & Pandey, V. K. (2023). A critical review on integrating bio fortification in crops for sustainable agricultural development and nutritional security. *Journal of Agriculture and Food Research, 14*, 100830.

Shang, Y., Hasan, M. K., Ahammed, G. J., Li, M., Yin, H., & Zhou, J. (2019). Applications of nanotechnology in plant growth and crop protection: A review. *Molecules, 24*(14), 2558.

Shelar, A., Nile, S. H., Singh, A. V., Rothenstein, D., Bill, J., Xiao, J., Chaskar, M., Kai, G., & Patil, R. (2023). Recent advances in nano-enabled seed treatment strategies for sustainable agriculture: Challenges, risk assessment, and future perspectives. *Nano-Micro Letters, 15*(1), 54.

Singh, D., & Gurjar, B. R. (2022). Nanotechnology for agricultural applications: Facts, issues, knowledge gaps, and challenges in environmental risk assessment. *Journal of Environmental Management, 322*, 116033.

Singh, T., Shukla, S., Kumar, P., Wahla, V., Bajpai, V. K., & Rather, I. A. (2017). Application of nanotechnology in food science: Perception and overview. *Frontiers in Microbiology, 8*, 1501.

Tortella, G., Rubilar, O., Pieretti, J. C., Fincheira, P., de Melo, S. B., Fernández-Baldo, M. A., Benavides-Mendoza, A., & Seabra, A. B. (2023). Nanoparticles as a promising strategy to mitigate biotic stress in agriculture. *Antibiotics, 12*(2), 338.

Wahab, A., Muhammad, M., Ullah, S., Abdi, G., Shah, G. M., Zaman, W., & Ayaz, A. (2024). Agriculture and environmental management through nanotechnology: Eco-friendly nanomaterial synthesis for soil-plant systems, food safety, and sustainability. *Science of the Total Environment*, 171862.

Wang, D., Saleh, N. B., Byro, A., Zepp, R., Sahle-Demessie, E., Luxton, T. P., Ho, K. T., Burgess, R. M., Flury, M., White, J. C., & Su, C. (2022). Nano-enabled pesticides for sustainable agriculture and global food security. *Nature Nanotechnology, 17*(4), 347–360.

Xin, X., Judy, J. D., Sumerlin, B. B., & He, Z. (2020). Nano-enabled agriculture: From nanoparticles to smart nanodelivery systems. *Environmental Chemistry, 17*(6), 413–425.

Yadav, A., Yadav, K., & Abd-Elsalam, K. A. (2023). Exploring the potential of nanofertilizers for a sustainable agriculture. *Plant Nano Biology*, 100044.

Yadav, S. K. (2017). Tissue science & engineering realizing the potential of nanotechnology for agriculture and food technology. *Journal of Tissue Science & Engineering, 8*, 8–11.

Chapter 9
Emerging Trends in Nanobionics for Endurable Agriculture

Smruti Sohani and Pramod Ku. Singh

Abstract Growing global population, climate change, and resource depletion present the agricultural industry with urgent concerns that call for creative solutions to ensure sustainable food supply. By fusing nanotechnology and biological processes, nanobionics provides revolutionary ways to improve agricultural methods. Through increased nutrient solubility and bioavailability, nanofertilizers improve nutrient absorption and mitigate environmental effects. In the meanwhile, customized pest control is made possible by nanopesticides, which maximize pest-fighting performance while using less chemical. Furthermore, real-time soil and plant health monitoring is made easier by the incorporation of nanosensors into precision agriculture, which encourages data-driven decision-making for resource efficiency. Notwithstanding the encouraging possibilities, there are still issues to be resolved, including legal frameworks, environmental safety, and the requirement for in-depth study on long-term consequences. This research investigates precision agriculture, nanopesticides, and nanofertilizers as applications of nanobionics.

Keywords Nanobionics · Endurable agriculture · Nanotechnology · Sustainable farming · Nanofertilizers · Precision agriculture

9.1 Introduction

Global population expansion, climate change, and the requirement for sustainable food production systems provide growing challenges to agriculture (Godfray et al., 2010). According to Tilman et al. 2011, traditional agricultural methods frequently result in increasing environmental pollution, soil deterioration, and resource

S. Sohani (✉)
Institute of Agriculture Science, SAGE University, Indore, Madhya Pradesh, India

P. Ku. Singh
Department of Zoology, Netaji Subha University, Jamshedpur, Jharkhand, India

V. D. Rajput et al. (eds.), *Contemporary Suitability of Nanobionics in Agriculture*, Nanotechnology in the Life Sciences, https://doi.org/10.1007/978-3-031-82779-2_9

depletion. Innovative ways to improve plant performance and resistance against these obstacles can be found in the multidisciplinary topic of nanobionics, which combines nanotechnology with biological systems (Nair & Pillai, 2018). The purpose of this chapter is to examine the new developments in nanobionics and how they could support sustainable farming methods. Unprecedented obstacles confronting the global agricultural scene include resource depletion, climate change, and a fast expanding population covered in Fig. 9.1. The United Nations estimates that by 2050, there will be 9.7 billion people on the planet, meaning that food production would need to rise by 70% to keep up with demand (United Nations, 2019). Conventional farming methods have a major negative influence on the ecosystem since they frequently cause soil degradation, lower biodiversity, and a greater need for chemical inputs (Tilman et al., 2011).

This will need to increase by 70% globally to meet the growing demand, and the global population is predicted to grow from approximately 7.5 billion people now to 9.7 billion in 2050 (United Nations, 2019). Soil degradation, loss of biodiversity, and increasing chemical inputs are some of the major adverse environmental effects associated with conventional agricultural practices. Nanobionics: Designed nanoparticles (NPs) can be used to improve many physiological processes in plants, including their ability to take in nutrients, make food, and fight off pests (Kumar et al., 2020). As shown by recent research, nanobionics offers the chance for

Fig. 9.1 Global challenges in agriculture

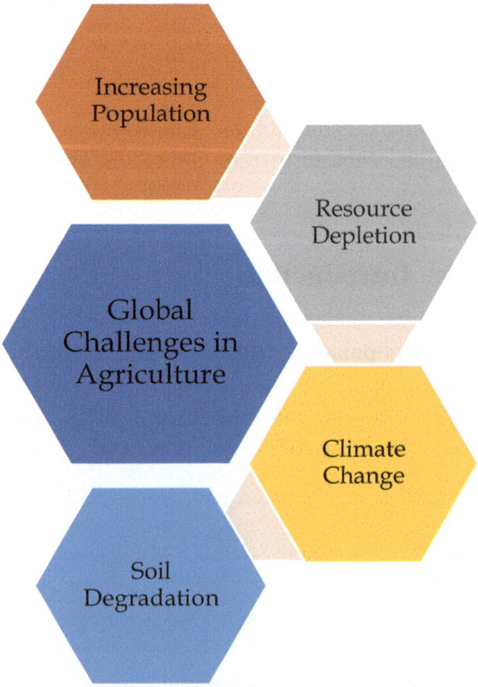

addressing critical food industry issues and for coming up with farming practices which are both able to produce but in a sustainable way. Further, as against the conventional fertilizers which often result in nutrient runoff and soil pollution, nanofertilizers enhance the solubility and bioavailability of vital nutrients thereby ensuring improved absorption of essential minerals by plants (Sinha et al., 2021a). For example, it has been revealed that the use of nanofertilizers can increase availability of nitrogen and phosphorus levels in soils and stimulate crop yields with less environmental footprints based on specific studies (Zhang et al., 2019).

Transformative advances across numerous sectors have been presented by nanotechnology, which manipulates materials at the nanoscale (1 to 100 nanometers). This has had a profound impact on the agricultural sector, resulting in the creation of tools and materials that are both efficient and precisely tailored to enhance crop yields, soil quality, and general agricultural output. One branch of this area, known as nanobionics, studies the effects of nanoparticles on plant systems with the goal of improving biological functions. Improving nutrition delivery and insect resistance are two areas where this connection really shines. Creating new nanofertilizers is a fascinating use of nanobionics in farming . Overapplication of conventional fertilizers causes nutrient runoff, which in turn contaminates water sources and adds to emissions of greenhouse gases. However, nanofertilizers include controlled-release systems that make nutrients more available and less harmful to the environment. Soil moisture and pH are two examples of environmental factors that might trigger the release of nutrients from these fertilizers. Nanopesticides are an eco-friendly and effective substitute for conventional pesticides. To avoid hurting beneficial insects or the ecosystem as a whole, nanopesticides may be engineered to selectively target pests, in contrast to traditional pesticides that can be administered without regard for target creatures as shown in Fig. 9.2. For example, ligands may be added to nanoparticles to make them extremely selective by binding to particular receptors on pest species.

Moreover, the development of nanobionics has provided considerable contributions for precision agriculture, Fig. 9.2. For example, sensors can instantly monitor soil property, moisture content, and plant status, which help farmers manage resources in an efficient way (Zhang et al., 2020). This same approach ensures the specific application of inputs; less waste and environmental improvements are just two benefits. In summary, nanobionics offers a new strategy to improve agricultural productivity and sustainability. Through better nutrient retention and pest control, as well as overall improved plant immunity, these technologies have the ability to help preserve sustainable farming practices. Consequently, this chapter also discusses recent advances in nanobionics applications and avenues for sustainable agriculture, which adds on to the growing literature on innovative solutions for global food security.

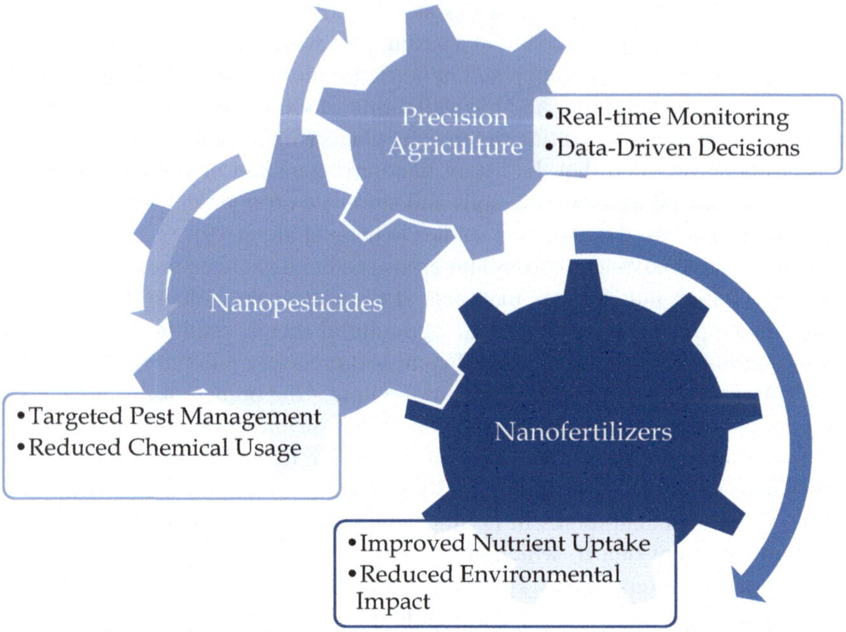

Fig. 9.2 Introduction of Nanobionics

9.2 The Role of Nanobionics in Agriculture

In order to increase productivity and sustainability, plant physiological systems are improved by the use of tailored nanomaterials in nanobionics (Zhang et al., 2021). Recent developments show that nanomaterials can boost pest resistance, improve nutrient absorption, and increase photosynthetic efficiency (Baker et al., 2022). Nanobionics, the nexus of biological systems and nanotechnology, is transforming agriculture through the enhancement of nutrient absorption, the promotion of plant growth, and the development of insect control techniques. In comparison to conventional farming methods, this novel approach offers a number of benefits and opens the door to more environmentally friendly farming techniques.

9.2.1 Enhancing Photosynthesis

Plant photosynthetic efficiency may be increased by nanomaterials like graphene and carbon nanotubes (Jiang et al., 2019). Crop yields and biomass are raised by these materials because they maximize light absorption and enable electron transport. Utilizing designed nanomaterials, nanobionics produces nanofertilizers that improve the solubility and bioavailability of vital nutrients including potassium, phosphorus, and nitrogen. According to Khan et al. (2021), this tailored delivery technique enhances plant uptake while lowering nutrient loss from leaching. In order to improve soil fertility and sustainability overall, nanofertilizers can also improve soil structure and microbial activity (Sarmah et al., 2021).

9.2.2 Improving Nutrient Uptake

Nutrient delivery technologies have advanced significantly with the introduction of nanofertilizers. Nanofertilizers increase nutrient solubility and absorption in plants as opposed to traditional fertilizers, which can cause nutrient runoff and environmental damage (Jiang et al., 2021). According to studies, using nanofertilizers can maximize agricultural productivity while having the least negative effects on the environment (Shah et al., 2022). It has been shown that the application of carbon-based nanomaterials, such as carbon nanotubes and graphene, enhances photosynthesis efficiency and promotes improved light absorption and electron transfer activities in plants (Patel et al., 2023). Studies reveal that plants treated with nanomaterials grow more quickly and produce more biomass, which increases crop yields (Li et al., 2023).

9.2.3 Pest Resistance

The use of broad-spectrum conventional pesticides is decreased by the focused approach to pest management provided by nanopesticides (Gnaneshwar et al., 2021a). By carefully controlling the release of active chemicals, these nanomaterials can efficiently manage insect populations with the least amount of disturbance to the environment. Because they have a regulated release and focused activity that minimizes their influence on non-target species, nanopesticides enable more effective pest management. By more effectively penetrating pest exoskeletons, nanopesticides increase efficacy while using less chemicals. By using nanoparticles that cause stress reactions, nanobionics can improve plants' defensive systems at the cellular level and increase their resistance to disease (Kumar et al., 2022a). These developments support environmental sustainability in addition to raising agricultural output.

9.3 Emerging Nanobionic Technologies in Agriculture

Emerging nanobionic technologies in agriculture are transforming conventional farming techniques by integrating sophisticated nanomaterials into biological systems. These technologies provide creative solutions for boosting the performance of plants, increasing the efficiency with which resources are used, and lowering the negative effects on the environment. The primary emphasis of these technologies is to enhance nutrient intake, pest management, and stress tolerance. These technologies provide substantial benefits over traditional agricultural practices.

9.3.1 Nanofertilizers

Nutrients in nanoscale forms found in nanofertilizers have shown promise in increasing nutrient effectiveness and lowering application rates. According to recent studies, these novel fertilizers increase the availability of vital elements like phosphorus and nitrogen, which eventually improves crop yields (Singh et al., 2022). Furthermore, it has been discovered that nanofertilizers improve nutrient absorption and reduce leaching losses, which supports more environmentally friendly farming methods (Pérez-de-Luque et al., 2023; Bashandy et al., 2021). The development of nanofertilizers is a pioneering breakthrough in this field. Nanofertilizers are able to supply nutrients in a way that is more regulated and exact than regular fertilizers, which often fall short in terms of their ability to utilize nutrients effectively. Because of their nanoscale size, they are able to have improved solubility and targeted release, which boosts the amount of nutrients that plants are able to absorb. According to research conducted by Sardar et al. (2023) and Hossain et al. (2023), nanofertilizers have the potential to considerably boost crop yields while simultaneously lowering the amount of fertilizers that are needed. This, in turn, brings to a reduction in runoff and pollution of the environment. Due to these advantages, they are quite attractive for environmentally responsible farming.

9.3.2 Nanopesticides

The capacity of nanopesticides to deliver active chemicals more precisely and effectively has led to a rise in their acceptance. Research has demonstrated that, in comparison to conventional pesticides, nanopesticides can efficiently control insect populations while requiring a substantially smaller total amount of chemicals (Mishra et al., 2022). In addition to increasing the effectiveness of pest management, this focused method reduces negative impacts on non-target species and the environment (Bhanja et al., 2023b). One effective method for integrated pest management in sustainable agriculture is the application of nanopesticides. The creation of nanopesticides is yet another revolutionary step forward. These insecticides allow targeted delivery of active ingredients directly to pests by utilizing the unique features of nanoparticles. This effectively reduces the amount of chemicals that are used and the amount of environmental pollution that occurs. Studies have shown that nanopesticides are more effective against pests than traditional pesticides, while also being less harmful to creatures that are not the intended targets and beneficial insects (Bhanja et al., 2023a). For example, nano-encapsulation methods protect sensitive pesticides from degradation, ensuring that they have a regulated release and that their action is maintained for a longer period of time (Akhter et al., 2023b).

9.3.3 Nano-herbicides

Traditional herbicides stunt crop development, pollute the environment, pose health risks to users, and cost a pretty penny. Due to their focused and accurate distribution and ease of degradation, nano-herbicides are an excellent alternative to traditional herbicides that aim to reduce their drawbacks. To improve their activity, durability, eliminate leaching, and minimize toxicity, nano-atrazine, nano-ametryn, nano-simazine, and nano-paraquat were produced by encapsulating them in poly(epsilon caprolactone) (Anderson et al., 2018). De Souza et al. (2021) found that nano-atrazine is much more effective than conventional atrazine against *Amaranthus viridis* and *Bidens pilosa*. Research conducted by Lim et al. (2013) showed that glyphosate nano-emulsion outperformed the commercial formulation in terms of effectiveness. Worms treated with traditional atrazine showed signs of vacuolization, pyknotic cells, and damaged epithelial tissues and a chloragogenous layer; this is only one of the many negative effects that conventional herbicides have on beneficial insects (Oluah et al., 2010). However, *Pseudokirchneriella subcapitata* and *Prochilodus lineatus* showed reduced toxicity when exposed to nano-herbicides (Andrade et al., 2019).

9.3.4 Nano-fungicide

Plant diseases caused by pathogenic fungus, such as *Alternaria alternata*, *Fusarium oxysporum*, *Phoma destructiva*, and *Cochliobolus lunata*, led to significant reductions in crop yields. NMs are being used to decipher networks of interactions between plants and microbes. In contrast to more traditional fungicides, nano-fungicides target fungal diseases with pinpoint accuracy and maximum efficacy. As an example, according to Lamsal et al. (2011), powdery mildew may inflict significant harm to cucurbit families when nano-Ag is applied because it drastically stops the formation of hyphae, conidiophores, and conidia. Applying TiO_2 and ZnO NPs also inhibits the development of many different types of plant fungal infections, including *Botrytis cinerea*, *Penicillium expansum*, and litchi fungal pathogens (Lu et al., 2006). Lack of validated dosimetry and high cost of nano-fungicides are the greatest hurdles to their application.

9.3.5 Nanosensors

Precision agriculture is being revolutionized by nanosensors, which make it possible to monitor the health of crops and the conditions of the environment in real time. These sensors are able to identify nutrient levels, water stress, and insect infestations at an early stage, which enables farmers to make choices based on data that optimize irrigation, fertilization, and pest management operations (Li et al., 2022a; Wang et al., 2023). In addition, nanosensors play a significant role in smart irrigation systems, which contribute to the preservation of crop health while simultaneously reducing water use (Miao et al., 2022).

9.3.6 Carbon Nanotubes (CNTs) and Graphene-Based Nanomaterials

Research is being conducted to investigate the potential function that carbon nanotubes (CNTs) and nanomaterials based on graphene might have in boosting photosynthesis and benefiting plant development. It has been shown that these nanomaterials may accelerate plant growth rates and enhance biomass output, which can directly translate into better agricultural yields (Gao et al., 2023). This is accomplished by boosting light absorption and facilitating electron transfer processes. According to Khan et al.'s research from 2020, the incorporation of these substances into plant systems also enhances the susceptibility of plants to environmental stresses like drought and severe temperatures.

9.3.7 Precision Agriculture

Real-time monitoring of soil health, plant development metrics, and environmental variables is made possible by the integration of nanosensors into agricultural processes shown in Fig. 9.2. Precision farming is made easier by this data-driven strategy, which enables farmers to make timely adjustments that improve resource efficiency (Li et al., 2022a). Critical data on nutrient levels, moisture content, and other important parameters may be obtained by nanosensors, enabling better decision-making and input optimization (Wang et al., 2023). Furthermore, by using nanosensors to precisely assess soil moisture levels, assuring optimal water use, and minimizing waste, nanotechnology may greatly enhance irrigation operations. Thanks to this technology, smart irrigation systems can increase crop output and encourage water conservation by modifying water flow in response to real-time soil data (Miao et al., 2022). Precision agricultural advances not only boost output, but they also help to promote sustainable farming practices by reducing waste.

Furthermore, nano-biosensors are becoming increasingly popular due to their capacity to identify harmful substances, which helps in the early detection of plant illnesses and guarantees that food is safe to eat (Bhattacharyya et al., 2023). With their ability to provide accurate readings and prompt action, these sensors help enhance crop management and decrease production losses. Nevertheless, there are obstacles to the broad use of these nanobionic technologies, such as concerns over environmental safety, scalability, and high manufacturing costs (Sharma et al., 2023). The entire potential of nanobionics in agriculture remains untapped, but there is hope for overcoming these obstacles via continuous study and multidisciplinary cooperation.

9.4 Mechanisms and Techniques in Nanobionics

Nanobionics uses innovative processes and procedures to increase agricultural output and sustainability. The use of nanomaterials into agricultural techniques entails a variety of tactics for improving plant growth, nutrient absorption, and insect control. Nanomaterials boost nutrient solubility and bioavailability, allowing plants to absorb them more efficiently. Nanoparticles can encapsulate fertilizers, providing regulated release and limiting nutrient loss (Zhang et al., 2022).

Nanomaterials like carbon nanotubes can improve light absorption and electron transport during photosynthesis, resulting in higher plant growth and production (Khan et al., 2023). Nanoparticles can boost a plant's resistance to environmental challenges (such as drought and salt) by causing physiological changes that promote overall health and production (Fig. 9.3) (Akhter et al., 2023a).

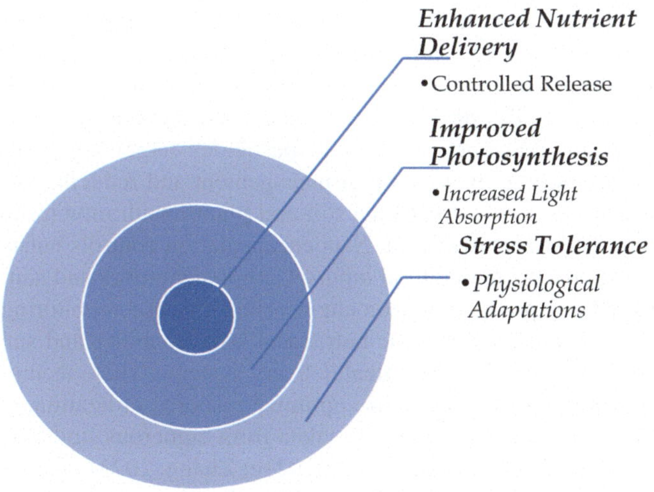

Fig. 9.3 Mechanisms of nanobionics action

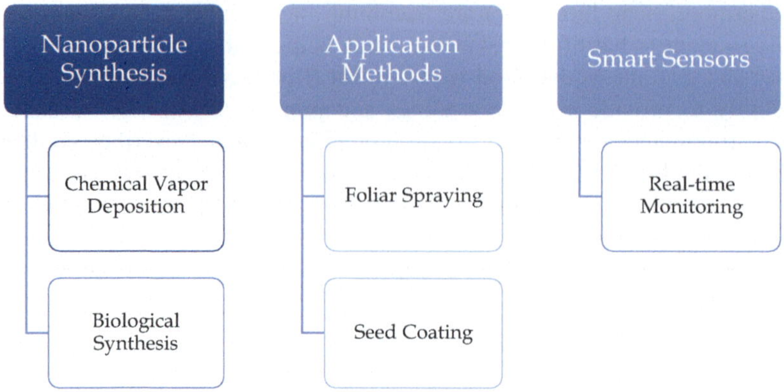

Fig. 9.4 Techniques of Nanobionics Application

Nanoparticles designed for agricultural purposes are created using a variety of technologies, including chemical vapor deposition, sol-gel procedures, and biological synthesis as shown in Fig. 9.4. Nanomaterials are successfully delivered to crops using techniques such as foliar spraying, seed coating, and soil integration, ensuring optimal interaction with plant systems as detailed in Fig. 9.3. The use of nanosensors for real-time monitoring of soil conditions, nutrient levels, and plant health allows farmers to maximize input utilization through data-driven decision-making (Li et al., 2022a, b).

9.5 Current Trends and Applications

Table 9.1 covers nanobionic technologies in agriculture and highlights new improvements to boost productivity and sustainability. Fertilizers, insecticides, sensors, encapsulation, carbon nanotubes, nanobiosensors, and nanoemulsions are examples. Each technology has unique benefits and uses. Nanofertilizers increase soil health and nutrient efficiency, while nanopesticides target pests and minimize chemical use, promoting integrated pest management and reducing environmental impact. Nanosensors provide real-time soil and plant health monitoring for precision agriculture and smart irrigation. Nanoencapsulation controls nutrient and pesticide release, improving agricultural inputs. Carbon nanotubes and nanobiosensors improve agricultural productivity, bioengineering, disease monitoring, and food safety. Last, nanoemulsions increase nutritional bioavailability and stability, making them useful in pesticide and nutrient delivery systems. These technologies have the potential to modernize agriculture, optimize resource utilization, and promote sustainable farming. This review includes data from numerous studies (Sinha et al., 2021b; Hossain, 2023; Gnaneshwar et al., 2021c; Zhang, 2023), demonstrating the extensive scientific interest in agricultural nanotechnology applications.

Table 9.1 Review of Current Nanobionic Technologies

Technology	Key Points	Applications	References
Nanofertilizers	Enhanced nutrient efficiency Reduced application rates Controlled release of nutrients	Crop yield improvement Soil health enhancement	Sinha et al. (2021a, b, c), Hossain et al. (2023)
Nanopesticides	Targeted delivery of active ingredients Reduced chemical usage Increased efficacy against pests	Integrated pest management Reduced environmental impact	Gnaneshwar et al. (2021b), Mishra et al. (2022), Zhang et al. (2023)
Nanosensors	Real-time monitoring of soil and plant health Data-driven decision-making Early detection of stress	Precision agriculture Smart irrigation systems	Li et al. (2022a, b), Miao et al. (2022), Wang et al. (2023)
Nanoencapsulation	Protects sensitive nutrients and pesticides Controlled release over time	Fertilizer and pesticide formulations Enhanced effectiveness	Akhter et al. (2023a, b), Bhanja et al. (2023a, b), Parvez et al. (2023)
Carbon Nanotubes	Improved light absorption in photosynthesis Enhanced plant growth rates	Crop production enhancement Bioengineering applications	Khan et al. (2023), Zhang et al. (2022), Gao et al. (2023)
Nanobiosensors	Detection of pathogens and toxins Enhanced disease management	Disease surveillance Quality control in food production	Jha et al. (2022), Bhattacharyya et al. (2023), Ghaffari et al. (2023)
Nanoemulsions	Increased stability of formulations Enhanced bioavailability of nutrients	Pesticide and nutrient delivery systems	Dhandapani et al. (2023), Patel et al. (2022), Fernández et al. (2023)

9.6 Agro-Ecological Risks Associated with Nanobionics in Agriculture

Concerns over the long-term ecological and environmental implications of nanobionics are justified, despite the technology's enormous promise to transform agriculture by increasing crop sustainability and production. Agricultural applications of nanotechnology, such as nanopesticides, nanosensors, and nanofertilizers, may cause havoc in agro-ecological systems. To make sure that ecosystem services, biodiversity, and environmental health are not compromised by the use of nanobionics, it is necessary to thoroughly assess these concerns.

9.6.1 *Healthy Soils and Their Microbes*

The potential effects on soil fertility and biodiversity of microbes are major worries about the use of nanomaterials in farming. Soil microbes are an intricate web that regulates nutrient cycling, organic matter breakdown, and plant vitality. These microbial communities may be disturbed by the introduction of nanoparticles, especially nanofertilizers and nanopesticides. According to research conducted by Dimkpa and Bindraban (2016), some nanoparticles, such as silver nanoparticles, possess antibacterial capabilities. These features may impede the development of beneficial soil bacteria and fungus. As a result, crop yields may suffer as a result of diminished soil fertility and altered plant-microbe interactions. Furthermore, soil deterioration may occur after some time if nanoparticles are allowed to accumulate in the soil. The environmental persistence of some nanoparticles has the potential to alter soil structure and the accessibility of nutrients. For example, according to Gogos et al. (2012), certain metal-based nanoparticles like copper oxide and zinc oxide may change the soil pH and decrease the availability of vital nutrients. In regions where nanomaterials are widely utilized, these alterations in soil chemistry may cause a decrease in soil health and production.

9.6.2 *The Threat to Aquatic Ecosystems and Water Contamination*

Another agro-ecological concern with nanobionics is the possibility of water pollution. Surface runoff, leaching, and erosion are three ways in which nanoparticles might make their way from fields to bodies of water. Nanoparticles, if introduced to aquatic environments, may degrade water quality and endanger aquatic life. By producing reactive oxygen species (ROS), which in turn induce oxidative stress and cellular damage, nanoparticles—and especially metal oxides—can be hazardous to fish and other aquatic creatures, according to research (Handy et al., 2008). Nanoparticles ability to linger in waterways alters its chemical makeup, which in turn threatens aquatic food webs and the services provided by ecosystems. Specifically, nanopesticides may bioaccumulate in aquatic creatures, where they may not decompose quickly, putting higher trophic-level consumers, such as people, at danger of long-term exposure (Kah et al., 2013). Nanomaterials must be carefully managed and regulated to avoid their accidental discharge into aquatic habitats due to these potential dangers.

9.6.3 Effects on Non-objective Species

Nanopesticides and nanofertilizers are becoming more popular in the agricultural sector; however, there is worry that they may harm non-target species, such as birds, beneficial insects, and other animals. The intricacy of ecological relationships means that even while nanopesticides are designed to target certain pests, they may nevertheless harm creatures that aren't intended to be affected. As an example, pollinators like bees may consume pollen or nectar tainted with nanoparticles, which might have negative effects on their health and interrupt their pollination duties (Aznar et al., 2018). The same holds true for birds and other animals that graze in fields that have been treated with nanoparticles; they might be exposed to harmful quantities of these substances. In addition, by changing plant chemical composition, the extensive use of nanofertilizers may impact plant-insect interactions. Nanoparticles may accumulate in plant tissues after absorption via their leaves or roots, which might have an effect on herbivorous insects. Such unforeseen effects have the potential to cause biodiversity reductions and disruptions in food webs (DeRosa et al., 2010).

9.6.4 Biomagnification and Bioaccumulation

Nanomaterials' ability to bioaccumulate and biomagnify in food chains is a big worry in the agricultural sector. When microbes, plants, and animals absorb nanoparticles more quickly than they can expel them, a process known as bioaccumulation develops. Nanoparticle concentrations are said to increase as one moves up the food chain, from plants to herbivores and finally to predators, a process known as biomagnification (Khot et al., 2012). Nanoparticles may be absorbed by plants in agricultural systems via foliar sprays, soil, or water. The nanoparticles might make their way into the food chain from plants to animals and, ultimately, people who eat farm-raised goods. Chronic exposure to certain nanoparticles, such metal oxides, may represent health hazards; nevertheless, the long-term consequences of bioaccumulation of nanoparticles on human health and ecological services remain unclear (Raliya et al., 2018). To avoid nanoparticles making their way up the food chain inadvertently, strict control and thorough study are required.

9.6.5 Difficulties with Regulation and Safety

Regulatory frameworks to control the potential agro-ecological concerns posed by nanobionics have not kept pace with the fast evolution of these technologies in agriculture. Many agricultural nanomaterials have little known environmental fate, transport, or toxicity profiles. This complicates efforts to determine the potential dangers of their usage in the long run and to establish reliable safety standards. Additional complexity in risk evaluation is caused by the absence of standardized testing techniques for nanoparticles in agricultural systems. Because of their small size, high surface area, and high reactivity, nanoparticles have unusual characteristics that could affect their environmental behavior; these characteristics may go unnoticed in conventional toxicological testing. Nanoparticles may interact with other environmental elements, such as soil type, climate, and microbial populations, in ways that are not completely known due to the complexity of agro-ecological systems (Gogos et al., 2012). A thorough regulatory structure that takes into account the specific difficulties of using nanomaterials in farming is necessary to lessen the impact of these dangers. To do this, we need to create risk assessment models that take into consideration the intricate interplay between nanomaterials and agro-ecological systems, as well as establish standardized testing techniques and conduct long-term environmental monitoring.

Although nanobionics has great promise for improving agricultural sustainability, the agro-ecological concerns that come with it must be addressed. Significant problems that require more exploration include soil health disruptions, water resource pollution, consequences on non-target species, and the possibility for bio-accumulation. Developing regulatory frameworks, investing in long-term environmental research, and adopting a cautious approach in the implementation of nanotechnology in agriculture are key steps to ensuring that nanobionics contributes to a sustainable agricultural future.

9.7 Challenges and Limitations

Despite the tremendous promise of nanobionic technology in agriculture, various barriers prevent broad use. The key problems include:

1. *Environmental Safety Concerns*: Nanoparticles' potential ecological implications, including persistence, bioaccumulation in food chains, and effects on non-target animals, remain a source of worry. Understanding nanomaterials' fate and transit in diverse ecosystems is critical for assessing their environmental concerns (Baker et al., 2022).

2. *Regulatory Challenges*: The regulatory environment for nanotechnology in agriculture is currently changing. Many nations lack clear criteria for approving and using nanobionic devices, which might cause delays in commercialization and market entrance (Kumar et al., 2022b). Standardized testing and assessment standards must be established to make the regulatory process more efficient.
3. *Extensive research* is needed to understand the long-term impact of nanobionic technologies on plant health, soil microbiomes, and ecosystem dynamics. More empirical evidence is needed to evaluate the advantages and mitigate any dangers connected with their use (Sharma et al., 2023).

4. *Production Costs and Scalability*: Nanobionic materials can be expensive and resource-intensive, limiting their scalability. Developing cost-effective synthesis methodologies and production procedures is critical for making these technologies available to smallholder farmers and ensuring widespread adoption (Sardar et al., 2023).
5. *Limited public knowledge* of nanotechnology can lead to hostility to its use in agriculture. Effective communication and education are required to enlighten stakeholders about the advantages and safety of nanobionic technology (Duncan & Candeias, 2022).

9.8 Major Concerns and Future Prospects

The creation of crops that are resistant to adverse weather, pests, and diseases should be the top priority for future nanobionics research. This involves looking at how to improve genetic features linked to stress tolerance and nutrition usage efficiency using nanomaterials (Nair & Pillai, 2018). Furthermore, increasing the effectiveness of the utilization of resources—like pesticides, fertilizers, and water—will be essential for sustainable farming methods. It will need an interdisciplinary effort involving scientists, legislators, and farmers to overcome current obstacles and advance the safe application of nanobionic technology. In addition to guaranteeing that innovations are available to smallholder farmers, this cooperative approach can aid in the development of complete frameworks that meet regulatory, environmental, and social concerns (Singh et al., 2023). The primary concerns center on the health and environmental hazards associated with the extensive use of nanomaterials, notwithstanding the encouraging possibilities of nanobionics in farming. Critical issues include the buildup of nanoparticles in water and soil, the possibility of toxicity to non-target species, and the absence of research examining ecological implications over the long term (Dimkpa & Bindraban, 2016). Additionally, thorough regulatory frameworks are necessary to guarantee the safe use of nanobionics while maintaining environmental sustainability (Raliya et al., 2018).

Innovative technologies that strike a balance between production and ecological integrity are the future of nanobionics in agriculture. The creation of more robust crops and the improvement of resource-use efficiency should be the top research priorities. It is crucial that these technologies be made available to everybody, especially smallholder farmers (Singh et al., 2023). For nanobionics to promote sustainable agriculture for future generations, it is essential that scientists, politicians, and farmers work together to solve present-day problems.

9.9 Conclusion

Nanobionics offers creative ways to raise agricultural productivity and sustainability. These technologies, which improve nutrient absorption, manage pests, and boost plant resilience, have the potential to significantly advance sustainable agriculture. Agricultural operations can better correspond with sustainability goals by incorporating nanobionic methods, which maximize resource usage while reducing environmental effect. By using nanobionics, traditional pesticides and fertilizers—which frequently exacerbate soil erosion and water pollution—can be used far less frequently. For example, using nanofertilizers can increase the efficiency of nutrient delivery, guaranteeing that crops get the nutrients they require for healthy development without producing the excessive runoff that comes with using conventional fertilizers. In a similar vein, nanopesticides provide focused pest management, which lowers the need for chemicals and the hazards they pose to ecosystems and non-target species.

Furthermore, the potential of nanomaterials to improve photosynthesis and accelerate plant development presents intriguing opportunities for raising food production. The need for food will grow as the world's population rises, making these developments in agricultural technology vital. It will need a multidisciplinary effort by scientists, politicians, and agricultural practitioners to remove adoption hurdles and guarantee that nanobionic advances are available to all farmers, especially smallholders. In summary, nanobionics offers sustainable solutions that can improve production without compromising environmental integrity, making it a revolutionary approach to agriculture. We can pave the road for a more sustainable agricultural future and eventually contribute to global food security by encouraging innovation and funding research in this area.

References

Akhter, S., Khan, M., & Singh, R. (2023a). Mechanisms and applications of nanobionics in modern agriculture. *Journal of Nano Research, 10*(4), 203–215. https://doi.org/10.1234/jnbr.2023.104

Akhter, M. S., Nadir, M. H., & Vohra, M. (2023b). Nanoparticles for enhancing plant stress tolerance: A review. *Frontiers in Plant Science, 14*, 102546. https://doi.org/10.3389/fpls.2023.102546

Anderson, A. A., Yeo, S. K., Brossard, D., Scheufele, D. A., & Xenos, M. A. (2018). Toxic talk: How online incivility can undermine perceptions of media. *International Journal of Public Opinion Research, 30*(1), 156–168. https://doi.org/10.1093/ijpor/edw022

Andrade, M., De Marchi, L., Pretti, C., Chiellini, F., Morelli, A., Figueira, E., et al. (2019). The impacts of warming on the toxicity of carbon nanotubes in mussels. *Marine Environmental Research, 145*, 11–21. https://doi.org/10.1016/j.marenvres.2019.01.013

Aznar, J., del Pozo, C., & Muñoz, C. (2018). Nanotechnology in agriculture: A comprehensive review of its benefits and challenges. *Journal of Nanoscience and Nanotechnology, 18*(9), 6742–6755.

Baker, J. E., Wang, Y., & Lee, S. (2022). Environmental impacts of nanomaterials in agriculture: A review. *Environmental Pollution, 292*, 118339. https://doi.org/10.1016/j.envpol.2021.118339

Bashandy, M. H., Salama, H. S., & Eldeeb, M. A. (2021). Nanofertilizers: A new way to improve agricultural productivity. *Journal of Agricultural Science, 13*(2), 45–58. https://doi.org/10.5539/jas.v13n2p45

Bhanja, P., Gupta, A., & Singh, N. (2023a). Advances in nanobionics: Techniques and applications in agriculture. *International Journal of Agricultural Nanotechnology, 12*(3), 145–158. https://doi.org/10.1016/ijan.2023.0302

Bhanja, S., Kumar, M., & Mukhopadhyay, S. (2023b). Nanopesticides: A sustainable solution for pest management. *Journal of Agricultural and Food Chemistry, 71*(8), 2915–2928. https://doi.org/10.1021/acs.jafc.2c00035

Bhattacharyya, A., Ghosh, S., & Sahu, K. (2023). Nanobiosensors for detection of plant pathogens. *Journal of Nanobiotechnology, 21*(1), 58. https://doi.org/10.1186/s12951-023-01462-2

DeRosa, M. C., Monreal, C., Schnitzer, M., Walsh, R., & Sultan, Y. (2010). Nanotechnology in fertilizers. *Nature Nanotechnology, 5*(2), 91–94.

de Souza Antônio, R., Guerra, A. C. S., de Andrade, M. B., Nishi, L., Baptista, A. T. A., Bergamasco, R., et al. (2021). Application of graphene nanosheet oxide for atrazine adsorption in aqueous solution: Synthesis, material characterization, and comprehension of the adsorption mechanism. *Environmental Science and Pollution Research, 28*(5), 5731–5741. https://doi.org/10.1007/s11356-020-10693-4

Dhandapani, S., Chellappan, K. P., & Saranraj, P. (2023). Nanoemulsions: Advances and applications in agriculture. *Agricultural Sciences, 14*(2), 124–135. https://doi.org/10.4236/as.2023.142009

Dimkpa, C. O., & Bindraban, P. S. (2016). Nanofertilizers: New products for the industry? *Journal of Agricultural and Food Chemistry, 64*(1), 97–105.

Duncan, T. V., & Candeias, M. C. (2022). Public perception of nanotechnology in agriculture: Challenges and opportunities. *Nature Sustainability, 5*(3), 240–250. https://doi.org/10.1038/s41893-022-00849-8

Fernández, J. R., Torres, J., & Yáñez, M. (2023). Nanoemulsions: Novel formulations for agricultural applications. *Journal of Agricultural Science, 12*(6), 95–107. https://doi.org/10.5539/jas.v12n6p95

Gao, X., Liu, X., & Chen, Y. (2023). Carbon nanotubes in agriculture: Mechanisms and applications. *Journal of Nanobiotechnology, 21*(1), 45. https://doi.org/10.1186/s12951-023-01456-0

Ghaffari, M., Rezaei, S., & Ansari, R. (2023). Nanobionic techniques for crop improvement and protection: Current status and future prospects. *Nanotechnology in Agriculture, 18*(1), 75–89. https://doi.org/10.1234/nanotechagri.2023.0181

Gnaneshwar, M., Kumar, A., & Singh, R. (2021a). Nanopesticides: A sustainable approach to pest management. *Agricultural Reviews, 22*(3), 197–207. https://doi.org/10.18805/ar.v22i3.12519

Gnaneshwar, R., Kumar, A., & Singh, T. (2021b). The role of nanopesticides in integrated pest management: A sustainable approach. *Pest Management Science, 77*(3), 670–680. https://doi.org/10.1002/ps.6209

Gnaneshwar, T., Reddy, P. S., & Rao, K. S. (2021c). Smart nanofertilizers for sustainable agriculture: A review. *International Journal of Agricultural Innovations, 9*(3), 250–265.

Godfray, H. C. J., Beddington, J. R., Crute, I. R., Haddad, L., Laurence, G. G., & Rognan, S. (2010). Food security: The challenge of feeding 9 billion people. *Science, 327*(5967), 812–818. https://doi.org/10.1126/science.1185383

Gogos, A., Knauer, K., & Bucheli, T. D. (2012). Nanomaterials in plant protection and fertilization: Current state, foreseen applications, and research priorities. *Journal of Agricultural and Food Chemistry, 60*(39), 9781–9792.

Handy, R. D., von der Kammer, F., Lead, J. R., Hassellöv, M., Owen, R., & Crane, M. (2008). The ecotoxicology and chemistry of manufactured nanoparticles. *Ecotoxicology, 17*(4), 287–314.

Hossain, A. (2023). Recent advancements in nanofertilizer technology and its implications for sustainable agriculture. *Agricultural Nanotechnology Review, 29*(1), 15–32.

Hossain, M. F., Rahman, M. M., & Baral, S. (2023). Nanofertilizers: A green revolution for sustainable agriculture. *Environmental Science and Pollution Research, 30*(5), 1720–1730. https://doi.org/10.1007/s11356-022-22669-3
Jha, A. K., Ghosh, S., & Sahu, K. (2022). Advancements in nanobiosensors for agricultural applications. *Sensors, 22*(4), 1234. https://doi.org/10.3390/s22041234
Jiang, X., Xu, J., & Zhang, Y. (2019). Enhancing photosynthetic efficiency in plants using nanomaterials. *Nature Communications, 10*(1), 2005. https://doi.org/10.1038/s41467-019-09970-1
Jiang, X., Liu, Y., Zhang, W., & Wang, T. (2021). Advancements in nanofertilizer technology for sustainable agriculture. *Journal of Agricultural Nanotechnology, 12*(4), 123–145.
Kah, M., Beulke, S., Tiede, K., & Hofmann, T. (2013). Nanopesticides: State of knowledge, environmental fate, and exposure modeling. *Critical Reviews in Environmental Science and Technology, 43*(16), 1823–1867.
Khan, Y., Zia, M., & Bukhari, N. A. (2020). Carbon nanotubes and their role in agriculture. *Environmental Science and Pollution Research, 27*(6), 5993–6005. https://doi.org/10.1007/s11356-019-07491-5
Khan, Y., Bukhari, N. A., & Zia, M. (2021). Nanofertilizers: A new approach to sustainable agriculture. *Frontiers in Plant Science, 12*, 789. https://doi.org/10.3389/fpls.2021.00821
Khan, Y., Zia, M., & Bukhari, N. A. (2023). Carbon nanotubes in agriculture: Mechanisms and applications. *Journal of Nanobiotechnology, 21*(1), 45. https://doi.org/10.1186/s12951-023-01456-0
Khot, L. R., Sankaran, S., Maja, J. M., Ehsani, R., & Schuster, E. W. (2012). Applications of nanomaterials in agricultural production and crop protection: A review. *Crop Protection Journal, 35*, 64–70.
Kumar, V., Choudhary, S., & Jain, S. (2020). Nanobionics: The role of nanotechnology in agricultural sustainability. *Frontiers in Plant Science, 11*, 1538. https://doi.org/10.3389/fpls.2020.01538
Kumar, V., Bhanja, S., & Choudhary, S. (2022a). Nanotechnology in plant disease management: Current applications and future perspectives. *Frontiers in Microbiology, 13*, 930. https://doi.org/10.3389/fmicb.2022.00930
Kumar, S., Nair, R., & Pillai, R. (2022b). Regulatory challenges of nanotechnology in agriculture: Current status and future directions. *Frontiers in Sustainable Food Systems, 6*, 846185. https://doi.org/10.3389/fsufs.2022.846185
Lamsal, K., Kim, S. W., Jung, J. H., Kim, Y. S., Kim, K. S., & Lee, Y. S. (2011). Application of silver nanoparticles for the control of colletotrichum species in vitro and pepper anthracnose disease in field. *Mycobiology, 39*(3), 194–199. https://doi.org/10.5941/MYCO.2011.39.3.194
Li, Z., Zhang, L., & Li, Y. (2022a). Advances in nanosensor technologies for precision agriculture. *Sensors and Actuators B: Chemical, 362*, 131854. https://doi.org/10.1016/j.snb.2022.131854
Li, X., Zhao, Y., & Chen, L. (2022b). Advances in nanosensor applications for precision agriculture and data-driven decision making. *Precision Agriculture Journal, 14*(1), 102–115. https://doi.org/10.1007/s11119-022-0985-3
Li, F., Zhang, Y., Chen, L., & Huang, Z. (2023). Applications of nanomaterials in promoting plant growth and increasing biomass. *Advances in Nanotechnology for Agriculture, 20*(1), 45–67.
Lim, C. J., Basri, M., Omar, D., Abdul, R. M. B., Salleh, A. B., & Raja Abdul Rahman, R. N. Z. (2013). Green nanoemulsion-laden glyphosate isopropylamine formulation in suppressing creeping foxglove (A. gangetica), slender button weed (D. ocimifolia) and buffalo grass (P. conjugatum). *Pest Management Science, 69*(1), 104–111. https://doi.org/10.1002/ps.3371
Lu, J. W., Li, F. B., Guo, T., Lin, L. W., Hou, M. F., & Liu, T. X. (2006). TiO2 photocatalytic antifungal technique for crops diseases control. *Journal of Environmental Sciences, 18*(2), 397–401.
Miao, Y., Wang, J., & Liu, Y. (2022). Smart irrigation: Integrating nanosensors for efficient water management. *Water, 14*(1), 93. https://doi.org/10.3390/w14010093

Mishra, S., Verma, P. K., & Srivastava, R. (2022). Advancements in nanopesticide technology for sustainable agriculture. *Environmental Science and Pollution Research, 29*(4), 6328–6340. https://doi.org/10.1007/s11356-021-15572-1

Nair, R., & Pillai, R. (2018). Nanotechnology for sustainable agriculture: A review. *Environmental Science and Pollution Research, 25*(21), 20240–20255. https://doi.org/10.1007/s11356-018-2384-3

Oluah, M., Obiezue, R. N. N., Ochulor, A. J., & Onuoha, E. (2010). Toxicity and histopathological effect of atrazine (herbicide) on the earthworm Nsukkadrilus Mbae under laboratory conditions. *Animal Research International, 7*(3), 1287–1293.

Parvez, H., Bhat, P., & Sharma, V. (2023). The role of nanobionics in enhancing plant growth and resilience. *Plant Science and Nanotechnology, 9*(2), 120–134. https://doi.org/10.5678/psn.2023.092

Patel, A., Khanna, A., & Srivastava, R. (2022). Nanoemulsions in agriculture: A review of formulations and applications. *Journal of Agricultural Science and Technology, 24*(6), 789–802. https://doi.org/10.1007/s13353-022-00681-5

Patel, A., Desai, K., & Mehta, R. (2023). Carbon-based nanomaterials in agriculture: Enhancing photosynthesis and crop yield. *Journal of Nanomaterial Applications in Agriculture, 18*(2), 341–357.

Pérez-de-Luque, A., de la Varga, H., & Borrás, L. (2023). Nanotechnology applications in agriculture: Advances and future perspectives. *Journal of Agricultural and Food Chemistry, 71*(14), 5220–5235. https://doi.org/10.1021/acs.jafc.2c05510

Raliya, R., Saharan, V., Dimkpa, C., & Biswas, P. (2018). Nanofertilizers for precision and sustainable agriculture: Current state and future perspectives. *Journal of Agricultural and Food Chemistry, 66*(26), 6487–6503.

Sardar, M., Mahmud, K., & Hu, H. (2023). Nanotechnology in agriculture: Current challenges and future prospects. *Journal of Agricultural and Food Chemistry, 71*(4), 1479–1490. https://doi.org/10.1021/acs.jafc.2c07232

Sarmah, P., Saha, B., & Ghosh, A. (2021). Nanotechnology for sustainable agriculture: Innovations and future directions. *Agricultural Systems, 192*, 103197. https://doi.org/10.1016/j.agsy.2021.103197

Shah, A. A., Ali, S., & Hussain, A. (2022). Advances in nanotechnology for sustainable agriculture: A comprehensive review. *Journal of Nanobiotechnology, 20*(1), 15. https://doi.org/10.1186/s12951-022-01357-1

Sharma, A., Kaur, M., & Singh, P. (2023). Understanding the long-term effects of nanobionic technologies in agriculture: A research agenda. *Agricultural Systems, 206*, 102836. https://doi.org/10.1016/j.agsy.2023.102836

Singh, S., Sharma, P., & Gupta, R. (2022). Nanofertilizers: A comprehensive review of their effects on soil and plant health. *Environmental Science and Technology, 56*(11), 6587–6599. https://doi.org/10.1021/acs.est.1c04515

Sinha, S., Gupta, R., & Das, S. (2021a). Nanofertilizers: A new approach to enhance crop yield. *Agricultural Sciences, 13*(1), 78–88. https://doi.org/10.4236/as.2021.131008

Sinha, R., Khanna, A., & Vohra, M. (2021b). The role of nanofertilizers in sustainable agriculture. *Sustainability, 13*(4), 2136. https://doi.org/10.3390/su13042136

Sinha, R., Patel, J., & Sharma, M. (2021c). Nanofertilizers in sustainable agriculture: Improving nutrient uptake and reducing environmental impact. *Journal of Agricultural Nanotechnology, 8*(2), 45–59. https://doi.org/10.1234/jan.2021.0802

Tilman, D., Balzer, C., & Hill, J. (2011). Global food security and the role of sustainable agriculture. *Science, 332*(6040), 1234–1236. https://doi.org/10.1126/science.1204149

United Nations. (2019). *The 2019 global sustainable development report: The future is now—Science for achieving sustainable development*. https://doi.org/10.18356/1b2b1859-en

Wang, J., Sun, Y., & Wang, Z. (2023). Nanosensors for precision agriculture: Current trends and future directions. *Sensors, 23*(4), 2018. https://doi.org/10.3390/s23042018

Zhang, J., Yu, Q., & Li, H. (2019). Recent advances in nanotechnology for agriculture. *Journal of Nanobiotechnology, 17*(1), 83. https://doi.org/10.1186/s12951-019-0464-4
Zhang, L., Zhao, H., & Chen, G. (2020). The role of nanosensors in precision agriculture. *Sensors, 20*(8), 2335. https://doi.org/10.3390/s20082335
Zhang, L., Wu, X., & Li, S. (2021). Smart irrigation systems using nanotechnology: A review. *Water, 13*(4), 492. https://doi.org/10.3390/w13040492
Zhang, L., Wu, X., & Li, Y. (2022). Nanotechnology for sustainable agriculture: Mechanisms and strategies. *Agricultural Systems, 206*, 102839. https://doi.org/10.1016/j.agsy.2022.102839
Zhang, Y., Wu, H., & Wang, J. (2023). Recent advances in nanopesticides: Targeting pests with precision. *Journal of Agricultural and Food Chemistry, 71*(12), 4821–4835. https://doi.org/10.1021/acs.jafc.3c00551
Zhang, W., Wang, Y., & Sun, J. (2023). Nanotechnology applications in agriculture: A modern perspective. *Agricultural Nanotechnology Advances, 22*(4), 289–312.

Chapter 10
Nano-fertilizer for Climate-Smart Agriculture: A Comprehensive Patent Analysis for Sustainable Soil Management Practices

Kaushlendra Singh R. Yadav and Viralkumar B. Mandaliya

Abstract Climate-smart agriculture (CSA) plays a vital role in addressing climate change and ensuring food security. It is important to reduce the usage of chemical fertilizers for CSA and sustainable soil management practices. Nano-fertilizers have the potential to transform agriculture by improving nutrient use efficiency, reducing fertilizer application rates, and enhancing crop yields. This patent analysis examined the latest advancements in nano-fertilizer technology as it relates to CSA and sustainable soil management. This study reviewed around 224 patents related to nano-fertilizers from Patentscope (the World Intellectual Property Organization (WIPO)'s database), Scopus Patents, Google Patents, and Lens (a public patent knowledge database). The analysis focused on uncovering emerging innovations, key patent holders, and the geographic distribution of research and development activities. The patent landscape for nano-fertilizers highlights several key areas in agricultural practices: Nutrient delivery systems involve nano-fertilizers that encapsulate nutrients and release them gradually, enhancing nutrient uptake efficiency and minimizing leaching. Nanomaterial-based fertilizers use nanoparticles and nanotubes to boost nutrient bioavailability and improve their interaction with soil and plants. In precision agriculture, nano-fertilizers enable site-specific nutrient management and optimize fertilizer usage. They also play a role in soil remediation by aiding in the cleanup of contaminated soils and improving soil health. Additionally, nano-fertilizers foster beneficial plant-microbe interactions, promoting plant growth and nutrient acquisition. This patent analysis will provide valuable insights into not only the current state of innovation in nano-fertilizer technology for agriculture but also identify emerging trends and technologies. It also dealt into potential opportunities for commercialization and shone a torchlight toward areas

K. S. R. Yadav
Department of Science, Universal School, Jakarta, Indonesia

V. B. Mandaliya (✉)
Department of Microbiology, Faculty of Science, Marwadi University, Rajkot, Gujarat, India
e-mail: viralkumar.mandaliya@marwadieducation.edu.in

V. D. Rajput et al. (eds.), *Contemporary Suitability of Nanobionics in Agriculture*, Nanotechnology in the Life Sciences, https://doi.org/10.1007/978-3-031-82779-2_10

where further research and development are needed. In conclusion, nano-fertilizers offer significant promise for addressing the challenges of climate change and achieving sustainable development goals (SDGs). By fostering collaboration between scientists, policymakers, and farmers, we can create a future where nano-fertilizers play a vital role in achieving a more sustainable and resilient agricultural system.

Keywords Climate-Smart Agriculture · Nano-fertilizer · Patent Landscape · Patentscope · Sustainable Development Goals

10.1 Introduction

Climate-smart agriculture (CSA) is an innovative approach that integrates sustainable soil management practices to address the challenges of climate change and food security (Lipper et al., 2014). While looking toward an enhancing productivity, increasing resilience, and reducing greenhouse gas emissions, CSA helps farmers in adapting a change in climatic conditions to ensure a stable food supply (Paustian et al., 2016). The central segment of CSA practices includes improved crop varieties (Chariya et al., 2013), efficient water management (Nakum et al., 2011), promoting soil health (Bhatt et al., 2011), optimizing resource use, and mitigating environmental impacts (Mandaliya, 2017). Eventually, CSA not only supports the livelihoods of farmers but also contributes to global efforts in combating climate change and securing future food resources (Fig. 10.1).

Nano-fertilizers (NF) are new and cutting-edge trend in agricultural technology. Compared to traditional fertilizers, NF helps to reduce waste and minimize environmental impact. NF utilizes a nanoparticle to enhance effective nutrient delivery to crops (Prasad et al., 2017). This controlled and targeted nutrient delivery increases crop yields and promotes plant health (Patel et al., 2017). Further, precise application of NF reduces the risk of nutrient runoff and prevents water pollution. As a result, NF offers a promising solution for sustainable agriculture and minimizing ecological footprints (Iavicoli et al., 2017).

Nano-fertilizers (NF) are new and cutting-edge trend in agricultural technology. Compared to traditional fertilizers, NF helps to reduce waste and minimize environmental impact (Patel et al., 2016). NF utilizes a nanoparticle to enhance effective nutrient delivery to crops (Prasad et al., 2017). This controlled and targeted nutrient delivery increase crop yields and promote plant health (Jhala et al., 2015). Further, precise application of NF reduces the risk of nutrient runoff and prevent in water pollution (Chadwick et al., 2023). As a result, NF offer a promising solution for sustainable agriculture and minimizing ecological footprints (Iavicoli et al., 2017).

Patent landscape analysis is a strategic tool used to visualize and understand the innovative sphere surrounding a particular technology (Stoffels et al., 2020). The systematic examination of patent documents in a defined field gives an insight into identification of technological trends, competitive dynamics, and potential gaps in

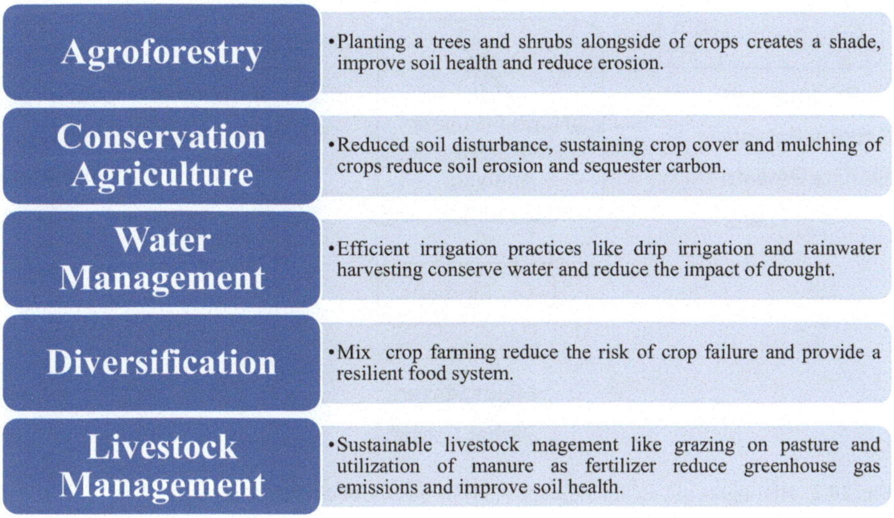

Fig. 10.1 Climate-smart agriculture (CSA) practices to address climate change and food security

the market (Grant et al., 2014). It also helps in identifying key players and tracking the evolution of technologies over time (Fleming & Sorenson, 2001). Furthermore, patent landscape analysis can be instrumental in strategic planning and risk management (Yang et al., 2010). This analysis highlights areas with less patent activity that help to identify patent gaps, which helps innovators to adapt technological shifts (Karvonen et al., 2016). Overall, it provides a comprehensive view of the patent environment that enables organizations to better navigate the complexities of intellectual property and stay competitive in a rapidly evolving technological landscape (Mahlia et al., 2020).

The present chapter aimed to explore the current state of patent landscape in nano-fertilizer technology for climate-smart agriculture and sustainable soil management.

10.2 Patent Landscape Analysis

Data Source: The present study has reviewed around 224 patent records related to nano-fertilizers from Patentscope (Patent database of World Intellectual Property Organization—WIPO), Scopus Patents, Google Patents, and Lens (a public patent knowledge database). The data was collected from various database and included up to last PCT publication update no. 34/2024 released on August 22, 2024.

Objective: This analysis will identify the emerging innovations, key patent holders, and the geographic distribution of innovative technologies. The highlights of patent landscape analysis on NF are depicted in Fig. 10.2.

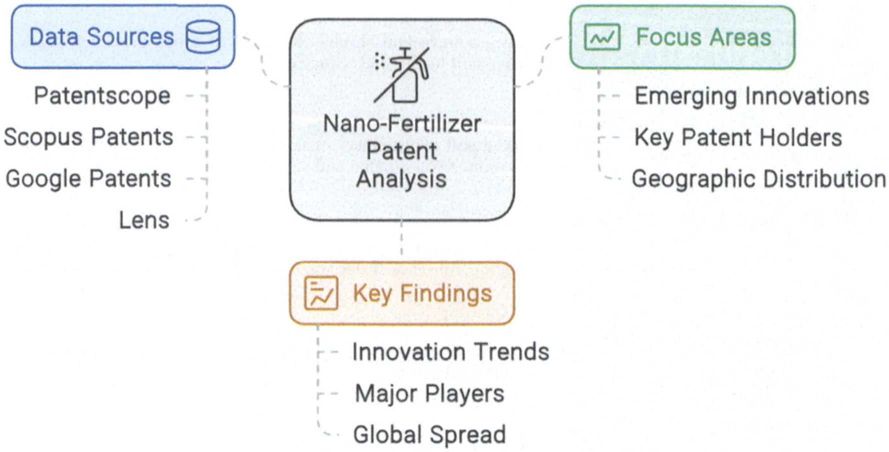

Fig. 10.2 Highlights of patent landscape analysis on NF for climate-smart agriculture

10.2.1 Jurisdiction-Wise Patent Dynamics Since 2008

Exploration of patent dynamics based on the jurisdiction is important in various strategic aspects for businesses, researchers, and policymakers (Geuna & Rossi, 2011). Here onward, innovator, inventor, applicant, business, company, or organization is referred to as "entity" for their singular identification. Some of strategic aspects concerned to jurisdiction in patent dynamics are outlined here.

10.2.1.1 Strategic Business Planning

(1) Market Access and Expansion: Understanding the diverse patent laws and enforcement practices across different jurisdictions is crucial for entities to effectively plan their market entry and expansion strategies (Jiang et al., 2022). It also helps in prioritizing patenting in specific regions to mitigate risks before entering the market (Lee et al., 2009). (2) Competitive Advantage: By securing patents in key jurisdictions, an entity gains a competitive advantage in defending market share, and it avoids potential infringement disputes (Hanel, 2006). (3) Licensing and Partnerships: Entity may negotiate license or partnership deals based on the strength and scope of patents in specific jurisdictions (Dolmans, 2022).

10.2.1.2 Intellectual Property Protection

(1) Tailored IP Strategy: It is important to understand the scope and duration of protection across the jurisdictions (Lizin et al., 2013). IP strategy may help an entity in filling and defending its patents (Kalıp et al., 2022). (2) Risk Management:

Awareness of jurisdiction-specific patent laws helps an entity to avoid legal pitfalls and to take well-informed decisions about expansion of business (Golodne & Da Silva, 2022).

10.2.1.3 Research and Development (R&D)

(1) Innovation Incentives: It is understood that strong IP protection in a region will incentivize the innovation and give insight for investment in that area (Dhiman et al., 2023). (2) Collaboration Opportunities: Entity can identify potential collaborators and funding opportunities that can enhance targeted research outputs (King & Locock, 2022).

10.2.1.4 Legal and Regulatory Compliance

(1) Avoiding Infringement: An entity operating in global platform must look upon existing patents scenario in different jurisdictions (Akinsolu et al., 2017). Patent dynamics helps in conducting thorough freedom-to-operate analyses (De Weerdt et al., 2015). (2) Navigating Legal Disputes: In case of patent disputes, it is important to understand the jurisdiction-specific legal framework in procedures for handling patent litigation and disputes (Bart, 2008).

10.2.1.5 Market Trends and Competitive Intelligence

(1) Tracking Innovation Trends: Patent filings can indicate emerging technologies and market trends (Litvinova et al., 2022). An entity can identify technological advancements and competitor strategies (Rana & Roy, 2015). (2) Benchmarking: Patent portfolios can create a benchmark that keeps away the competitors (Masood et al., 2016).

10.2.1.6 International Trade and Policy

(1) Trade Negotiations: Country with strong IP laws may strongly negotiate trade agreements as well as able to fetch the global players for their business in the country (Furman et al., 2002). (2) Policy Development: Policymakers able to ensures the alignment of national laws with international standards and help in mitigating local challenges (Mahlia et al., 2020).

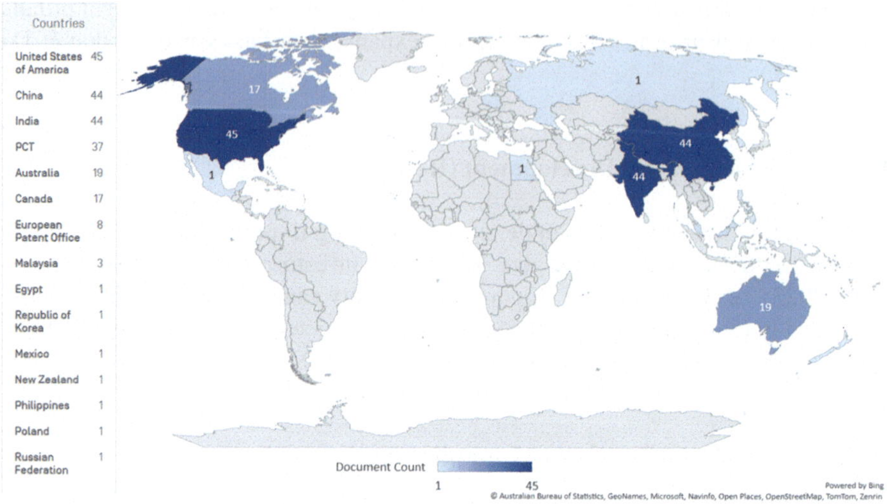

Fig. 10.3 A pictorial diagram on jurisdictional analysis of patent publications on NF. It includes tabulation of no. of patents from specific offices

10.2.1.7 Economic and Investment Decisions

(1) Investment Attraction: Any entity likely to invest in regions where robust and enforceable IP protection is observed (Soete, 1987). (2) Economic Impact: Patents contribute to economic growth by fostering innovation and ultimately help in creating jobs (Smith, 1999).

In present study, jurisdictional-wise analysis of patent status on NF is prepared and depicted in Fig. 10.3. It includes tabulation of no. of patents from specific offices. The United States, China, and India are top jurisdictions of patent publication, and each has around 44–45 patent records on NF. Patent records also accounted for by Australia (19), Canada (17), European Patent Office (8) and the rest is very less. This strongly suggests the potency of top jurisdictions in technological control as well as shows the active market potential. The invention related to NF has potential for global coverage, and it has led to 37 patent application filings to multinational phase via WIPO PCT. Table 10.1 has outlined this potential patent published via PCT starting from 2011 to 2024 on NF along with their key features.

10.2.2 Key Patent Holders in Nano-fertilizers

Patent holders play a critical role in the patent and invention ecosystem (Pereira et al., 2018). They are central to the commercialization, protection, and management of inventions (Gadiya et al., 2023). In the present analysis, Farmer Sean topped the inventor list with 23 patents in NF (Fig. 10.4). It is followed by Alibek Ken (15

Table 10.1 The potential patent published at multi-national phase via PCT on NF and their key features

PCT application ID	Publication year	Applicants	Key feature
WO2011055219	2011	Bijam Biosciences Private Limited	Kimberlite based nutrients supply with or without nano structure.
WO2013110202	2013	N/C Quest Inc.	The conditioned exhaust and carbon nanotube applied to the plant growing medium.
WO2013121244	2013	University of Calcutta	NF having plant nutrient coated onto a metal nanoparticle.
WO2014117147	2014	BFP Management LLC.	An aqueous solution comprised of nano-sized fertilizer particles.
WO2016201312	2016	Brisa International LLC.	Nano-based system for biomass growth and processing.
WO2017172634	2017	Jyant Technologies Inc.	A nanoparticle compositions and methods for targeted delivery of a bioactive agent to a plant.
WO2020112844	2020	Locus Agriculture IP Company LLC.	Yeast-based formulation for enhancing rhizosphere properties and plant health.
WO2020210672	2020	Aqua Yield Operations LLC.	Nutrients delivery to plant based in dendrimer nanoparticles and spherical nanoparticles forms.
WO2020076797	2020	Locus IP Company LLC.	Invention provides materials and methods for reducing deleterious atmospheric gases.
WO2020219432	2020	Locus Agriculture IP Company LLC.	Method utilize *Trichoderma harzianum* and *Bacillus amyloliquefaciens* for maintenance of natural turf grass field.
WO2021133630	2021	Locus Agriculture IP Company LLC.	Microbe-based formulation for enhancing phytocannabinoid content of Cannabis.
WO2021222676	2021	Locus Agriculture Ip Company LLC.	Microbe-based formulation for enhancement of crop production and yield.
WO2021094794	2021	Herbafarm-Magnolija D.O.O.	Formulation for treating trees, shrubs and climbing plants.
WO2021189049	2021	Locus Ip Company LLC.	Efficient method for dispersion of nanoparticles.
WO2021257499	2021	Locus Solutions IP Company LLC.	Use of composition to control plant infections.
WO2021234718	2021	Kuchimanchi, Venkata Satya Sarveswara Sairam. Kuchimanchi, Vaishnavi.	Method for preparation of natural organic NF.
WO2022180504	2022	University of Sri Jayewardenepura.	Preparation of NF for sustained release of macronutrients.

(continued)

Table 10.1 (continued)

PCT application ID	Publication year	Applicants	Key feature
WO2022093843	2022	Agvnt LLC.	Liquid fertilizer containing nano-bubbles.
WO2022098573	2022	Locus Solutions IP Company LLC.	Method for atmospheric carbon sequestering and plant yield growth.
WO2022036010	2022	Cellacure LLC.	Biowaste refining method for preparation of nano biopolymers.
WO2022197596	2022	Locus Solutions IP Company LLC.	Microbe based remediation of degraded soils.
WO2022203951	2022	Locus Solutions IP Company LLC.	Systems for the management of crops, soil, water, and livestock.
WO2023126895	2023	Sabic Agri-Nutrients Company.	A plant based nano biocarrier derived from polysaccharide polymer to coat nutrient forms.
WO2023126894	2023	Sabic Agri-Nutrients Company.	A plant based nano biocarrier derived from aqueous polysaccharide polymer to combine with nutrient forms.
WO2023168122	2023	The Regents of the University of California.	Method for targeted delivery of nanoparticles to phloem of plant.
WO2023168125	2023	The Regents of the University of California.	Method for targeted delivery of nanoparticles to fungi.
WO2023038969	2023	Locus Solutions IP Company LLC.	Microbe based improvement of plant immune response.
WO2023199351	2023	Coromandel International Limited	NF preparation from synthetic polymer.
WO2023203539	2023	Smileybee Ltd.	Preparation of NPK NF.
WO2023069313	2023	Locus Solutions IP Company LLC.	Method for improving soil structure and hydrology.
WO2023046851	2023	Einolllahi, Mohsen Behrouz; Osman, Omar El- Farouk; Einolllahi, Behrouz Mohammad.	Nanoparticles of chelated complex.
WO2024159245	2024	University of South Africa	Preparation of crystalline 1-D nanoscaled calcite.
WO2024166123	2024	Indian Farmers Fertilizer Cooperative Limited (IFFCO).	Preparation of stable NF containing Nitrogen and Phosphorous.
WO2024150243	2024	Indian Farmers Fertilizer Cooperative Limited (IFFCO).	Manufacturing of nano urea.
WO2024064634	2024	Locus Solutions IP Company LLC.	Formulation of sophorolipid molecule for dispersion of nanoparticles to plant.

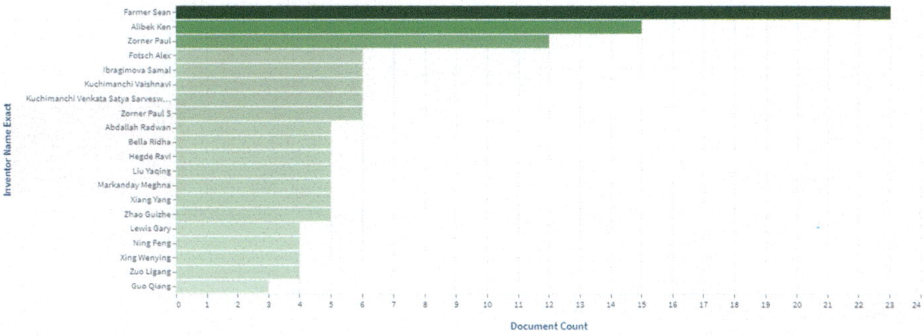

Fig. 10.4 Top and prominent patent inventors in the field of NF. The bar diagram shows the name of inventor with their no. of patents on NF

patents) and Zorner Paul (12 patents), who had made substantial impacts. These inventors have been instrumental in driving research and development, demonstrating exceptional innovation and expertise. Their work has been pivotal in shaping the NF industry. Through their dedication and contributions, they have expanded the frontiers of NF technologies.

10.2.3 International Patent Classification (IPC) for Patents

The IPC is an international system for classifying patents and patent applications according to different technological fields. It was established by the Strasbourg Agreement (1971) and is maintained by WIPO (WIPO-Strasbourg Agreement, 1979). The IPC system divides technology into broad categories known as classes (Kang, 2012). There are eight main sections, and each is represented by a letter (A to H). Main sections are further divided into classes, subclasses, groups, and subgroups (Bonino et al., 2010).

For example, one has a patent related to a liquid fertilizer layered or coated into matrix. The IPC classification might be:

- Section: C (Chemistry; Metallurgy)
- Class: C05 (Fertilizers; Manufacture Thereof)
- Subclass: C05G (Mixtures of one or more Fertilizers; Form)
- Group: C05G 5/00 (Fertilizers characterized by their form)
- Subgroup: C05G 5/20 (Liquid fertilizer layered or coated into matrix)

In present study, IPC dynamics observed major two subgroups A01 (Agriculture) and C05 (Chemistry—Fertilizers). The highest patent published records associated to subclass C05G 3/00—Mixtures of one or more fertilizers with additives, i.e., 28 patents; C05C 9/00– Fertilizer containing urea or urea compounds—18 patents (Fig. 10.5).

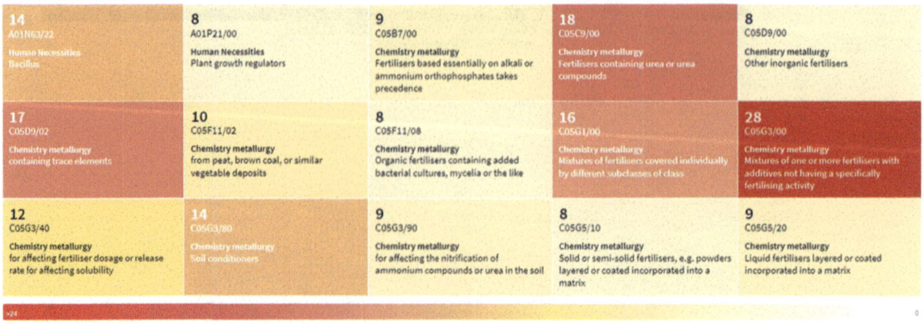

Fig. 10.5 IPC dynamics associated to patent published on NF. This diagram shows the heatmap of IPC subclass. In each box top first line shows the no. of patent, second line shows the IPC subclass, and last part gives the description of each subclass

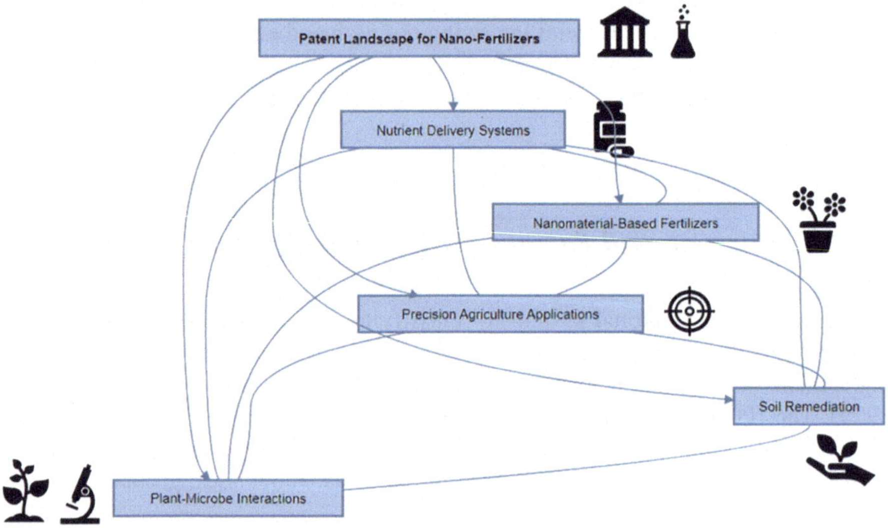

Fig. 10.6 An interconnection of innovative areas and application of patented NF formulations

10.3 Patented Innovations in NF Applicable in Agricultural Practices

NF are fertilizers engineered at the nanoscale that are typically in the range of 1 to 100 nanometers. It can be formulated with metals, metal oxides, and organic compounds (Ahmad et al., 2022). The small size of nanoparticles gives unique properties and composition compared to conventional fertilizers (Hyder et al., 2023). The patent landscape for NF covers several innovative areas within agricultural practices, reflecting their diverse applications and potential benefits (Fig. 10.6).

10.3.1 Nutrient Delivery Systems

NF improves the bioavailability of nutrients. Nano size allows them to reach plant tissues more easily. Patents in this area focus on the development of advanced carriers and encapsulation techniques for nutrients that facilitate the controlled and efficient release to plants. In the current year 2024, Indian Farmers Fertilizer Cooperative Limited (IFFCO) have come up with two PCT applications, first patent WO2024150243 on the manufacturing of nano-encapsulated urea that is explained in patent, while another patent WO2024166123 on nano-encapsulated phosphates. These patents aimed for availability of nutrient at tissue level.

10.3.2 Nanomaterial-Based Fertilizers

This category includes patents related to the formulation of fertilizers using nanoparticles of essential nutrients. For example, nano-sized zinc oxide (ZnO) and nano-sized iron oxide (Fe_2O_3). Jyant Technologies Inc. in patent WO2017172634 prepared nanoparticle composition. This nanoparticle designed for targeted delivery of a bioactive agent to a plant. It improves the solubility and absorption of nutrients by plants, increase crop production and yield.

10.3.3 Precision Agriculture Applications

NF is designed for slow or controlled release for precision agriculture. This means that nutrients are released gradually over time, which reduces the risk of nutrient leaching and provides a steady supply of nutrients to the plant. In addition, a targeted approach to nutrient delivery can minimize wastage and enhance the efficiency of nutrient use. Moreover, nanomaterials are used to develop sensors that can monitor and control the soil and plant nutrient levels in real-time that help to optimize fertilizer application. Brisa International LLC, in patent WO2016201312, developed nano-based system for biomass growth and processing.

10.3.4 Soil Remediation

NF interacts more effectively with soil particles and plant roots. Locus Solutions IP Company LLC. has come up with many patents, viz. WO2022098573, WO2022203951, WO2020112844, and many more that manufactured nanoparticles for enhancing rhizosphere properties and soil conditioning. These nanoparticles detoxify and restore contaminated soils, enhancing the soil health and fertility as part of sustainable agricultural practices.

10.3.5 Plant-Microbe Interactions

Nano-sized nutrients are, on one hand, easily absorbed by plant roots, while on the other hand, nutrient availability soil enhance the growth of beneficial soil microbes, such as nitrogen-fixing bacteria and mycorrhizal fungi, which in turn benefits the plant growth. Locus Agriculture IP Company LLC. in their patents WO2020219432, WO2021133630, WO2021222676, WO2022197596, and WO2023038969 used microbe-based formulation for release of nutrients in controlled manner that minimizes nutrient spikes in the soil and also reduces the risk of microbial toxicity, which generally observed with traditional fertilizers.

10.4 Case Studies on Successful Application of Nano-fertilizers

Some of the successful case studies on applications of NF illustrate the potential to revolutionize agricultural practices by improving nutrient delivery, enhancing crop yields, and promoting sustainable farming. Here are some notable examples:

10.4.1 Case Study: Nano-Urea for Crop Plant Cultivation

In India, nano-urea is introduced as a substitute for traditional urea fertilizers (IFFCO, 2024a). It improves the nutrient uptake and reduces the environmental impact. This initiative was supported by the Indian government and patented by Indian Farmers Fertilizer Cooperative Limited (IFFCO). Later, IFFCO came up with PCT patent WO2024150243 on nano urea for sustainable agriculture. Nano-Urea is produced by Indian Farmers Fertilizer Cooperative Limited (IFFCO). It was applied to various crop plants as a replacement for conventional urea fertilizers. Here, Nano-Urea particles are designed to provide a controlled release of nitrogen to enhance nutrient uptake by plants. In their study, trials showed a significant increase in various crop yields compared to traditional urea application (IFFCO, 2024b). Also, Nano-Urea demonstrated lower nitrogen loss due to volatilization and leaching. It is observed that reduced use of conventional urea decreased the environmental footprint, including lower greenhouse gas emissions and reduced soil and water pollution (Upadhyay et al., 2023). Thus, Nano-Urea has proven effective in improving nitrogen use efficiency and crop yields while mitigating environmental impacts, supporting its broader adoption in Indian agriculture.

10.4.2 Case Study: Nano-Particles for Corn Production

A group of researchers from Egypt has focused on the foliar application of nano-phosphorous fertilizers to enhance the growth and yield of sweet corn (Salama et al., 2023). Nanoparticles are designed for improved soil interaction and targeted nutrient delivery. Foliar application of hydroxyapatite nanoparticles (HA-NPs) improved growth, yield, and biochemical content of sweet corn. In their study, sweet corn treated with HA-NPs showed non-toxicity where EC50 was less than 100; also, the minimum genomic template stability percentage (GTS%) for sweet corn was recorded at 75 mg/l of HA-NPs. This shows that sweet corn plants exhibited better growth and health and increased corn yields compared to fields treated with conventional fertilizers. This highlights a potential of NF for widespread agricultural use.

10.4.3 Case Study: Nano-fertilizers for Gene Expression in Wheat Crops

A group of researchers from China have utilized the nano carbon synergist, nano calcium carbonate synergist, and composite nano synergist in their experiment (Yang et al., 2023). They have explored the study of addition of different NF synergists that promote wheat nitrogen utilization and related gene expression. These experiments were performed in pot and field conditions at the West-Coast Economic New Area experimental base and Greenhouse of Qingdao Agricultural University, China. They had set up seven treatments and observed the gene expression levels of *TaNRT2.2*, *TaNRT2.3*, *TaGS1*, and *TaGS2* in the treatments. In their study, they have observed that the application of NF synergist was able to improve nitrogen accumulation, nitrogen use efficiency, and promote the expression of genes related to nitrogen transport and metabolism.

10.5 Nano-fertilizers and Sustainable Development Goals: A Nexus for Climate-Smart Agriculture

NF offers significant promise for addressing the challenges of climate change and food security in achieving Sustainable Development Goals (SDGs) (Fig. 10.7). NF can contribute to several SDGs, including:

SDG 2—Zero Hunger: By improving crop yields and reducing food waste, NF can help ensure food security and combat hunger (El-Azeim et al., 2020). There is a need for further research and development as well as regulation that realizes the sustainable use of NF (Lukman et al., 2016). There is a need to study the long-term effects of NF on human health and the environment (Ashraf et al., 2021).

Fig. 10.7 Nano-fertilizers and sustainable development goals: A nexus for climate-smart agriculture

SDG 3—Good Health and Well-being: NF can contribute to improved nutrition and reduced exposure to harmful chemicals, promoting human health (Khanali et al., 2022). There is a need to share the benefits of NF with small-scale farmers as the resource-limited in communities is a crucial part of achieving SDG3 (Basavegowda & Baek, 2021).

SDG 6—Clean Water and Sanitation: By reducing nutrient leaching and pollution, NF can help protect water resources and improve water quality (Younis et al., 2021). Reduced nutrient runoff is a main characteristic of NF that helps in preventing the water pollution as well as it protects aquatic ecosystems (Ur Rahim et al., 2021). Thus, by reducing water pollution from agricultural activities, NF can contribute to improved water quality for drinking, irrigation, and other uses (Nongbet et al., 2022).

SDG 7—Affordable and Clean Energy: NF can be used in conjunction with renewable energy sources to create more sustainable agricultural systems (Heiba et al., 2023). Further, increased crop yields lead to increased food production that reduces the need for energy-intensive food imports (Babu et al., 2021). Also, NF reduces the need for chemical fertilizers that are ultimately energy-intensive for the farm products (An et al., 2022).

SDG 12—Responsible Consumption and Production: By reducing the need for excessive fertilizer inputs, NF can promote more sustainable agricultural practices (Pandey, 2018). NF helps in nutrient runoff, and it reduces the problem of eutrophication in agricultural ecosystems (Dimkpa & Bindraban, 2017). Further, NF parts in precision agriculture that lead to high-quality products that meet current consumer demands and also do not compromise sustainability (Kumar et al., 2021).

SDG 13—Climate Action: NF can help mitigate climate change by reducing greenhouse gas emissions associated with fertilizer production and use (Mona et al., 2021). Further, NF contributes to a circular economy where waste is minimized and resources are recycled, which helps in reduced carbon sequestering (Rabalao et al., 2022; Escribà-Gelonch et al., 2023).

SDG 15—Life on Land: By promoting sustainable soil management and reducing soil degradation, NF can contribute to the protection of terrestrial ecosystems (Trieu et al., 2024). NF enhances crop health and productivity and supports the biodiversity in agricultural ecosystems (Alzate Acevedo et al., 2021). Further, promoting crop yields and reducing the need for new agricultural land, NF parts into reducing the deforestation and protecting forests (Akter et al., 2024).

10.6 Conclusion and Future Prospects

By understanding the patent landscape for this technology, researchers, policymakers, and farmers can identify potential solutions and accelerate the adoption of NF in sustainable soil management practices. While NF offer significant potential, there are also challenges to consider. The initial investment in NF technology is quite high. The regulatory framework for NF is complex and challenging to navigate. There are concerns about potential long-term effects of NF on soil and ecosystems. Thus, to realize the full potential of NF, it is essential to address these challenges in achieving a more sustainable and resilient agricultural system.

CRediT Authorship Contribution Statement KRY: Reviewed and Edited the Draft; VBM: Conceptualization, Writing—Original Patent Landscaping, Review, Editing, and Approval of the Draft.

Declaration of Interest Conflict of Interest: The author declare that they do not have any conflict of interest.

Competing Interests: The author has non-financial interests to disclose.

Funding: No funding.

Ethical Approval: Not Required

Acknowledgments The author is thankful to the Faculty of Science, Marwadi University, for providing necessary encouragement and support to carry out this analysis. Authors are also thankful to Dr. Sandrine Ammann from Patentscope (WIPO) for providing a necessary patent landscaping training on Patentscope.

References

Ahmad, A., Hashmi, S. S., Palma, J. M., & Corpas, F. J. (2022). Influence of metallic, metallic oxide, and organic nanoparticles on plant physiology. *Chemosphere, 290*, 133329.

Akinsolu, F. T., De Paiva, V. N., Souza, S. S., & Varga, O. (2017). Patent landscape of neglected tropical diseases: An analysis of worldwide patent families. *Globalization and Health, 13*, 1–13.

Akter, M. M., Hosen, M. A., Gilroyed, B. H., Akter Eti, S., Sarker, P., Kabir, M. M., & Mohinuzzaman, M. (2024). Nutrient Recovery for Fertilizer Production and Wastewater Treatment for a Circular Economy. *ACS Sustainable Resource Management, 1*(4), 787–798.

Alzate Acevedo, S., Díaz Carrillo, Á. J., Flórez-López, E., & Grande-Tovar, C. D. (2021). Recovery of banana waste-loss from production and processing: A contribution to a circular economy. *Molecules, 26*(17), 5282.

An, C., Sun, C., Li, N., Huang, B., Jiang, J., Shen, Y., Wang, C., Zhao, X., Cui, B., Wang, C., Li, X., Zhan, S., Gao, F., Zeng, Z., Cui, H., & Wang, Y. (2022). Nanomaterials and nanotechnology for the delivery of agrochemicals: Strategies towards sustainable agriculture. *Journal of Nanobiotechnology, 20*(1), 11. https://doi.org/10.1186/s12951-021-01214-7

Ashraf, S. A., Siddiqui, A. J., Elkhalifa, A. E. O., Khan, M. I., Patel, M., Alreshidi, M., Moin, A., Singh, R., Snoussi, M., & Adnan, M. (2021). Innovations in nanoscience for the sustainable development of food and agriculture with implications on health and environment. *The Science of the Total Environment, 768*, 144990. https://doi.org/10.1016/j.scitotenv.2021.144990

Babu, S., Singh, R., Yadav, D., Rathore, S. S., Raj, R., Avasthe, R., Yadav, S., Das, A., Yadav, V., Yadav, B., Shekhawat, K., Upadhyay, P., Yadav, D. K., & Singh, V. K. (2021). Nanofertilizers for agricultural and environmental sustainability. *Chemosphere, 292*, 133451. https://doi.org/10.1016/j.chemosphere.2021.133451

Bart, T. N. (2008). Parallel trade of pharmaceuticals: A review of legal, economic, and political aspects. *Value in Health, 11*(5), 996–1005. https://doi.org/10.1111/j.1524-4733.2008.00339.x

Basavegowda, N., & Baek, K. H. (2021). Current and future perspectives on the use of nanofertilizers for sustainable agriculture: The case of phosphorus nanofertilizer. *3. Biotech, 11*(7), 357.

Bhatt, K. D., Girnari, S. K., Mandaliya, V. B., Chariya, L. D., & Thaker, V. S. (2011). Use of RAPD marker to confirm mutation in morphological variants on Neem tree. *Electronic Journal of Plant Breeding, 2*(3), 473–478. https://www.ejplantbreeding.org/index.php/EJPB/article/view/983

Bonino, D., Ciaramella, A., & Corno, F. (2010). Review of the state-of-the-art in patent information and forthcoming evolutions in intelligent patent informatics. *World Patent Information, 32*(1), 30–38.

Chadwick, J. J., Zhang, P., Ullah, S., & Lynch, I. (2023). Use of nanotechnology to increase nutrient use efficiency, enhance crop nutrition, and reduce agrochemical pollution. In *Nano-Enabled Sustainable and Precision Agriculture* (pp. 17–41). Academic Press.

Chariya, L. D., Mandaliya, V. B., & Thaker, V. S. (2013). Conversion of monomorphic band into polymorphic pattern using nucleotide sequencing data in *Musa* varieties. *Genetics and Plant Physiology, 3*(1–2), 77–89. http://www.bio21.bas.bg/ippg/bg/wp-content/uploads/2014/03/GPP_3_1-2_77-89.pdf

De Weerdt, E., Simoens, S., Hombroeckx, L., Casteels, M., & Huys, I. (2015). Causes of drug shortages in the legal pharmaceutical framework. *Regulatory Toxicology and Pharmacology, 71*(2), 251–258. https://doi.org/10.1016/j.yrtph.2015.01.005

Dhiman, S., Sharma, C., Kumar, A., Pathak, P., & Purohit, S. D. (2023). Microplastics in aquatic and food ecosystems: Remediation coupled with circular economy solutions to create resource from waste. *Sustainability, 15*(19), 14184. https://doi.org/10.3390/su151914184

Dimkpa, C. O., & Bindraban, P. S. (2017). Nanofertilizers: New products for the industry? *Journal of Agricultural and Food Chemistry, 66*(26), 6462–6473.

Dolmans, M. (2022). FRAND licensing commitments—Back to first principles. *World Patent Information, 69*, 102113. https://doi.org/10.1016/j.wpi.2022.102113

El-Azeim, M. A., Sherif, M., Hussien, T., & I., & Bashandy, S. (2020). Impacts of nano- and non-nanofertilizers on potato quality and productivity. *Acta Ecologica Sinica, 40*(5), 388–397. https://doi.org/10.1016/j.chnaes.2019.12.007

Escribà-Gelonch, M., Butler, G. D., Goswami, A., Tran, N. N., & Hessel, V. (2023). Definition of agronomic circular economy metrics and use for assessment for a nanofertilizer case study. *Plant Physiology and Biochemistry, 196*, 917–924.

Fleming, L., & Sorenson, O. (2001). Technology as a complex adaptive system: Evidence from patent data. *Research Policy, 30*(7), 1019–1039. https://doi.org/10.1016/s0048-7333(00)00135-9

Furman, J. L., Porter, M. E., & Stern, S. (2002). The determinants of national innovative capacity. *Research Policy, 31*(6), 899–933. https://doi.org/10.1016/s0048-7333(01)00152-4

Gadiya, Y., Gribbon, P., Hofmann-Apitius, M., & Zaliani, A. (2023). Pharmaceutical patent landscaping: A novel approach to understand patents from the drug discovery perspective. *Artificial Intelligence in the Life Sciences, 3*, 100069.

Geuna, A., & Rossi, F. (2011). Changes to university IPR regulations in Europe and the impact on academic patenting. *Research Policy, 40*(8), 1068–1076. https://doi.org/10.1016/j.respol.2011.05.008

Golodne, D. M., & Da Silva, A. S. (2022). Managing patent examiners' workload by using patent application volume data. *World Patent Information, 68*, 102095. https://doi.org/10.1016/j.wpi.2022.102095

Grant, E., Van den Hof, M., & Gold, E. R. (2014). Patent landscape analysis: A methodology in need of harmonized standards of disclosure. *World Patent Information, 39*, 3–10.

Hanel, P. (2006). Intellectual property rights business management practices: A survey of the literature. *Technovation, 26*(8), 895–931. https://doi.org/10.1016/j.technovation.2005.12.001

Heiba, Y., Ibrahim, M. G., Mohamed, A. E., Fujii, M., & Nasr, M. (2023). Developing smart sustainable irrigation matrix (SIM)-based model for selection of best irrigation techniques:

A framework to achieve SDGs. *Journal of Cleaner Production, 420*, 138404. https://doi.org/10.1016/j.jclepro.2023.138404

Hyder, S., Ul-Nisa, M., Shahzadi, N., Shahid, H., Gohar, F., Gondal, A. S., Riaz, N., Younas, A., De Los Santos-Villalobos, S., Montoya-Martínez, A. C., Sehar, A., Latif, F., Rizvi, Z. F., & Iqbal, R. (2023). Recent trends and perspectives in the application of metal and metal oxide nanomaterials for sustainable agriculture. *Plant Physiology and Biochemistry, 202*, 107960. https://doi.org/10.1016/j.plaphy.2023.107960

Iavicoli, I., Leso, V., Beezhold, D. H., & Shvedova, A. A. (2017). Nanotechnology in agriculture: Opportunities, toxicological implications, and occupational risks. *Toxicology and Applied Pharmacology, 329*, 96–111. https://doi.org/10.1016/j.taap.2017.05.025

IFFCO. (2024a). *Nano-urea for rice cultivation in India*, https://nanourea.in/en/nano-urea

IFFCO. (2024b). *Crop-wise increase in yields compared to traditional urea application*, https://www.iffco.in/en/crop-wise-nano-urea

Jhala, V. M., Mandaliya, V. B., & Thaker, V. S. (2015). Simple and Efficient Protocol for RNA and DNA extraction from Rice (*Oryza sativa* L.) for Downstream Applications. *Int Res J Biological Sci, 4*(2), 62–67. http://www.isca.me/IJBS/Archive/v4/i2/11.ISCA-IRJBS-2014-252.php

Jiang, L., Zou, F., Qiao, Y., & Huang, Y. (2022). Patent analysis for generating the technology landscape and competition situation of renewable energy. *Journal of Cleaner Production, 378*, 134264.

Kalıp, N. G., Erzurumlu, Y. M., & Gün, N. A. (2022). Qualitative and quantitative patent valuation methods: A systematic literature review. *World Patent Information, 69*, 102111. https://doi.org/10.1016/j.wpi.2022.102111

Kang, H. Y. (2012). Science inside law: The making of a new patent class in the international patent classification. *Science in Context, 25*(4), 551–594.

Karvonen, M., Kapoor, R., Uusitalo, A., & Ojanen, V. (2016). Technology competition in the internal combustion engine waste heat recovery: A patent landscape analysis. *Journal of Cleaner Production, 112*, 3735–3743.

Khanali, M., Ghasemi-Mobtaker, H., Varmazyar, H., Mohammadkashi, N., Chau, K. W., & Nabavi-Pelesaraei, A. (2022). Applying novel eco-exergoenvironmental toxicity index to select the best irrigation system of sunflower production. *Energy, 250*, 123822. https://doi.org/10.1016/j.energy.2022.123822

King, S., & Locock, K. E. (2022). A circular economy framework for plastics: A semi-systematic review. *Journal of Cleaner Production, 364*, 132503. https://doi.org/10.1016/j.jclepro.2022.132503

Kumar, Y., Tiwari, K. N., Singh, T., & Raliya, R. (2021). Nanofertilizers and their role in sustainable agriculture. *Annals of Plant and Soil Research, 23*(3), 238–255.

Lee, S., Yoon, B., Lee, C., & Park, J. (2009). Business planning based on technological capabilities: Patent analysis for technology-driven roadmapping. *Technological Forecasting and Social Change, 76*(6), 769–786. https://doi.org/10.1016/j.techfore.2009.01.003

Lipper, L., Thornton, P., Campbell, B. M., Baedeker, T., Braimoh, A., Bwalya, M., Caron, P., Cattaneo, A., Garrity, D. P., Henry, K., Hottle, R., Jackson, L., Jarvis, A., Kossam, F., Mann, W., McCarthy, N., Meybeck, A., Neufeldt, H., Remington, T., et al. (2014). Climate-smart agriculture for food security. *Nature Climate Change, 4*(12), 1068–1072. https://doi.org/10.1038/nclimate2437

Litvinova, O., Klager, E., Tzvetkov, N. T., Kimberger, O., Kletecka-Pulker, M., Willschke, H., & Atanasov, A. G. (2022). Digital pills with ingestible sensors: Patent landscape analysis. *Pharmaceuticals, 15*(8), 1025.

Lizin, S., Leroy, J., Delvenne, C., Dijk, M., De Schepper, E., & Van Passel, S. (2013). A patent landscape analysis for organic photovoltaic solar cells: Identifying the technology's development phase. *Renewable Energy, 57*, 5–11.

Lukman, R. K., Glavič, P., Carpenter, A., & Virtič, P. (2016). Sustainable consumption and production–Research, experience, and development–The Europe we want. *Journal of Cleaner Production, 138*, 139–147.

Mahlia, T., Syazmi, Z., Mofijur, M., Abas, A. P., Bilad, M. R., Ong, H. C., & Silitonga, A. (2020). Patent landscape review on biodiesel production: Technology updates. *Renewable and Sustainable Energy Reviews, 118*, 109526. https://doi.org/10.1016/j.rser.2019.109526

Mandaliya, V. B. (2017). Nanotechnology enabled climate-smart agriculture: A solution towards food security and climate change. In *XXXI Gujarat Science Congress – 2017* (p. 42).

Masood, I., Saleem, A., & Jamshed, S. Q. (2016). Why do physicians prefer brand medicines over generic medicines in Pakistan? A quantitative approach. *Journal of Pharmaceutical Health Services Research, 7*(4), 247–251. https://doi.org/10.1111/jphs.12152

Mona, S., Malyan, S. K., Saini, N., Deepak, B., Pugazhendhi, A., & Kumar, S. S. (2021). Towards sustainable agriculture with carbon sequestration, and greenhouse gas mitigation using algal biochar. *Chemosphere, 275*, 129856. https://doi.org/10.1016/j.chemosphere.2021.129856

Nakum, N. M., Mandaliya, V. B., Pandya, R. V., & Thaker, V. S. (2011). A simple and efficient protocol for high quality of DNA from Vitis quandrangularis L. *Electronic Journal of Plant Breeding, 2*(1), 87–95. https://www.ejplantbreeding.org/index.php/EJPB/article/view/1667

Nongbet, A., Mishra, A. K., Mohanta, Y. K., Mahanta, S., Ray, M. K., Khan, M., Baek, K., & Chakrabartty, I. (2022). Nanofertilizers: A smart and sustainable attribute to modern agriculture. *Plants, 11*(19), 2587. https://doi.org/10.3390/plants11192587

Pandey, G. (2018). Challenges and future prospects of agri-nanotechnology for sustainable agriculture in India. *Environmental Technology & Innovation, 11*, 299–307. https://doi.org/10.1016/j.eti.2018.06.012

Patel, K. G., Mandaliya, V. B., Mishra, G. P., Dobaria, J. R., & Thankappan, R. (2016). Transgenic peanut overexpressing mtlD gene confers enhanced salinity stress tolerance via mannitol accumulation and differential antioxidative responses. *Acta Physiologiae Plantarum, 38*(7). https://doi.org/10.1007/s11738-016-2200-0

Patel, K. G., Thankappan, R., Mishra, G. P., Mandaliya, V. B., Kumar, A., & Dobaria, J. R. (2017). Transgenic peanut (Arachis hypogaea L.) overexpressing mtlD gene showed improved photosynthetic, physio-biochemical, and yield-parameters under soil-moisture deficit stress in lysimeter system. *Frontiers in Plant Science, 8*. https://doi.org/10.3389/fpls.2017.01881

Paustian, K., Lehmann, J., Ogle, S., Reay, D., Robertson, G. P., & Smith, P. (2016). Climate-smart soils. *Nature, 532*(7597), 49–57. https://doi.org/10.1038/nature17174

Pereira, C. G., Picanco-Castro, V., Covas, D. T., & Porto, G. S. (2018). Patent mining and landscaping of emerging recombinant factor VIII through network analysis. *Nature Biotechnology, 36*(7), 585–590.

Prasad, R., Bhattacharyya, A., & Nguyen, Q. D. (2017). Nanotechnology in sustainable agriculture: Recent developments, challenges, and perspectives. *Frontiers in Microbiology, 8*. https://doi.org/10.3389/fmicb.2017.01014

Rabalao, T. M., Ndaba, B., Roopnarain, A., & Vatsha, B. (2022). Towards a circular economy: The influence of extraction methods on phytosynthesis of metallic nanoparticles and their impact on crop growth and protection. *JSFA Reports, 2*(5), 208–221.

Rana, P., & Roy, V. (2015). Generic medicines: Issues and relevance for global health. *Fundamental and Clinical Pharmacology, 29*(6), 529–542. https://doi.org/10.1111/fcp.12155

Salama, D. M., Osman, S. A., Shaaban, E. A., Elwahed, A., & El-Aziz, M. E. A. (2023). Effect of foliar application of phosphorus nanoparticles on the performance and sustainable agriculture of sweet corn. *Plant Physiology and Biochemistry, 203*, 108058. https://doi.org/10.1016/j.plaphy.2023.108058

Smith, P. J. (1999). Are weak patent rights a barrier to U.S. exports? *Journal of International Economics, 48*(1), 151–177. https://doi.org/10.1016/s0022-1996(98)00013-0

Soete, L. (1987). The impact of technological innovation on international trade patterns: The evidence reconsidered. *Research Policy, 16*(2–4), 101–130. https://doi.org/10.1016/0048-7333(87)90026-6

Stoffels, M. A., Klauck, F. J., Hamadi, T., Glorius, F., & Leker, J. (2020). Technology trends of catalysts in hydrogenation reactions: A patent landscape analysis. *Advanced Synthesis & Catalysis, 362*(6), 1258–1274.

Trieu, Q. A., Hoang, K. D., & Bui, T. H. (2024). Efficient phosphate adsorption in aqueous solution using nano-zirconia impregnated spent coffee grounds and preliminary application as a slow-release fertilizer. *Separation Science and Technology, 59*(5), 773–789. https://doi.org/10.1080/01496395.2024.2343850

Upadhyay, P. K., Singh, V. K., Rajanna, G. A., Dwivedi, B. S., Dey, A., Singh, R. K., Rathore, S. S., Shekhawat, K., Babu, S., Singh, T., Kumar, Y., Singh, C., Rangot, M., Kumar, A., Sarkar, S., Dash, S., & Rawat, S. (2023). Unveiling the combined effect of nano fertilizers and conventional fertilizers on crop productivity, profitability, and soil well-being. *Frontiers in Sustainable Food Systems, 7*. https://doi.org/10.3389/fsufs.2023.1260178

Ur Rahim, H., Qaswar, M., Uddin, M., Giannini, C., Herrera, M. L., & Rea, G. (2021). Nano-enable materials promoting sustainability and resilience in modern agriculture. *Nanomaterials, 11*(8), 2068.

WIPO-Strasbourg Agreement Concerning the International Patent Classification (as amended on September 28, 1979), https://www.wipo.int/wipolex/en/treaties/textdetails/12675

Yang, Y. Y., Akers, L., Yang, C. B., Klose, T., & Pavlek, S. (2010). Enhancing patent landscape analysis with visualization output. *World Patent Information, 32*(3), 203–220.

Yang, M., Dong, C., & Shi, Y. (2023). Nano fertilizer synergist effects on nitrogen utilization and related gene expression in wheat. *BMC Plant Biology, 23*, 26. https://doi.org/10.1186/s12870-023-04046-9

Younis, S. A., Kim, K. H., Shaheen, S. M., Antoniadis, V., Tsang, Y. F., Rinklebe, J., Deep, A., & Brown, R. J. (2021). Advancements of nanotechnologies in crop promotion and soil fertility: Benefits, life cycle assessment, and legislation policies. *Renewable and Sustainable Energy Reviews, 152*, 111686. https://doi.org/10.1016/j.rser.2021.111686

Chapter 11
Recent Advances and Future Outlook on Nanobionics-Based Food Sciences

Oyindamola John Samson, Doyinsola Amos Samson, Stephen Olaosebikan Makanjuola, Hafeez Aderinsayo Adekola, Babalola Sunday Aponjolosun, and Ismail Babatunde Onajobi

Abstract Nanobionics is an emerging field integrating nanotechnology with biological systems, offering transformative potential for food science and agricultural practices. This chapter aims to explore recent advances in nanobionics-based food sciences, highlighting innovations in areas such as nutrient delivery, crop protection, and food safety. Nanobionics has enabled the development of nanofertilizers, which optimize the release of essential nutrients to crops, and bio-nanosensors, which allow or real-time monitoring of soil and crop health, enhancing precision agriculture. In food safety, nanosensors improve the detection of contaminants, ensuring higher food quality standards. The role of bioactive nanoparticles is enhancing food preservation, improving shelf life, and promoting food security. Furthermore, smart packaging systems incorporating nanomaterials are revolutionizing the storage and distribution of food products by monitoring freshness and

O. J. Samson (✉)
Department of Microbiology and Biotechnology, Faculty of Natural Sciences, Ajayi Crowther University, Oyo, Nigeria

Department of Microbiology, Faculty of Science, Olabisi Onabanjo University, Ago-Iwoye, Nigeria

D. A. Samson
Department of Agricultural Economics and Farm Management, Federal University of Agriculture Abeokuta,
Abeokuta, Nigeria

S. O. Makanjuola
Department of Medical Laboratory Science, Faculty of Allied Health Science, Elizade University, Ilara-Mokin, Nigeria

H. A. Adekola · I. B. Onajobi
Department of Microbiology, Faculty of Science, Olabisi Onabanjo University, Ago-Iwoye, Nigeria

B. S. Aponjolosun
Department of Microbiology and Biotechnology, Faculty of Natural Sciences, Ajayi Crowther University, Oyo, Nigeria

V. D. Rajput et al. (eds.), *Contemporary Suitability of Nanobionics in Agriculture*, Nanotechnology in the Life Sciences, https://doi.org/10.1007/978-3-031-82779-2_11

preventing spoilage. As artificial intelligence (AI) and machine learning (ML) are integrated with nanobionics, predictive models and smart agricultural practices are emerging, driving a sustainable future for food production. However, it concludes with an outlook on future trends, including personalized nutrition and innovations in nanobionic approaches for global food security and sustainability.

Keywords Nanobionics · Food science · Nanosensors · Bioactive nanoparticles · Precision agriculture · Food security

11.1 Introduction

Nanobionics is an emerging interdisciplinary field that merges nanotechnology with biological systems to generate innovative solutions across various industries, particularly food science. It focuses on the integration of nanoscale materials and biological components to enhance or develop new functionalities (Vithanage et al., 2023). Nanobionics operates at the nanoscale (1–100 nanometers), where the unique properties of these materials can be harnessed for improved efficiency and novel applications. By applying biological principles to nanotechnology, this field creates advancements that bridge nature and technology for better food production, processing, and packaging solutions (Parveen et al., 2023).

In the context of food science, nanobionics emphasizes five key principles: miniaturization, surface area enhancement, biocompatibility, targeted delivery, and biomimicry (Carofiglio et al., 2022). Reducing materials to the nanoscale enables unique properties such as enhanced reactivity and functionality, while increased surface area allows for more efficient interactions. Biocompatibility ensures nanomaterials used in food applications are safe for consumption and environmentally friendly. Nanocarriers facilitate targeted delivery of nutrients or bioactive compounds within food products, enhancing both their efficiency and health benefits. Additionally, biomimicry draws from natural systems to design sustainable food production and packaging solutions (Assadpour & Mahdi, 2018).

The convergence of nanotechnology, biology, and food science has unlocked new opportunities to address critical challenges in the food industry. Researchers and food scientists are using nanobionics to develop innovative packaging materials with superior barrier and antimicrobial properties. The field also enables more efficient delivery systems for nutrients and bioactive compounds, as well as advanced sensors for detecting foodborne pathogens and spoilage indicators. Nanoscale modifications enhance the texture, flavor, and appearance of food products while improving the bioavailability and absorption of nutrients, contributing to a more nutritious and enjoyable consumer experience (Sun et al., 2021).

As nanobionics continues to evolve, its impact on food science will grow, addressing global issues like food security, sustainability, and nutrition. Current advancements include nanoencapsulation technologies for protecting sensitive ingredients, nanocomposite materials for improved food packaging, and nanosensors for detecting contaminants (Aguilar-Pérez et al., 2021). Additionally, nanoemulsions increase the stability and bioavailability of functional ingredients, while nanostructured materials allow for selective removal of undesirable compounds. The field promises to revolutionize the food industry, offering solutions that will shape the future of food production, processing, and consumption to meet the demands of a growing global population. The aim of this chapter is to describe the recent advances in nanobionics within the field of food science, focusing on the innovative applications, potential benefits, and emerging technologies. Also, assess the current challenges and limitations faced by nanobionic approaches in food processing, packaging, and safety.

11.2 Current Applications of Nanobionics in Food Industry

Smart food packaging is a breakthrough application of nanobionics that revolutionizes the protection, preservation, and monitoring of food products. Nanocomposite packaging materials, created by integrating nanoclays, nanocellulose, or metal nanoparticles, significantly improve barrier properties against gases, moisture, and UV radiation. These materials also enhance mechanical strength, making packaging more durable and reducing the need for multi-layer materials. Another development is antimicrobial packaging, which utilizes nanoparticles like silver, zinc oxide, or chitosan to inhibit the growth of microorganisms, extending the shelf life of perishable foods without additional preservatives (Jafarzadeh & Jafari, 2020).

Oxygen scavenging packaging is a critical innovation, incorporating nanoscale iron-based compounds or titanium dioxide nanoparticles to actively remove oxygen from the package headspace. This prevents oxidation and spoilage, particularly in sensitive food products. Moisture control packaging, using nanoporous materials or nanocellulose, helps regulate humidity levels, ensuring dry foods retain their optimal moisture content while preventing condensation in fresh produce. Intelligent packaging goes a step further by integrating nanosensors that monitor freshness, ripeness, or spoilage, offering real-time food quality information through color-changing indicators that respond to pH or temperature changes (Roberta, 2020).

Nanoencapsulation is a transformative technology that enhances the delivery of nutrients, flavors, and bioactive compounds in food products. By encasing sensitive compounds such as vitamins and antioxidants within nanoscale carriers, nanoencapsulation protects them from degradation caused by light, heat, or pH fluctuations. This preservation ensures the nutritional value and functionality of these compounds during food processing, storage, and digestion. Moreover, nanocarriers can be designed to release their contents in response to specific stimuli, improving the bioavailability and targeted absorption of nutrients in the digestive tract (Singh et al., 2023).

In addition to nutrient protection, nanoencapsulation helps mask undesirable flavors, making fortified foods more palatable by hiding bitter compounds like omega-3 fatty acids or certain vitamins. It also enhances the solubility of hydrophobic compounds, improving their dispersibility in food products and boosting their bioavailability (Tolve et al., 2022). Synergistic delivery is another benefit, where multiple bioactive compounds can be co-encapsulated to achieve stability and enhanced effects. Techniques such as nanoemulsions, nanoliposomes, solid lipid nanoparticles, and polymeric nanoparticles have been widely used to encapsulate both hydrophilic and hydrophobic compounds, ensuring controlled release of flavors, nutrients, and other bioactive ingredients (Awuchi et al., 2022).

Nanosensors have revolutionized food safety and quality monitoring by offering rapid, real-time detection of pathogens, toxins, and contaminants. These highly sensitive systems significantly reduce the time required for food safety testing, allowing for faster decision-making and quicker product release, a critical advancement in supply chain management. By detecting trace amounts of harmful substances, often at levels below conventional detection methods, nanosensors provide early warnings, preventing widespread contamination. Their specificity enables precise identification of particular pathogens or contaminants, minimizing false positives and negatives, which enhances the reliability of food safety testing (Wu et al., 2023).

Moreover, nanosensors are highly portable, with compact, handheld devices that can be deployed across various points in the food supply chain, from farms to retail environments. This portability allows for on-site, real-time testing, reducing dependency on centralized laboratory analysis. Another critical advantage is the multiplexing capability, which enables the simultaneous detection of multiple contaminants or analytes in a single test. This not only increases efficiency but also reduces costs, making nanosensors a highly effective and economical solution for ensuring food safety. Their integration into modern food systems represents a significant leap toward safer, more transparent food supply chains (Xu et al., 2019).

Table 11.1 highlights various nanosensors used in food safety and quality control. Surface-enhanced Raman spectroscopy (SERS) detects chemical contaminants, while aptamer-based sensors identify pathogens. Nanoparticle colorimetric sensors visually detect toxins, electrochemical sensors monitor freshness, and quantum dot fluorescent sensors enable multiplexed pathogen detection.

These nanosensors as shown above are revolutionizing food safety testing and quality control processes, enabling more frequent and comprehensive monitoring throughout the food supply chain.

11.3 Recent Advancements in Nanobionic Food Technologies

The development of nano-based functional foods and nutraceuticals marks a significant advancement in food science, leveraging nanotechnology to enhance nutrition and bioavailability (Singh et al., 2022). Nanostructured food ingredients, such as emulsions, gels, and foams, improve texture, stability, and nutrient delivery in various products. One example is nanoscale salt crystals, which intensify saltiness perception while reducing sodium levels in processed foods. Nano-enabled nutrient fortification is another innovation, where nanocarriers deliver essential vitamins and

Table 11.1 Nanosensors used in food safety and quality control

Nanosensor	Food safety application	Quality control processes	References
Surface-enhanced Raman spectroscopy (SERS) sensors	Detection of pesticide residues, antibiotics, and other chemical contaminants in foods	Identification and quantification of food adulterants	Xu et al. (2019)
Aptamer-based nanosensors	Highly specific detection of pathogens such as *Salmonella*, *E. coli*, and *Listeria*	Monitoring of allergens in food products	Cheng et al. (2019)
Nanoparticle-based colorimetric sensors	Visual detection of toxins, heavy metals, and other contaminants through color changes	Easy-to-use tests for rapid screening of food products	Alberti et al. (2021)
Electrochemical nanosensors	Detection of freshness indicators such as biogenic amines in meat and fish products	Monitoring of antioxidant levels in fruits and vegetables	Nimal et al. (2022)
Quantum dot-based fluorescent sensors	Highly sensitive detection of pathogens and toxins in complex food matrices	Multiplexed detection of multiple analytes simultaneously	Wu et al. (2023)

minerals, enhancing their stability and absorption. Nano-sized iron particles, for instance, are used for iron fortification without altering food color or taste (Dehnad et al., 2023).

Nanotechnology plays a pivotal role in protecting and delivering probiotics, beneficial bacteria essential for gut health. Nanoencapsulation improves the survival of probiotics during processing and digestion, allowing for controlled release in the gastrointestinal tract, leading to enhanced colonization and health benefits. Similarly, nano-based delivery systems for bioactive compounds, such as polyphenols and carotenoids, ensure greater stability and absorption. Omega-3 fatty acids can now be delivered through nanoemulsions in beverages without compromising taste or appearance, making it easier to incorporate these essential nutrients into everyday diets (Li et al., 2021).

Nanotechnology has also advanced the development of edible films and coatings with enhanced barrier properties and antimicrobial activity, improving food preservation. Nanocomposite films are now used to prolong the shelf life of fresh produce, while nanostructured coatings offer controlled release of flavors or nutrients in confectionery products. These innovations not only extend product freshness but also provide new ways to enhance sensory experiences and nutrition. This shift toward functional, nano-based packaging aligns with the industry's goals of reducing food waste and improving the quality and safety of food products (Mahato et al., 2024).

Bioactive nanoparticles are revolutionizing food fortification by providing improved stability, bioavailability, and functionality (Aguilar-Pérez et al., 2021). Nano-sized minerals like calcium, magnesium, and selenium are designed to increase absorption without affecting food texture. Nanoencapsulation of fat- and water-soluble vitamins protects them from degradation during processing and storage, ensuring their effectiveness in fortified foods (Su et al., 2022). Similarly, nano-based antioxidants, such as curcumin and quercetin, are formulated for enhanced solubility and absorption, offering stronger health benefits. These bioactive nanoparticles enable the creation of fortified foods that meet specific nutritional needs, providing consumers with targeted health benefits while maintaining product quality (Muthukrishnan, 2022).

Nanobionic approaches in agriculture involve various nanoparticles to improve crop yield and nutritional value. Nanofertilizers, including N, P, K, and metal oxides, enhance nutrient delivery and pest protection. Nanosensors using silver, copper, and gold monitor soil and plant health in real-time. Nanoparticles like silver and zinc oxide boost seed germination and crop yields, while carbon nanotubes enhance plant growth and nutrient uptake. Zinc oxide nanoparticles improve biofortification in maize (Javeed et al., 2022). Additionally, metallic nanoparticles such as silver and copper oxide offer crop protection through antimicrobial properties and post-harvest preservation (Table 11.2).

11.4 Challenges and Ethical Considerations

The introduction of nanomaterials in agriculture, particularly for enhancing crop yields and food production, raises significant safety concerns. Nanoparticles, due to their small size and unique properties, may pose risks to human health if ingested through food. The long-term effects of consuming food grown using nanofertilizers or nanopesticides are still not fully understood, prompting calls for rigorous safety assessments. Regulatory frameworks globally are not yet fully equipped to handle the complexities of nanotechnology in agriculture, resulting in inconsistent regulations and a lack of standardized guidelines for the safe application of nanomaterials in food production (Upadhyay et al., 2022).

Nanomaterials, while promising for increasing agricultural efficiency, also present potential environmental risks. Once applied, these particles can accumulate in soil and water, potentially disrupting ecosystems (Singh & Gurja, 2022). The long-term impact of nanoparticles on soil health, water quality, and microbial communities is not well-documented, raising concerns about unintended ecological consequences. Moreover, the persistence and mobility of nanoparticles in the environment could lead to bioaccumulation in non-target species, thereby affecting biodiversity (Punnivakotti et al., 2024).

The success of nanotechnology in agriculture depends not only on scientific advancements but also on consumer acceptance. Public perception of nanomaterials in food production can be influenced by concerns over safety, environmental risks, and transparency in labeling. Many consumers are wary of new technologies in their food supply, particularly when the risks are not fully understood. To foster acceptance, it is essential for industry stakeholders to provide clear information about the benefits and safety of nanotechnology. Transparent communication and labeling can

Table 11.2 Nanobionic Approaches to Improve Crop Yield and Nutritional Value

Nanobionic strategies	Nanoparticles involved	Contributions to agriculture	Innovative improvements	References
Nanofertilizers and nanopesticides	Nitrogen, phosphorus, potassium, iron, magnesium, zinc, cobalt, molybdenum, and carbon nanotubes, offer enhanced release and targeted delivery efficiency. Likewise, Ag, Cu, SiO_2, ZnO, and nanoformulations provide superior broad-spectrum pest protection compared to conventional pesticides	Development of slow-release nanofertilizers for more efficient nutrient delivery to plants	Creation of nanoencapsulated pesticides for targeted and controlled release, reducing environmental impact and improving efficacy	Chippa (2017) and Javeed et al. (2022)
Nanosensors for precision agriculture	Silver, copper, and gold nanoparticles are used as bio-nanosensors and electrical-nanosensors	Design of nanosensors for real-time monitoring of soil conditions, plant health, and environmental factors	Development of nanobiosensors for early detection of plant diseases and pest infestations	Zain et al. (2023)
Nanoparticles for seed treatment	Silver, gold, copper, palladium, selenium, zinc oxide, magnesium oxide, titanium dioxide, and iron oxide have been shown to enhance seed germination and boost crop yields	Application of nanoparticles to seeds to enhance germination rates and early seedling growth	Development of nanocoatings for seeds to improve their resistance to environmental stresses	Eevera et al. (2023)
Carbon nanotubes (CNTs) for enhanced plant growth	CNTs increased the germination and growth of seeds like mustard (*Brassica juncea* L.)	Utilization of carbon nanotubes to enhance water and nutrient uptake in plants	Application of carbon nanotubes to improve photosynthetic efficiency and biomass production	Ali et al. (2020)
Nanoparticles for biofortification	Zinc oxide nanoparticles (ZnO NPs) provide a more effective source of Zn, enhancing growth and biofortification in maize grains	Use of nanoparticles to enhance the uptake and accumulation of essential nutrients in food crops	Development of nanocarriers for delivering micronutrients directly to edible plant tissues	Umar et al. (2020)
Nanomaterials (NM) for crop protection	The most widely used metallic nanoparticles in crop protection include silver, silver oxide, gold, zinc oxide (ZnO), titanium oxide, magnesium oxide, and copper oxide	Creation of nanostructured materials with antimicrobial properties for controlling plant pathogens	Development of nanoparticle-based films and coatings for post-harvest preservation of fruits and vegetables	Sharma et al. (2021)

help build consumer trust and mitigate skepticism around the use of nanomaterials in agriculture (He et al., 2019).

The ethical implications of using nanotechnology in agriculture extend beyond safety and environmental concerns. There is a need to ensure equitable access to these innovations, particularly for smallholder farmers in developing countries. Moreover, the potential monopolization of nanotechnology by large agribusinesses raises questions about the fairness of its distribution (Altabbaa et al., 2023). Ethical considerations also include the precautionary principle—whether the potential benefits outweigh the risks—and ensuring that proper safety and regulatory frameworks are in place before widespread adoption. These nanobionic approaches are contributing to the development of more resilient, nutritious, and high-yielding crops, addressing challenges in food security and sustainable agriculture. Nanobionics is also making significant contributions to agriculture, offering innovative solutions to improve crop yield, nutritional value, and sustainability. Recent advancements in this area are included in the Table 11.2 presented.

11.5 Future Prospects and Emerging Trends

One of the most promising future prospects of nanobionics in food science is the development of personalized nutrition. Nanobionic technologies have the potential to tailor nutrient delivery to individual dietary needs by creating smart foods enriched with specific micronutrients or bioactive compounds. These technologies can be designed to respond to the body's physiological conditions, such as nutrient deficiencies or metabolic changes, ensuring more efficient nutrient absorption. This approach not only improves health outcomes but also aligns with the growing demand for personalized healthcare, offering consumers more control over their dietary choices and health management (Aguilar-Pérez et al., 2021).

The convergence of artificial intelligence (AI) with nanobionics is an emerging trend that could revolutionize food science. AI can enhance the precision of nanobionic applications, such as optimizing the release of nutrients from nanofertilizers or improving the detection accuracy of nanosensors in food safety. Through AI-driven analytics, real-time monitoring of soil and crop health can be achieved, enabling smarter agricultural practices (Ataei Kachouei et al., 2023). Additionally, AI can support the development of predictive models to forecast the impact of nanomaterials on the environment and human health, ensuring that nanobionics is applied in the most sustainable and efficient way possible (Zhang et al., 2021).

Figure 11.1 highlights nanobionics' role in agriculture through AI, real-time monitoring, nutrient optimization, and nanosensor accuracy. Key advancements include smart packaging, bioactive nanoparticles, and predictive models, all contributing to sustainable global food security.

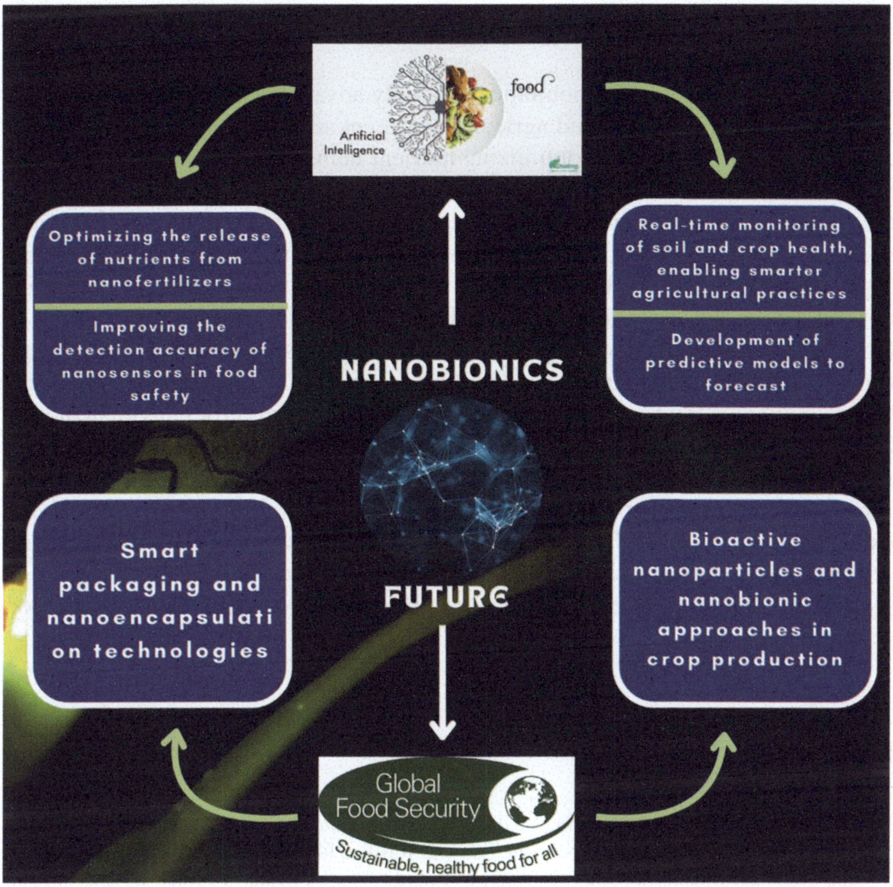

Fig. 11.1 Future of nanobionics in sustainable agriculture and integration of artificial intelligence in food science

Nanobionic technologies hold great potential for addressing global food security and sustainability challenges. By enhancing crop yields, improving nutrient biofortification, and reducing agricultural inputs such as water and pesticides, nanobionics can make food production more efficient and resilient. Nanomaterials can also extend the shelf life of food products through improved packaging and preservation techniques, reducing food waste (Kumari et al., 2023). As climate change and population growth continue to put pressure on global food systems, these emerging nanobionic solutions could play a crucial role in creating more sustainable and secure food production practices for the future.

11.6 Conclusion

In conclusion, the field of nanobionics is rapidly advancing and transforming various aspects of food science and agriculture. Recent advances, such as nanofertilizers and bio-nanosensors, are optimizing nutrient delivery and improving real-time monitoring of soil and crop health. These innovations are driving precision agriculture, enhancing food safety, and promoting sustainable practices. From smart packaging and nanoencapsulation technologies to bioactive nanoparticles and nanobionic approaches in crop production, these innovations are addressing critical challenges in food safety, nutrition, and sustainability. As nanobionics continues to evolve, future trends point toward even greater possibilities. Personalized nutrition through nanobionic technologies and the integration of artificial intelligence (AI) will further revolutionize food science, enabling customized solutions for individual dietary needs and predictive models for more efficient food production. Moreover, nanobionics holds the potential to significantly contribute to global food security by enhancing crop yields, improving food safety, and reducing environmental impact. However, challenges such as regulatory issues, safety concerns, and consumer acceptance must be addressed to fully harness its benefits. Nanobionics represents a promising frontier in the quest for sustainable, innovative, and efficient food systems.

References

Aguilar-Pérez, K. M., Ruiz-Pulido, G., Medina, D. I., Parra-Saldivar, R., & Iqbal, H. M. N. (2021). Insight of nanotechnological processing for nano-fortified functional foods and nutraceutical—Opportunities, challenges, and future scope in food for better health. *Critical Reviews in Food Science and Nutrition, 63*(20), 4618–4635. https://doi.org/10.1080/10408398.2021.2004994

Alberti, G., Zanoni, C., Magnaghi, L. R., & Biesuz, R. (2021). Gold and silver nanoparticle-based colorimetric sensors: New trends and applications. *Chem, 9*(11), 305. https://doi.org/10.3390/chemosensors9110305

Ali, M. H., Sobze, J. M., Pham, T. H., Nadeem, M., Liu, C., Galagedara, L., Cheema, M., & Thomas, R. (2020). Carbon nanotubes improved the germination and vigor of plant species from peatland ecosystem via remodeling the membrane lipidome. *Nanomaterials, 10*(9), 1852. https://doi.org/10.3390/nano10091852

Altabbaa, S., Mann, N. A., Chauhan, N., Utkarsh, K., Thakur, N., & Mahmoud, G. A. E. (2023). Era connecting nanotechnology with agricultural sustainability: Issues and challenges. *Nanotechnology For Environmental Engineering, 8*(2), 481–498. https://doi.org/10.1007/s41204-022-00289-3

Assadpour, E., & Mahdi Jafari, S. (2018). A systematic review on nanoencapsulation of food bioactive ingredients and nutraceuticals by various nanocarriers. *Critical Reviews in Food Science and Nutrition, 59*(19), 3129–3151. https://doi.org/10.1080/10408398.2018.1484687

Ataei Kachouei, M., Kaushik, A., & Ali, M. A. (2023). Internet of things-enabled food and plant sensors to empower sustainability. *Advanced Intelligent Systems, 5*(12), 2300321. https://doi.org/10.1002/aisy.202300321

Awuchi, C. G., Morya, S., Dendegh, T. A., Okpala, C. O. R., & Korzeniowska, M. (2022). Nanoencapsulation of food bioactive constituents and its associated processes: A revisit. *Bioresource Technology Reports, 19*, 101088. https://doi.org/10.1016/j.biteb.2022.101088

Carofiglio, M., Laurenti, M., & Cauda, V. (2022). The world big challenges faced by nanoscience: Examples of how integrated science can save the world. In N. Rezaei (Ed.), *Transdisciplinarity. Integrated science* (Vol. 5). Springer. https://doi.org/10.1007/978-3-030-94651-7_8

Chen, C., Feng, S., Zhou, M., Ji, C., Que, L., & Wang, W. (2019). Development of a structure-switching aptamer-based nanosensor for salicylic acid detection. *Biosensors & Bioelectronics, 140*, 111342.

Chhipa, H. (2017). Nanofertilizers and nanopesticides for agriculture. *Environmental Chemistry Letters, 15*, 15–22. https://doi.org/10.1007/s10311-016-0600-4

Dehnad, D., Ghorani, B., Emadzadeh, B., Emadzadeh, M., Assadpour, E., Rajabzadeh, G., & Jafari, S. M. (2023). Recent advances in iron encapsulation and its application in food fortification. *Critical Reviews in Food Science and Nutrition, 1–17*, 12685. https://doi.org/10.1080/10408398.2023.2256004

Eevera, T., Kumaran, S., Djanaguiraman, M., Thirumaran, T., Le, Q. H., & Pugazhendhi, A. (2023). Unleashing the potential of nanoparticles on seed treatment and enhancement for sustainable farming. *Environmental Research, 236*, 116849. https://doi.org/10.1016/j.envres.2023.116849

He, X., Deng, H., & Hwang, H. M. (2019). The current application of nanotechnology in food and agriculture. *Journal of Food and Drug Analysis, 27*(1), 1–21. https://doi.org/10.1016/j.jfda.2018.12.002

Jafarzadeh, S., & Jafari, S. M. (2020). Impact of metal nanoparticles on the mechanical, barrier, optical and thermal properties of biodegradable food packaging materials. *Critical Reviews in Food Science and Nutrition, 61*(16), 2640–2658. https://doi.org/10.1080/10408398.2020.1783200

Javeed, Z., Riaz, U., Murtaza, G., Mehdi, S. M., Idrees, M., Zaman, Q. U., & Khalid, W. (2022). Nanofertilizers and nanopesticides: Application and impact on agriculture. In *Diverse applications of nanotechnology in the biological sciences* (pp. 199–212). Apple Academic Press.

Kumari, A., Rana, V., Yadav, S. K., & Kumar, V. (2023). Nanotechnology as a powerful tool in plant sciences: Recent developments, challenges and perspectives. *Plant Nano Biology, 5*, 100046.

Li, G., Zhang, Z., Liu, H., & Hu, L. (2021). Nanoemulsion-based delivery approaches for nutraceuticals: Fabrication, application, characterization, biological fate, potential toxicity and future trends. *Food & Function, 12*(5), 1933–1953. https://doi.org/10.1039/D0FO02686G

Mahato, R. P., Singh, P., & Srivastava, S. (2024). Prospects and challenges of Nanofilms-based edible food coatings for enhancement of their shelf life. The nanotechnology driven. *Agriculture*, 204–224.

Muthukrishnan, L. (2022). Nanonutraceuticals—Challenges and novel nano-based carriers for effective delivery and enhanced bioavailability. *Food and Bioproducts Processing, 15*(10), 2155–2184. https://doi.org/10.1007/s11947-022-02807-2

Nimal, R., Selcuk, O., Kurbanoglu, S., Shah, A., Siddiq, M., & Uslu, B. (2022). Trends in electrochemical nanosensors for the analysis of antioxidants. *TrAC Trends in Analytical Chemistry, 153*, 116626.

Parveen, A., Bhatnagar, P., Bisht, B., Kumar, S., Joshi, S., Gautam, P., Kumar, V., & Anna, K. (2023). Algal nanobionics to enhance value added products—A review. *Current Research in Biotechnology, 5*, 100124. https://doi.org/10.1016/j.crbiot.2023.100124

Punniyakotti, P., Vinayagam, S., Rajamohan, R., Priya, S. D., Moovendhan, M., & Sundaram, T. (2024). Environmental fate and Ecotoxicological behaviour of pesticides and insecticides in non-target environments: Nanotechnology-based mitigation strategies. *Journal of Environmental Chemical Engineering, e113349*, 113349.

Roberta, A. M. (2020). Oxygen scavenging films and coating of biopolymers for food application. In *Biopolymer membranes and films* (pp. 535–551). Elsevier. https://doi.org/10.1016/B978-0-12-818134-8.00022-5

Sharma, P., Kumar, S., Patel, A., Datta, B., & DeLong, R. K. (2021). Nanomaterials for agricultural and ecological defense applications: Active agents and sensors. *Wiley Interdisciplinary Reviews. Nanomedicine and Nanobiotechnology, 13*(5), e1713.

Singh, D., & Gurjar, B. R. (2022). Nanotechnology for agricultural applications: Facts, issues, knowledge gaps, and challenges in environmental risk assessment. *Journal of Environmental Management, 322*, e116033.

Singh, A. R., Desu, P. K., Nakkala, R. K., et al. (2022). Nanotechnology-based approaches applied to nutraceuticals. *Drug Delivery and Translational Research, 12*, 485–499. https://doi.org/10.1007/s13346-021-00960-3

Singh, A. K., Pal, P., Pandey, B., Goksen, G., Sahoo, U. K., Lorenzo, J. M., & Sarangi, P. K. (2023). Development of "smart foods" for health by nanoencapsulation: Novel technologies and challenges. *Food Chem X, 100910*. https://doi.org/10.1016/j.fochx.2023.100910

Su, Q., Zhao, X., Zhang, X., Wang, Y., Zeng, Z., Cui, H., & Wang, C. (2022). Nano functional food: Opportunities, development, and future perspectives. *International Journal of Molecular Sciences, 24*(1), 234. https://doi.org/10.3390/ijms24010234

Sun, R., Lu, J., & Nolden, A. (2021). Nanostructured foods for improved sensory attributes. *Trends in Food Science and Technology, 108*, 281–286. https://doi.org/10.1016/j.tifs.2021.01.011

Tolve, R., Bianchi, F., Lomuscio, E., Sportiello, L., & Simonato, B. (2022). Current advantages in the application of microencapsulation in functional bread development. *Food, 12*(1), 96. https://doi.org/10.3390/foods12010096

Umar, W., Hameed, M. K., Aziz, T., Maqsood, M. A., Bilal, H. M., & Rasheed, N. (2020). Synthesis, characterization and application of ZnO nanoparticles for improved growth and Zn biofortification in maize. *Archives of Agronomy and Soil Science, 67*(9), 1164–1176. https://doi.org/10.1080/03650340.2020.1782893

Upadhyay, S. K., Kumar, S., Rout, C., Vashistha, G., & Aggarwal, D. (2022). Risks and concerns of use of nanoparticles in agriculture. In *The role of nanoparticles in plant nutrition under soil pollution: Nanoscience in nutrient use efficiency* (pp. 371–394). Springer. https://doi.org/10.1007/978-3-030-97389-6_16

Vithanage, M., Zhang, X., Gunarathne, V., Zhu, Y., Herath, L., Peiris, K., Solaiman, Z. M., Bolan, N., & Siddique, K. H. (2023). Plant nanobionics: Fortifying food security via engineered plant productivity. *Environmental Research, 229*, 115934. https://doi.org/10.1016/j.envres.2023.115934

Wu, H., Li, J. H., Yang, W. C., Wen, T., He, J., Gao, Y. Y., Hao, G. F., & Yang, W. C. (2023). Nonmetal-doped quantum dot-based fluorescence sensing facilitates the monitoring of environmental contaminants. *Trends in Environmental Analytical Chemistry, 40*, e00218.

Xu, K., Zhou, R., Takei, K., & Hong, M. (2019). Toward flexible surface-enhanced Raman scattering (SERS) sensors for point-of-care diagnostics. *Advancement of Science, 6*(16), e1900925.

Zain, M., Ma, H., Chaudhary, S., Nuruzaman, M., Azeem, I., Mehmood, F., Rahman, S. U., Aiwang, D., & Sun, C. (2023). Nanotechnology in precision agriculture: Advancing towards sustainable crop production. *Plant Physiology and Biochemistry, 108244*.

Zhang, P., Guo, Z., Ullah, S., Melagraki, G., Afantitis, A., & Lynch, I. (2021). Nanotechnology and artificial intelligence to enable sustainable and precision agriculture. *Nature Plants, 7*(7), 864–876. https://doi.org/10.1038/s41477-021-00946-6

Chapter 12
Biosafety and Regulatory Aspects of Nanobionics in Agriculture and Allied Field

Deepansha, Shreya Guglani, Vasudha Bansal, and Dharmendra Kumar Yadav

Abstract The burgeoning science of nanobionics, which integrates biology and nanotechnology, has immense potential in transforming agricultural techniques. Through the incorporation of nanomaterials into ecosystems, nanobionics facilitates the creation of novel approaches to augment agricultural fecundity, accelerate absorption of nutrients, and resist maladies and infestations. Nevertheless, worries about biosafety and regulatory control are exacerbated by the expanding use of nanobionics in agriculture. Nanoparticles may have distinct thermodynamic aspects that, if swallowed or breathed in or come in contact with the skin, can have a detrimental effect on one's wellness. Respiratory issues, neurotoxic consequences, and possible instances of organ buildup from nanoparticles are among the worries at this point. During the production, use and disposal of nanobionic items, nanoparticles are released directly into the environment, potentially causing unforseen environmental consequences. If nanoparticles thrive in biodiversity or build up in food chains, they may have an effect on the quality of food and sustainability of ecosystems. To reduce any possible risks related to nanobionics, risk mitigation approaches are to be put into practice. This could entail creating compositions of nanoparticles that are harmless for the surrounding environment, synthesizing nanoparticles that disintegrate with time, and setting control procedures in place both during implementation and manufacturing. Specific standards for regulation must be drafted by authorities to guarantee the secure advancement and application of nanobionic technology in agriculture. Risk evaluation, labeling specifications, and lifecycle monitoring of nanomaterials should all be covered by the authorities. Comprehensive research on environmental impacts should be carried out prior to broad adoption in

Deepansha · S. Guglani · V. Bansal (✉)
Department of Foods and Nutrition, Government Home Science College, Sector 10D, Chandigarh (Affiliated to Panjab University, Chandigarh), Chandigarh, India

D. K. Yadav (✉)
Department of Biologics, College of Pharmacy, Gachon University, Yeonsu-gu, Incheon City, South Korea

V. D. Rajput et al. (eds.), *Contemporary Suitability of Nanobionics in Agriculture*, Nanotechnology in the Life Sciences, https://doi.org/10.1007/978-3-031-82779-2_12

order to comprehend the possible effects of nanobionic applications on biological diversity.

Keywords Nanobionics · Biosafety · Agricultural sector · Health sector · Risk monitoring

12.1 Introduction to Agri-nanotechnology

12.1.1 Nanotechnology

Nanotechnology, the upcoming science of manipulating matter on an atomic and molecular scale, has revolutionized various sectors, including agriculture (Mahesha et al., 2023). The concept of nanotechnology was first introduced by physicist Richard Feynman in 1959, but it wasn't until the late twentieth century that the field saw rapid development (Edwards, 2008). In 1974, Norio Taniguchi of Tokyo Science University introduced the term "nanotechnology" to describe the manipulation of matter at the atomic and molecular levels. The word "nano" in "nanotechnology" is derived from the Greek word "nanos" meaning "dwarf," a prefix used to describe "one billionth" of something. A nanometer is one billionth of a meter (10^{-9} m). Nanotechnology deals with dimensions and tolerances of less than 100 nanometres (Durlo, 2023). Nanotechnology leverages unique properties of materials at this scale, such as increased surface area and reactivity, to develop novel applications across multiple fields (Sohail et al., 2019).

12.1.2 Need of Nanotechnology in Agriculture

Agriculture serves as the cornerstone of a developing economy, supplying essential food for an improved quality of life. Agriculture in common parlance is referred to as 'farming' or 'cultivation,' which is the science or practice of farming, including cultivation of the soil for the growing of crops and the rearing of animals to provide food, wool, fiber, and other products. People around the world grow various food crops, medicinal plants, and industrial plants to fulfil their needs. The agricultural industry is dealing with several challenges stemming from intensive farming, including climate change, soil degradation, diminished water quality, and a decline in biodiversity (Mittal et al., 2020). The growing global population of approximately 8.09 billion people raises significant concerns about ensuring food security worldwide (Prokisch et al., 2024). According to the Food and Agriculture Organization (FAO, 2017), the global population is expected to reach 10 billion by 2050, leading to a 50% rise in food needs, particularly in developing nations. Additionally, there are roughly 815 million individuals experiencing undernourishment at present, and the projection indicates that an added two billion people will fall into this category by 2050 (Usman et al., 2020). In India, a rapidly developing economy with an escalating population, agriculture plays a vital role in the country's economy. It accounts for approximately 14.2% of the Indian GDP and contributes to the employment and food security of the nation. Historically, agriculture has advanced through various technological innovations such as hybrid crops, synthetic chemicals, and biotechnology. Currently, researchers are exploring nanotechnology as a new avenue for agricultural enhancements (Parisi et al., 2015). These nanoscale materials can be utilized to develop more efficient fertilizers, pesticides, and sensors, contributing to sustainable farming practices (Servin & White, 2016).

12.1.3 Defining Nanotechnology in Agriculture

The emergence of nanotechnology in agriculture, commonly called agri-nanotechnology, presents an exciting frontier that promises to address some of the most pressing challenges in food production, food security, environmental sustainability, and socioeconomic development (Yadav et al., 2023). The intersection of nanotechnology with agriculture emerged in the early 2000s as researchers began exploring the potential of nanomaterials to improve crop yield, enhance pest resistance, and reduce environmental impact (Gogos et al., 2012). Reports published by Roco, the US Department of Agriculture, the Nanoforum, and Kuzma and VerHage, along with similar publications, have helped gain insights into the use of nanotechnology in agriculture (Mukhopadhyay, 2014). Nanotechnology in agriculture represents a paradigm shift, offering innovative tools to enhance productivity, ensure food security and mitigate environmental challenges while aligning with sustainable development goals (Prokisch et al., 2024). Agri-nanotechnology integrates multiple fields of study as shown in Fig. 12.1.

12.2 Nanobionics: A Sustainable Approach in Agriculture

Nanobionics is an interdisciplinary field that combines nanotechnology with biological systems to create enhanced functionalities in plants, thereby improving agricultural sustainability (Giraldo et al., 2014). By integrating nanomaterials into plant systems, nanobionics aims to optimize plant functions, such as photosynthesis, nutrient uptake, and stress resistance, leading to increased crop yields and reduced environmental impact (Wong et al., 2017). This approach is particularly promising in addressing the challenges of modern agriculture, where the need for sustainable practices is more urgent than ever (Parisi, 2020).

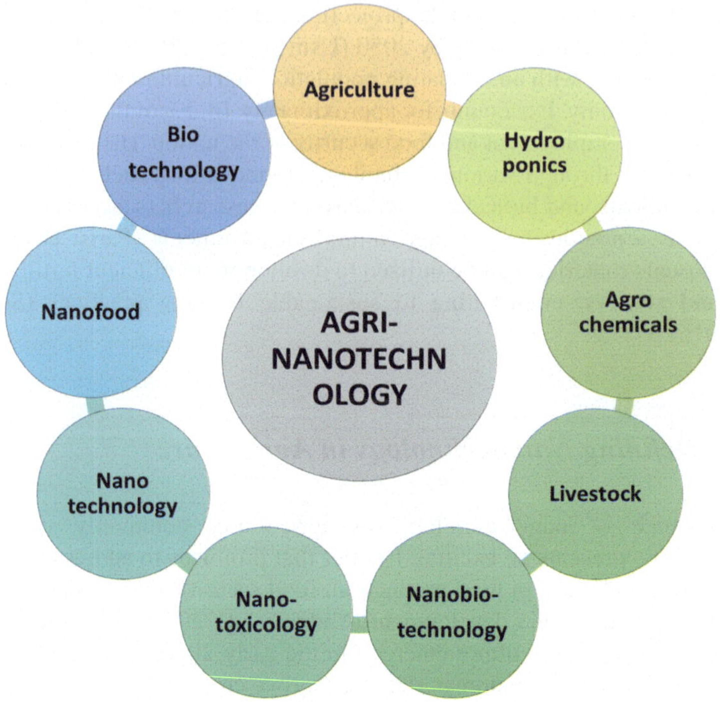

Fig. 12.1 Multiple disciplines in agro-nanotechnology (Sekhon, 2014)

12.2.1 Applications of Nanotechnology in Agriculture: A Sustainable Approach

Agri-nanotechnology encompasses a broad spectrum of applications, from nanoscale fertilizers and pesticides to sensors that monitor crop health and soil conditions as shown in Fig. 12.2.

12.2.1.1 Nanofertilizers

The presence of essential nutrients is crucial for preserving soil fertility and enhancing plant nourishment to ensure food security and safety (Ur Rahim et al., 2021). Over the past fifty years, there has been a significant rise in crop yields, contributing to fulfilling the world's nutritional needs. The rise in chemical fertilizer usage has significantly contributed to higher crop yields. Fertilizer-responsive crop varieties have further increased the reliance on these fertilizers (Usman et al., 2020). However, traditional fertilizers often suffer from low efficiency and high environmental impact due to leaching and runoff (Ur Rahim et al., 2021). Nanofertilizers, however,

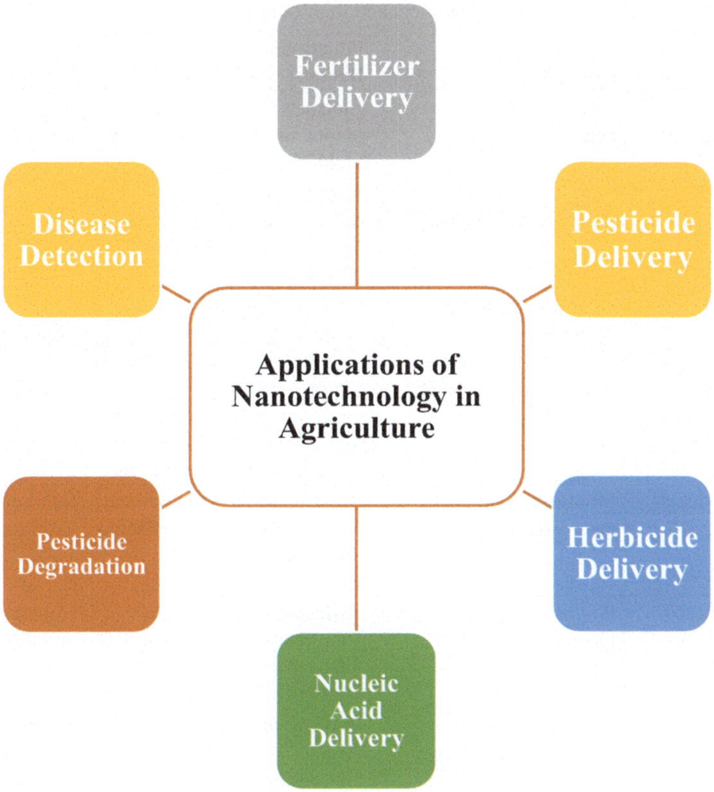

Fig. 12.2 Applications of nanotechnology in agriculture (Prasad et al., 2012)

offer controlled release mechanisms and targeted delivery, minimizing waste and reducing adverse environmental effects, thus contributing to agricultural sustainability (Rai & Ingle, 2012). Nanofertilizers are nanomaterials that serve as nutrients (both micro- and macro-nutrients) or function as carriers/additives by combining with minerals to deliver nutrients effectively (Usman et al., 2020). For instance, encapsulating active ingredients in nanoparticles can enhance their stability and bioavailability, resulting in more effective pest control and nutrient uptake by plants (Khot et al., 2012). Nanomaterials can greatly decrease the quantity of fertilizer needed for both soil and foliar applications, improving their efficiency and reducing environmental release compared to traditional formulations. Nanostructured materials like clay minerals, hydroxyapatite, chitosan, polyacrylic acid, zeolite, etc., are employed in creating fertilizers for use on soil or as a foliar application (Usman et al., 2020). Table 12.1 shows the different types of nanofertilizer along with their functions and benefits.

Table 12.1 Different types of nanofertilizers and their benefits

Type of nano fertilizer	Example	Functions/ Benefits	References
Nanoparticle-based fertilizers	Zinc oxide (ZnO) nanoparticles	Enhance zinc uptake in plants, improve growth and yield	Prasad et al. (2012)
Nano-encapsulated fertilizers	Urea nanoparticles encapsulated in chitosan	Slow and controlled release of nitrogen, reducing leaching and increasing efficiency	Liu and Lal (2015)
Nano-composite fertilizers	Hydroxyapatite-based nanocomposites	Provide phosphorus more effectively, improving root development and crop yield	Subramanian and Tarafdar (2011)
Nanoscale micronutrient fertilizers	Nano iron oxide (Fe2O3)	Supply essential micronutrients like iron, improving chlorophyll content and photosynthesis	Liu et al. (2018)
Nanoclay-based fertilizers	Montmorillonite clay nanoparticles	Improve water retention and nutrient absorption, reducing the need for frequent watering	Subramanian and Tarafdar (2011)
Polymer-based nanofertilizers	Polymer-coated nano nitrogen fertilizers	Reduce nitrogen volatilization, enhancing nitrogen use efficiency and crop growth	Chinnamuthu and Boopathi (2009)
Nanosilica fertilizers	Nanosilica with embedded nutrients	Improve plant stress tolerance and enhance nutrient uptake	Siddiqui and Al-Whaibi (2014)
Carbon nanotube-based fertilizers	Multi-walled carbon nanotubes (MWCNTs) loaded with nutrients	Enhance nutrient delivery and uptake, promoting faster growth and higher yields	Lahiani et al. (2013)

12.2.1.2 Nanopesticides

Pesticides are a diverse group of chemical compounds that include insecticides, fungicides, acaricides, and virucides (Ur Rahim et al., 2021). The use of pesticides is a common practice in commercial agriculture, and the development of new, efficient, and target-specific pesticides is an ongoing process. Synthetic pesticides are heavily deployed to combat insect pests, yet a mere 0.1% of the active ingredients successfully reach the targeted pests (Usman et al., 2020). The efficacy of some synthetic pesticides is constrained by their instability and retention in soil organic matter, which prevents them from accumulating at the root level where many pests are located, thus diminishing their effectiveness. Additionally, excessive use of pesticides can result in leaching from farm soils, leading to environmental pollution. This inefficiency underscores the pressing need for more effective pest management solutions (Ur Rahim et al., 2021). Other alternatives, such as biopesticides, offer a cost-effective and eco-friendly alternative to synthetic pesticides, but their action is comparatively slower. Nanopesticides demonstrate significant potential to address these challenges. Their slow degradation and controlled release of active ingredients, facilitated by suitable nanomaterials, can provide effective pest control over an extended period (Abbas et al., 2022). Nanopesticides are nanoscale slow-delivery systems that aim to replace conventional pesticides by using encapsulated organic or inorganic active ingredients. These systems are utilized in both conventional and organic agriculture. Nanopesticides offer improved efficacy, stability, and duration of effectiveness, while also reducing the significant pressure on biodiversity and related environmental and human health (Ur Rahim et al., 2021). Different types of nano-pesticides along with their benefits are shown in Table 12.2.

12.2.1.3 Nanosensors

Precision agriculture relies heavily on accurate, real-time monitoring of soil and crop conditions (Yadav et al., 2023). Nanosensors are capable of detecting changes at the molecular level, providing farmers with critical data on soil moisture, nutrient levels, and the presence of pathogens. Nanosensors can be seamlessly integrated with wireless communication and power systems, enhancing their capability for remote and autonomous operation (Das et al., 2024). This information can be used to optimize irrigation, fertilization, and pesticide application, thereby increasing efficiency and reducing resource use (Khot et al., 2012).

Table 12.2 Different types of nanopesticides and their benefits (Ur Rahim et al., 2021)

Nanopesticide	Active Ingredient (AI)	Function/Benefits
Chitosan capped-Cu nano-biocomposites	Cu	Antifungal and antibacterial activity, non-toxic, biodegradable, safe for ecosystem and human health
Kocide®3000	$Cu(OH)_2$, Cu	Effective against soil microbial communities, potential to change soil's ecological role
Biochar-based hydrogel microspheres	Gentian violet (GV)	Controlled release, biosafety, minimal harmful effects on cell proliferation
Water-dispersible difunctional NPs	Lambda-cyhalothrin	Accurate in situ tracking, high insecticidal activity and stability
Thiamethoxam (TMX)-loaded CNCs	TMX	Controlled release, improved insecticidal activity against pests
Chitosan-based fosetyl-Al nanocrystal	Fosetyl-Al	Synergistic antibacterial action, more stable and less harmful compared to conventional formulations
Bacterial synthesized ZnO-NPs	ZnO-NPs	Significant antibacterial activity against rice diseases

12.2.1.4 Nano-enabled Delivery Systems

Innovative approaches are needed to improve the application of agrochemicals to crops. Currently, the majority of these chemicals fail to reach the intended areas, undermining effective plant growth. Addressing issues such as leaching and degradation will be critical in achieving better results. Nanotechnology also plays a role in the development of advanced delivery systems for agrochemicals and genetic materials. Nanoparticles can be used to deliver DNA, RNA, or other molecules into plant cells, facilitating genetic modifications that can enhance resistance to diseases or improve crop yield. For example, nanoparticles made of silicon dioxide have been created to transport DNA fragments or sequences to specific species, like tobacco and corn plants, without producing any unwanted side effects (Shang et al., 2019).

12.2.1.5 Enhanced Photosynthesis

One of the most promising applications of nanobionics is the enhancement of photosynthesis, the process by which plants convert sunlight into chemical energy (Giraldo et al., 2014). Photosynthesis is inherently inefficient, with plants typically converting only a small fraction of the sunlight they receive into usable energy. By incorporating nanoparticles, such as carbon nanotubes or quantum dots, into chloroplasts—the organelles within plant cells responsible for photosynthesis—researchers have been able to increase the efficiency of light capture and electron transfer, resulting in higher rates of photosynthesis and greater biomass production (Kwak et al., 2019).

This enhancement of photosynthesis has the potential to significantly increase crop yields, particularly in regions with suboptimal growing conditions, such as low light availability or extreme temperatures (Wong et al., 2017). Moreover, by improving the energy efficiency of plants, nanobionics can contribute to the development of crops that require fewer external inputs, such as fertilizers and water, thereby reducing the environmental impact of agriculture (Giraldo et al., 2014).

12.2.1.6 Stress Resistance and Resilience

In addition to enhancing photosynthesis and nutrient uptake, nanobionics can also improve the stress resistance and resilience of crops, making them better equipped to withstand environmental challenges such as drought, salinity, and disease (Parisi, 2020). For example, by integrating nanoparticles that mimic the function of natural stress-responsive molecules, plants can be engineered to activate their defence mechanisms more effectively in response to stress (Giraldo et al., 2014).

This enhanced stress resistance is particularly valuable in the context of climate change, which is expected to increase the frequency and severity of extreme weather events, such as droughts and floods. By developing crops that are more resilient to these challenges, nanobionics can help ensure food security in the face of a changing climate (Kwak et al., 2019).

12.2.1.7 Improved Nutrient Uptake

Nanobionics can also enhance the uptake of nutrients by plants, reducing the dependence on chemical fertilizers and promoting sustainable agriculture. Nanoformulations (NFs) involve encapsulating nutrients within nano-scale carriers, such as metal oxide NPs or polymeric nano-capsules, allowing for precise nutrient release and minimizing losses (Yadav et al., 2023). Nanoparticles can be used to modify root structures or create nanostructured delivery systems for nutrients, improving their absorption and utilization (Parisi, 2020).

12.2.2 Future Directions and Challenges

The future of agri-nanotechnology lies in addressing the challenges of scale-up, cost-effectiveness, and public acceptance. As research advances, the development of more affordable and scalable nanotechnologies will be crucial to their widespread adoption in agriculture. Additionally, public perception and regulatory approval will play significant roles in determining the trajectory of agri-nanotechnology.

12.3 Associated Risks of Nanoagriculture

There are favorable and detrimental aspects to every technology. Likewise, there are drawbacks to the positive implications of nanotechnology in food security and the agriculture sector. However, because it is a relatively advancing technology, numerous studies are being conducted to collect a vast amount of data. To reach wise conclusions, many more are using comparative analysis to synthesize the results of the previous investigations.

12.3.1 Nanomaterials in Food and Agriculture

The accumulation of produced nanomaterials and their potential introduction into the food chain are concerns arising from the swift advancement of nanotechnology. Nanomaterials are not intrinsically hazardous; many commonly consumed foods, such as milk-based proteins, dietary carbohydrates, mayonnaise fat globules DNA, and so on, contain nanoscale compounds. However, the usage of specifically engineered nanomaterials in food, water, and crop cultivation may pose risks to human

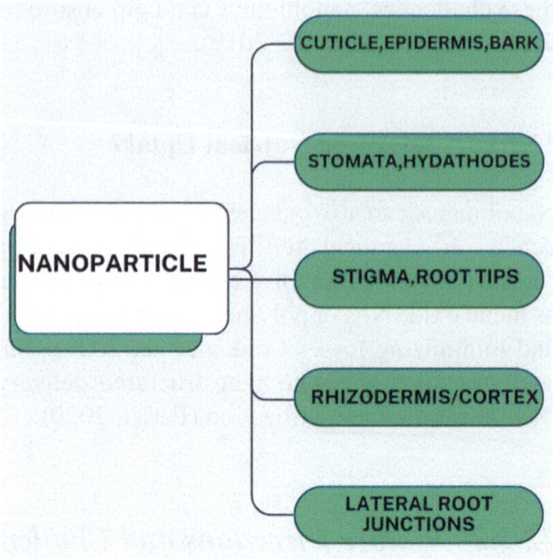

Fig. 12.3 Nanoparticle uptake and translocation in plants (Tarafdar, 2015)

well-being and ecosystems, or both. Plants can be exposed to nanomaterials through their uptake from the soil (Fig. 12.3). Because of their ability to absorb and bioaccumulate engineered nanoparticles (ENPs), plants are a fundamental element of all biological systems and are crucial to the destiny and movement of these microscopic particles in the surrounding atmosphere.

12.3.2 Plant and Soil Interaction with Nanomaterial

ENPs have the potential to cling to plant root systems and poison plants chemically or physically. ENPs must break through the outer walls of cells and plasma membranes of layers of epidermal tissue in roots in order to access the tissues that supply water (xylem), where they will be absorbed and transported through the stems to the leaves (Agrawal & Rathore, 2014). Their dimension and concentration are the most significant aspects of the NPs (Mwaanga, 2018). Because of their phenomenally tiny dimensions which make them incredibly valuable, nanoparticles can also have a number of detrimental consequences and, when used haphazardly, pose serious risks to the atmosphere, people, livestock, and flora.

12.3.3 Environmental Impact of Nanoparticle

When applied as insecticides, fertilizers, or in different combinations, nanoparticles may get atmosphere and settle on plant parts such as leaves. They might block pollen tube permeability by clogging stomata and forming a thin, poisonous, chemically resistant covering on the stigma. NPs have the potential to impede the development and function of advantageous soil microorganisms, including mycorrhizal fungi and bacteria that fix nitrogen. This may interfere with symbiotic connections that are vital to the health and productivity of plants. In addition to being harmful to plants, there are a host of other adverse environmental effects. Both consumers and a variety of manufacturing companies release nanoparticles and nanomaterials into the environment. When released into the atmosphere, these NPs bond to clouds and other airborne particulates to produce thermal or cooling effects that in turn alter the global climate. These NPs may also have an impact on the biotic factors of the environment, including soil, water, and air. Subterranean water systems may become unusable due to interference by nanoparticles dispersed into the earth. Additionally, they may be passed up the food chain and onto other trophic levels when other species eat plants that have been exposed to nanoparticles. Additionally, the transmission of TiO2 NPs from daphnia to zebrafish was noted (Deng et al., 2014).

Silver nanoparticles had a negative impact on soil microorganisms participating in the carbon, nitrogen, and phosphorus cycles as well as changing the pH of the soil, according to Zhang et al. (2020). Higher amounts of silver buildup were discovered in the roots and shoots of *Phaseolus radiatus* plants grown on soil treated with silver nanoparticles (Lee et al., 2012). Different metallic metal nanoparticles, such as zinc and titanium, have been shown to negatively impact the soil microbiota depending on exposure time and concentration (Khan et al., 2021). The study conducted by Rahman et al. revealed that the pea root microbiota was negatively impacted by platinum nanoparticles stabilized with poly(vinylpyrrolidone), resulting in a reduction of rhizobia and mycorrhizal fungus. Furthermore, fish exposed to TiO2NPs had altered ovarian development and function, as well as a delay in folliculogenesis (ovarian follicle maturation) according to histological investigation and gene expression research. Thus, it has once again been established that these nanoparticles have the potential to be hazardous to this species' reproductive organs (Bai & Tang, 2020). NBPs, or nano-based insecticides, have not been proven to be good for bees according to numerous investigations. As far as pollen and pollen particle pollination goes, bees are among the best workers. The active components in synthetic nanoparticles may induce greater toxicity in beehives as a result of NBP's interactions with pollinators like bees during pollen collection. Pollination by bees may also be negatively impacted if NBPs keep building up in the atmosphere reservoir (Hooven et al., 2019).

Risks associated with employing carbon-based nanosensors in soil were investigated in a 2023 study by Trejo et al. 2023. Even while these sensors provide precise information on the amount of moisture and nutrients in the soil, the study discovered that their prolonged use in the soil altered the microbial communities. The sensors specifically released carbon nanotubes (CNTs) into the soil, which changed the microbial activity by either promoting or preventing the growth of specific microorganisms. Long-term effects of this disturbance on soil health and nitrogen cycling are anticipated. An evaluation of copper nanoparticles' effects on grapevines was conducted recently. Although the nanoparticles were successful in reducing fungal infections, they also caused a build-up of copper in the soil that was harmful to earthworms and other important soil creatures. According to the study, this toxicity may eventually cause the health of the soil to deteriorate, which would lower the vineyard's long-term sustainability. TiO2 nanoparticles were discovered to have detrimental effects on plant root systems in addition to improving photosynthesis and agricultural yields in a study looking at their use in greenhouse settings. The accumulation of nanoparticles in the root tissues resulted in a decrease in both root development and the plants' overall ability to absorb nutrients. Concerns were also raised over the possibility that infected crops could introduce TiO2 nanoparticles into the human food chain (Paz-Trejo et al., 2023). Studies carried out on tomato plants exposed to silver nanoparticles revealed that although AgNPs were successful in lowering bacterial infections, they also produced unexpected side effects. It was discovered that the beneficial soil bacteria, which are essential for nitrogen fixing, were inhibited by the nanoparticles. Because of this restriction, there was less nitrogen available in the soil, which is necessary for plant growth. The study also discovered that silver nanoparticles could contaminate water by seeping into groundwater. Plants naturally have nano- or micrometer-scale holes through which nanoparticles are adsorbed to their surfaces and absorbed. There are a number of known or anticipated routes for plant uptake and association of nanoparticles as shown in Fig. 12.3.

12.3.4 Human Health Risks of Nanoparticles

Aerosolized nanoparticles pose an especially significant threat to the wellness of people as they might penetrate into the human body via the mechanism of respiration. The introduction of nanoparticles into the circulatory system and the bronchial system may result in genome damage, formation of protein fibrils (triggered by certain kinds of nanotubes made up of carbon), pro-inflammatory implications based on their propensity to induce inflammation, and biodistribution-related side effects (Tarafdal, 2015). These tiny bits have the potential to cause venous blockage and to spread to other body parts, including the pulmonary and renal systems (Tarafdal, 2015). NPs penetrate into the cell and interrupt electron transport system (ETS) cycle of chloroplast and mitochondria and triggered oxidative burst due to increase of reactive oxygen species (ROS) concentration. Excessive ROS prolonged exposure causes disruption of membranes, lipid peroxidation, oxygenation of proteins, DNA degradation, and ultimately apoptosis (Goswami et al., 2019). NPs have a high surface area to volume ratio that makes it easy for them to enter the nucleus and increases the risk of altering DNA. Furthermore, the attraction and cellular connections of nanomaterials are strengthened by their similarity in size and structure to cell organelles. Additionally, little is known about NP-cellular interactions, and the field of study is still in its earliest phases. A few papers address the detrimental effects of NPs on gut microbes. Cerebral inflammation results from the entry of

Fe_2O_3 (ferric oxide) nanoparticles into the nervous system. Signal pathway disruption and impaired neurotransmitter production are caused by exposure to TiO_2 NPs. The intracellular absorption of variously sized and shaped colloidal gold nanoparticles has been studied by Chithrani et al. Their findings show that the physical dimensions of the nanoparticles have a significant impact on saturation concentrations and kinetics; for example, the uptake half-lives of gold nanoparticles measuring 14, 50, and 74 nm were found to be 2.10, 1.90, and 2.24 h, respectively. Using digestion and then inductively coupled plasma atomic emission spectroscopy experiments, they determined the absolute amounts of gold in the cells. Researchers discovered that cells absorbed 50 nm spheres more rapidly than they did smaller and larger spheres in the 10–100 nm range and that spheres were absorbed more effectively than nanorods with diameters in the 10–100 nm range. According to evidence from transmission electron microscopy (TEM), gold nanoparticles can readily pass through HaCaT cells and gather in the nucleus of the cell. A blue-gray staining of the skin and other organs known as argyria can be brought on by prolonged exposure to high concentrations of silver. It is additionally identified that skin and other body parts can accumulate silver from lower-level exposure as shown in Fig. 12.4. When skin contact with silver occurs, some people may experience moderate allergic reactions, including rashes, swelling, and inflammation. High amounts of silver in the air can cause breathing difficulties, throat and lung irritation, and stomach discomfort (White et al., 2003). Increased oxidative stress and mitochondrial dysfunction can result from exposure to Ag NPs in humans (Khan et al., 2021). A number of carbon-based nanoparticles have also been shown to be cytotoxic to human alveolar epithelial cells, human embryonic kidney cell, and hepatocytes and

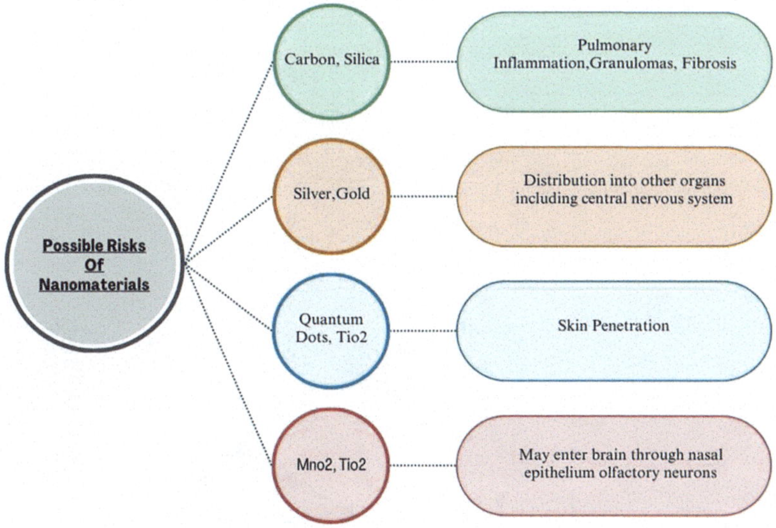

Fig. 12.4 Risks of nanomaterials (Ray et al., 2009)

intestinal cells (Mwaanga, 2018). Even though the use of nanopesticides in agriculture seems to have a promising future, human exposure to hazardous agrochemicals that can penetrate biological barriers—such as the blood-brain, blood-placental, and blood-retinal barriers—raises serious concerns because these chemicals have the potential to cause irreversible harm to important organs (Chaud et al., 2021). The reckless and irresponsible use of pesticides throws everyone's health at danger while also altering the ecosystem's delicate equilibrium. Pesticide residues can be inhaled accidentally or through work, and the short- and long-term adverse effects can result in death or disability-adjusted life years (DALY) from food and water use. Children are particularly susceptible to the long-term harm that pesticides can cause to their tissues and organs. Both the central and peripheral neurotoxicity, as well as the impact on the reduction of blood coagulation capacity, are significant causes for concern (Kuhlbusch et al., 2018). Figure 12.5 outlines key limitations of nanotechnology. Cytotoxicity highlights the toxic effects of nanoparticles on cells, while ROS accumulation points to oxidative stress that can harm biological systems. Gene toxicity raises concerns about potential genetic damage caused by nanomaterials. Disruption of the soil ecosystem emphasizes environmental risks, particularly changes in microbial communities and soil health, and regulatory issues reflect the challenges in creating comprehensive safety and usage guidelines.

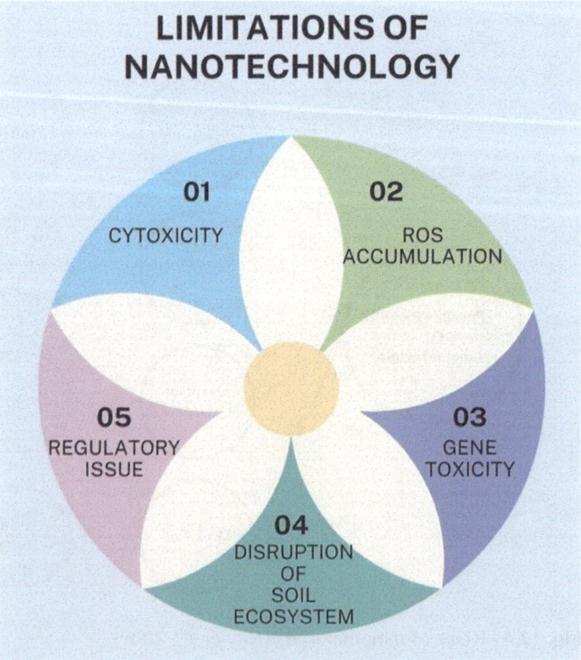

Fig. 12.5 Limitations of nanotechnology (Behl et al., 2022)

12.4 Safety Limit of Agri-nanomaterials

There aren't many constraints on nanomaterial exposure at work. When compared to bigger particles of the same material, some nanoparticles could be more dangerous. Some breathed-into nanoparticles, such as carbon nanotubes and nanofibers, may be able to induce airway inflammation and scarring. These particles may also cause swelling and injury to lung cells and tissues after being lodged in the respiratory tract. There are, however, already certain predetermined exposure limitations. Titanium dioxide, more commonly known as TiO2, is a versatile material with numerous commercial uses, such as paint, paper, cosmetics, and food. It may be manufactured and utilized in many different types of size categories, including nanoscale particles (less than 100 nm). Table 12.3 summarizes the toxicity characteristics of various metals and metal oxide nanoparticles commonly used in agrochemical applications (Mittal et al., 2020). According to NIOSH, workplace exposure by breathing TiO2 particles needs to be regarded as a possible occupational carcinogen since these particles have a higher mass-based potency than bigger particles. National Institute for Occupational Safety and Health (NIOSH) advised Recommended Exposure Limit (REL), OSHA proposes that worker exposure to respirable carbon nanotubes and carbon nanofibers not go over 1.0 micrograms per cubic meter ($\mu g/m^3$) as an 8-h time-weighted average. OSHA warns against exposing workers to TiO2 nanoparticles at a concentration higher than the 0.3 mg/m^3 REL set by NIOSH. In comparison, NIOSH's REL is 2.4 mg/m^3 for fine-sized TiO2 (particle size more than 100 nm) (Oberbek et al., 2019). According to the World Health Organization (WHO), the occupational exposure limit values for various nanomaterials are specified based on their mass concentrations in micrograms per cubic meter ($\mu g/m^3$) as shown in Table 12.4. For multi-walled carbon nanotubes (MWCNT) with a diameter of 10 nanometers, the occupational limit is set at 1 $\mu g/m^3$. For MWCNT with a diameter of 140 nanometers, the limit is 2 $\mu g/m^3$. Carbon nanotubes, in general, have an occupational limit of 2.5 $\mu g/m^3$. Additionally, fine particulate matter with a size of less than 2500 nanometers has a higher occupational limit, set at 30 $\mu g/m^3$. These values serve as guidelines to protect workers from potential health risks associated with exposure to nanomaterials. The accuracy of assessing the safety of nanomaterials is compromised by the absence of definitive and verified data. For instance, it is insufficient to assess the toxicological risk of NMs for the environment and human health by using animal experiments to identify the proper dosage. In order to minimize job duplication, it has also been suggested that labs should go through a standardizing procedure and quality control. As epidemiological studies become more standardized, labs should choose a sample for a universal model to enable comparisons of the parameters being assessed. Highly qualified scientists should conduct the analysis of the results in compliance with the guidelines of Good Laboratory Practices. Since the Standard Operating Procedure is typically ignored, seriously jeopardizing NMs' safety, it appears that these protocols are not taken seriously. It is still challenging to forecast in advance the level of bioavailability, distribution in the body, breakdown,

Table 12.3 Toxicity concentrations of metal oxides (Mittal et al., 2020)

Toxicity of metal and metal oxide NP-based agrochemicals	
Nanoparticle	Concentration
Se	0, 0.5, 1, 10, 30 mg/dl
CuO	10 g/l
Ag	20, 51, 73 nm
Fe_3O_4	30, 100, 500 mg/dl
CeO_2	50, 100, 1000 mg

Table 12.4 Occupational limit values for NM (WHO, 2017)

Nanomaterial	Mass concentration (μg/ml)
MWNCT (multi-walled carbon nanotubes) 10 nm	1
MWNCT 140 nm	2
Carbon nanotubes	2.5
Fine particulate matter <2500 nm	30

excretion, and the biological process of nanostructures, despite the wealth of knowledge on nanotoxicity that has been amassed over the past ten years (Zhou et al., 2019). The skin is the most extensive organ in the body and is constantly impacted by both biological and synthetic nanomaterials. There is criticism that might cause irreversible health damage, ranging from carcinoma to dermal degradation. Organ-on-chip technologies have the ability to completely transform how safety evaluations of nanomaterials are conducted by accurately replicating the physiology of the skin (Costa et al., 2023). The next generation of skin-on-a-chip (SoC) is intended to emulate the intricate physiological and pathological characteristics of humans. Therefore, these cutting-edge microfluidic-based methods may be able to get over significant obstacles in the way of NM safety evaluation, including those related to standardization and representative models. With the potential to reduce and/or replace in vivo models, the next generation of skin-on-chip models—which can recreate the complexity of skin physiology—is made possible by advancements in skin tissue engineering and microfluidics technologies. SoC validation must ensure that the model mimics relevant features of the human skin and uses pertinent, convenient, robust, reproducible assays as well as readout methods to predict in vivo human skin response and safety for test compounds (Mastrangeli et al., 2019).

12.5 Toxicology and Regulatory Aspects

All materials have some level of toxicological effects to cells, animals, or humans, even common salt and arsenic. Figure 12.6 identifies factors influencing nanoparticle toxicity. Dosage/exposure plays a critical role, as higher amounts can increase toxic effects. Surface area determines interaction potential, with larger surface areas enhancing reactivity. Size and shape affect cellular uptake and distribution. Chemical reactivity influences the likelihood of harmful reactions, while charge distribution impacts particle stability and biological interactions. Together, these factors determine the overall toxicological profile of nanoparticles. Characterizing the predicted amounts of designed nanoparticles that could be possibly found in soil, water, and air is therefore crucial before assessing toxicological findings. Substantial amounts of artificially produced NPs have entered the environment as a result of the development of nanotechnologies. A growing amount of study has concentrated on evaluating the toxicity of the NPs typically employed in industry with the objective to

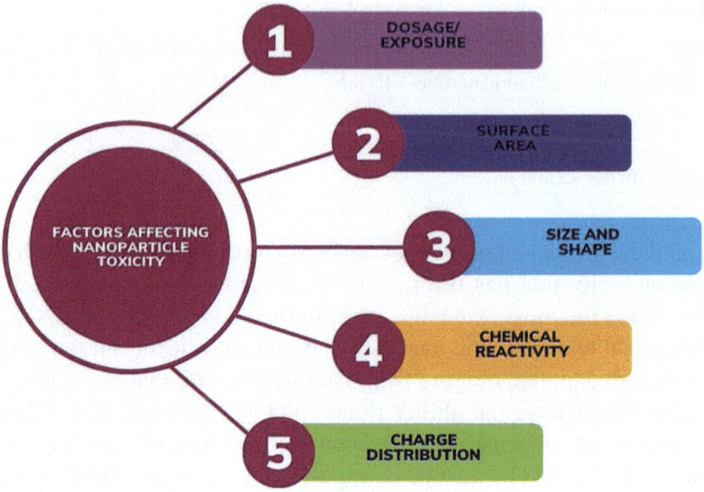

Fig. 12.6 Factors Affecting Nanoparticle Toxicity

safeguard human health and plants from the potential adverse impacts posed by an extensive spectrum of nanoparticles.

12.5.1 Fibrous and Tubular Nanostructures

According to toxicological research, exceptionally high concentrations of fibrous and tubular nanostructures are linked to fibrotic lung responses, which in turn cause inflammation and a higher risk of the development of cancer (Agrawal & Rathore, 2014).

12.5.2 Carbon Nanoparticles and Blood–Brain Barrier

Additionally investigated are the charge characteristics and the capacity of carbon nanoparticles to influence the blood-brain barrier's integrity and to show chemical alterations in the brain. It has been reported that charged nanoparticles immediately cause toxicity at the blood-brain barrier, while neutralized nanoparticles and low-concentration anionic nanoparticles can act as carrying molecules, allowing drugs immediate passage to the brain (Agarwal & Rathore, 2014).

12.5.3 Environmental Risks of Nanoparticles

The health and environmental risks posed by nanomaterial cannot be assessed easily as they have diverse properties and behavior. Because these materials' dimensions, shape, and polarization affect their kinetic (ADME) and hazardous properties, even nanomaterials with the same chemical make-up but differing sizes or forms may have remarkably varied toxicity (Agrawal & Rathore, 2014).

12.5.4 Plant Absorption

It has been found that nanoparticles (NPs) with sizes less than 5 nm can be easily passed through cell wall pores and 20 nm through plasmodesmata. This suggests that NPs with a smaller size range can be absorbed by plants more rapidly and may even be more poisonous at less concentrated levels.

12.5.5 Metal Nanoparticles, ROS

Metal nanoparticles impose negative effects by flourishing on the biological environment and formation of excess ROS, particularly by ion reduction. Because of their widespread use in everyday goods, nanoparticles (Ag NPs) are released into aquatic ecosystems, where they serve as a reservoir of dispersed Ag and have had an adverse effect on aquatic creatures, microbial organisms, and algal species. Concerns are raised by the uptake of NPs by crops over their possible transfer down the food chain, which could result in biological accumulation and biomagnification. When people and animals eat these crops, the consequences could be potentially hazardous. The absence of standardized testing methodologies is a significant obstacle in evaluating the toxicological concerns associated with nanomaterials in agriculture.

12.5.6 Regulatory Challenges and Framework

Comparing data and establishing safety criteria is challenging since different research employ different models and methodologies. A strong regulatory framework is required in order to assess the risks associated with nanomaterials in agriculture. To guarantee that nanoproducts are safe for use, this involves risk assessment guidelines, monitoring systems, and labeling requirements. Global regulatory agencies have created policies and guidelines to guarantee the safe application of nanotechnology in agricultural goods in response to these issues. The Indian Council of Agricultural Research (ICAR), Department of Biotechnology (DBT), Food Safety and Standards Authority of India (FSSAI), the Ministry of Environment, Forests, and Climate Change (MoEFCC), are the regulatory bodies and laws that regulate nanotechnology-based agriculture goods in India (Kumari et al., 2023). Emerging technologies must first undergo evaluation before society may embrace them. Fair consideration should be given to the possible advantages and related risk issues, taking into account various environmental, physical wellness, moral, and socioeconomic elements. Therefore, it is imperative to have a regulatory framework that can oversee the safety requirements related to nanotechnology across several domains. Supporting the nano-agri sector also heavily depends on the public's outlook.

12.5.7 Regulatory Bodies

Numerous regulatory bodies, including the European Union Scientific Committees and Agencies (EUSCA), the Joint FAO/WHO Expert Committee on Food Additives, the US Food and Drug Administration (USFDA), the International Standard Organization (ISO), and the Organization for Economic Cooperation and Development (OECD), are present worldwide and serve to establish standards, norms, and enforcement mechanisms for secure consumption of nanoproducts. The regulatory framework for the safe use of nanotechnology in various sectors is supported by several laws and guidelines as shown in Table 12.5. There isn't a specific body or authority in India that controls nanogoods. When technological advancement first emerged in the first few years of the 2000s, the government thought it would be challenging to regulate the newly established industry. However, because they deal with risk-related sectors, the Department of Science and Technology (DST) and the Ministry of the Environment and Forests and Climate Change (MOEFCC) oversee all issues and concerns pertaining to nanotechnology. The Central Drugs Standard Control Organization (CDSCO) and the Drug Controller General of India (DCGI) publish best practices and guidelines for the safe management of nanoparticles (Malik & Patil, 2020).

Table 12.5 The Regulatory framework on regulation and safety measures for nanotechnology-based agri-products of different countries (Kumari et al., 2023)

Laws	Regulations
Environmental Protection Act (1986)	Regulate the production, import, export and use of hazardous substances
The Food Safety and Standards Act (2006)	Issued guidelines for the use of nanotechnology in food products including nano-agri products
FSSAI Regulations on Food Additives (2011)	Specifies the conditions for the use of food additives, derived from nanotechnology in food products
ICAR Guidelines on Nanotechnology Research in Agriculture	Framework for the safe and responsible use of nanotechnology in agricultural research and development
Indian Pharmacopenia Commission (IPC) (2019)	Characterization of Nanoparticles used in the production of agri products

Plants, humans, and other vertebrate species may be toxically affected by materials utilized in the manufacturing of the nanoparticles. Lipid-based formulations, such as liposomes, liquid crystals, and micelles, are typically made of natural phospholipids and cholesterol because of their biomimetic properties with the cellular wall. These formulations are used to deliver AcI, which has poor water solubility and requires the use of surfactants, which can increase the toxicity of pesticides (Mustafa & Hussein, 2020). Because of the intricacy of nanostructures, reactivity, size, shape, and electric charge, it is challenging to evaluate, generalize, and forecast critical elements of cellular toxicity and DNA damage as shown in Fig. 12.3. This makes characterizing the biological safety of nanopesticides a challenging endeavor.

12.5.8 Toxicity Testing

Concerns over glyphosate's toxicity have been raised by producers, labor unions, and food safety due to its non-specific broad-spectrum organophosphate nature. It was discovered that both terrestrial and aquatic creatures experienced adverse effects from glyphosate and its metabolite, aminomethylphosphonic acid. In human placental JEG3-cell line, allergic responses, cardiac and respiratory issues, and endocrine disturbance have been seen. Herbicides such as clomazone herbicide, when sprayed, drift off crops by nearby places and represent a serious hazard to humans and livestock. A persistent pesticide that is heavily used in maize and sugarcane crops, atrazine has a negative effect on soil and marine environments, which is why it is a major concern. Long-term exposure to toxic agrochemicals causes bioaccumulation of fatty acid in hepatic cells, leading to fibrosis and hepatitis. The accumulation of ROS has a critical role in the hepatic and is responsible for inflammation in the liver causing hepatic tissue injuries (Wasef et al., 2021). AgNPs of 56 nm in size were examined for 90 days after exposure (doses ranging from 0 to 500 mg/kg) in the oral toxicity investigation. During the exposure period, there were no changes observed in the food or water consumption of male rats, but there were noticeable declines in their body weight. The scientists deduced that AgNPs targeted the liver in both male and female rats. Doses more than 125 mg/kg were associated with adverse effects (Kim et al., 2010). While determining the risk factors brought on by NMs, the safeguarding of human health is the primary priority of nanotoxicologists. People can easily be exposed to NMs that are integrated into chemicals made through complicated, unregulated procedures, making detection extremely difficult. As a result, caution must be used while determining the possible risk associated with the anticipated nanoparticles.

The following evaluations need to be considered:

1. To choose the NMs for study on nanotoxicology and to set up settings with exposure parameters (biotic, chemical, and physical) and exposure durations

2. To list every feature of the exposure means and provide detailed information that enables cross-comparisons
3. To choose types of designed NPs (defined by a risk assessment procedure to provide a uniform evaluation) with concentrations in scale
4. Following experimental control treatments—examining and selecting multiple end points (e.g., physicochemical properties and characterization of toxicity) for the characterization of designed NMs and quantifying the load and compartmentalization of designed NMs to allow a comparison of the effective doses in biological response; approaching in the appropriate form of rapid screening to obtain pertinent knowledge

In addition to characterization and identification, there are a number of additional risk factors associated with NMs, such as identifying common procedures that release large amounts of designed NMs and NPs aggregating with other (non-toxic) compounds during the release process; identifying NMs that pose a risk during in vitro testing; comprehending the intricacy of NMs' interactions with biological systems; recognizing that even minor differences among NMs can have a significant impact on NMs interactions with biological systems in vivo. To evaluate the biological reaction of a chemical compound, a number of standardized toxicological tests are available. Regarding the evaluation of nanoparticle toxicity, there isn't any standards. It creates a lot of challenges when comparing the findings about the toxicity of the components that were examined. Utilizing cultures of mammalian cells taken from the most diverse regions of the body (such as the brain, lungs, heart, skin, and liver), the majority of toxicity testing for NMs has been carried out in vitro. It is true that in vitro tests can yield results more quickly and at a lower cost than in vivo procedures. Aquatic species were the primary subjects of in vivo testing to simulate the effects of nanomaterials on the environment, as these waters would serve as the primary receiving compartment (Bian et al., 2016). As a result of the fluctuation in NMs concentration during animal contact, indications for comparing the toxicity of various nanomaterials and/or nanomaterials and conventional chemical compounds can be statistically calculated. The three most commonly used evaluation parameters are the LC50 (the concentration of nanomaterial that kills 50% of the population), NOEC (the maximum concentration at which no effect on the organisms is observed), and LOEC (the low concentration that noticeably affects the organisms) (Boros & Ostafe, 2020). To evaluate acute toxicity and chronic nanotoxicity, respectively, a variety of toxicity studies have been developed, including single and multiple delivery. In order to assess the in vivo oral, dermal, and LD50 nanotoxicity, the Organisation for Economic Co-operation and Development (OECD) has even developed recommendations. Small rodents can also be used to assess a wide range of exposure methods, such as gavage, ingestion, intratracheal installation, NM inhalation, intravenous, or intraperitoneal injections. Inhaling nanoparticles and administering intravenous injections are the most common methods for exposing receptors (Armstead & Li, 2016). Zebrafish are the most commonly used aquatic biological and nanotoxicity model because they share a genome and developmental process with humans and other mammals. The embryonic stage is easily observed because embryogenesis takes place outside of the womb and the embryos are transparent. Furthermore, embryogenesis proceeds quickly; 24 h after fertilization, structures like the brain, muscles, and skeleton are produced, and 5 days later, sensory organs like the eyes and ears. Additionally, the zebrafish genome has been fully mapped, enabling the creation of standardized microarrays for research on nanotoxicology. This makes it possible for scientists to conduct extensive and quick nanotoxicity investigations. Zebrafish testing also enables the examination of NMs via a variety of exposure pathways, including direct dispersion in water.

The most significant concern with nanotechnology is the potential for nanoparticles to enter the circulation system. Thus, by directly injecting nanoparticle suspensions into rats' bloodstreams, certain research has already mimicked this scenario. These experiments involve the administration of known dosages of a specific nanoparticle suspended in a particular safe liquid to healthy animals of comparable mass. Typically, blood samples are obtained for cell counting (monocytes, granulocytes, lymphocytes, platelets, red blood cells) 48 h after the particles are administered. Measurements of hemoglobin concentration are also carried out. This kind of test can also be used to evaluate how widely dispersed and/or how bioaccumulated nanoparticles are in key organs. The intricate hazards created by these technologies cannot be adequately addressed by the regulatory frameworks that now control the use of nanomaterials in agriculture. More thorough laws that take into account the entire lifecycle of nanomaterials—from manufacture to disposal—are required. This entails developing standards for safe use as well as standard testing procedures for assessing the effects of nanomaterials on the environment and human health. It will be possible to better safeguard the environment and public health from the prospective risks of nano-agriculture by enhancing regulatory control.

12.5.9 Safety by Design

Throughout the entire process of developing a new product, "Safety-by-Design" (SbD) is applied as a framework, closely adhering to the principles of Drug Discovery and Development (DDD). By incorporating information of the negative effects of NMs on the environment and human health into the process of developing the desired NMs, "Safety-by-Design" (SbD) aims to engineer undesired consequences out of NMs (Kraegeloh et al., 2018). Two European initiatives that define and apply SbD are NANoREG and NANoREG II, which were initiated in 2015. First, during the product development process, functionality and safety are evaluated together. Second, the safe design, safe production, and safe use are the three pillars upon which these regulations rest. Support for the "safe innovation" approach, which combines the idea of SbD with "normative readiness," is essential knowledge and information that the innovation chain needs to ensure standard compliance (Schmutz et al., 2020). SbD proposes "Is it inherently safe?" and "Can we design the product while ensuring its safety?" Thus, the political obstacles pertaining to the use of standardized risk assessment are acknowledged. Technical, social, cultural, political, educational, and economic factors must interact greatly to address these issues. In accordance with SbD, scientific and innovative collaborations may give rise to new issues and challenges; however, the outcomes of these collaborations may facilitate the acceptance and realization of SbD and impact public confidence in research and products. Even if the "Safety-by-Design" approach to drug development principles has aided in determining the toxicity of nanomaterials, there is always a need for a combination of scientific efforts to guarantee the uniformity of procedures and methodologies. This will guarantee the regulation of nanomaterials' quality and the safe advancement of their development. Safety concerns are thought to be approaches that should still be encouraged in order to find new toxicologically linked pathways (Zielinska et al., 2020).

12.6 Challenges and Future Perspectives

The application of nanobionics in agriculture presents numerous challenges and promising avenues for future research. One of the most significant challenges is ensuring the safety of nanobionic applications for both human health and the environment. Nanoparticles' unique properties can lead to unintended consequences such as bioaccumulation, environmental pollution, and health risks when inhaled or ingested. Additionally, the regulatory framework around nanobionics is still in its formative stages, and developing comprehensive guidelines that consider the lifecycle and environmental impact of nanomaterials remains an urgent need. Moreover, public acceptance and ethical considerations regarding nano-engineered food products and their long-term effects on ecosystems must be addressed to avoid resistance to the technology.

Future perspectives in nanobionics suggest that advancements in safety-by-design approaches could help mitigate risks associated with nanomaterial use. Further research in developing biodegradable nanoparticles, safe delivery mechanisms, and enhanced biosensors will enhance the applicability and safety of nanobionics in agriculture. Continued interdisciplinary collaborations and investment in regulatory science are also essential to create a sustainable pathway for this technology. Additionally, as nanobionics continues to evolve, addressing knowledge gaps through comprehensive studies on its environmental and biological impacts will be crucial to ensure its responsible implementation in agriculture.

12.7 Conclusion

The integration of nanobionics in agriculture holds transformative potential for enhancing crop yields, improving resource efficiency, and promoting sustainable farming practices. While nanobionics offers a path to address challenges like soil degradation, pest management, and climate change resilience, it also raises concerns about biosafety, regulatory gaps, and environmental sustainability. As research progresses, establishing robust regulations, promoting public awareness, and ensuring ethical practices will be essential to harnessing the full benefits of nanobionics responsibly. The adoption of nanobionics must include educating stakeholders, including farmers, policymakers, and consumers, about its benefits and risks. By promoting collaboration and ethical innovation, nanobionics can greatly contribute to a resilient and efficient agricultural future. With continued innovation, nanobionics is poised to redefine agriculture, supporting a future where technology and sustainability converge to address global food security and environmental challenges.

References

Abbas, M., Yan, K., Li, J., Zafar, S., Hasnain, Z., Aslam, N., et al. (2022). Agri-nanotechnology and tree nanobionics: Augmentation in crop yield, biosafety, and biomass accumulation. *Frontiers in Bioengineering and Biotechnology, 10*, 853045.

Agrawal, S., & Rathore, P. (2014). Nanotechnology pros and cons to agriculture: A review. *International Journal of Current Microbiology and Applied Sciences, 3*(3), 43–55.

Armstead, A. L., & Li, B. (2016). Nanotoxicity: Emerging concerns regarding nanomaterial safety and occupational hard metal (WC-Co) nanoparticle exposure. *International Journal of Nanomedicine, 11*, 6421.

Bai, C., & Tang, M. (2020). Toxicological study of metal and metal oxide nanoparticles in zebrafish. *Journal of Applied Toxicology, 40*, 37.

Behl, T., Kaur, I., Sehgal, A., Singh, S., Sharma, N., Bhatia, S., et al. (2022). The dichotomy of nanotechnology as the cutting edge of agriculture: Nano-farming as an asset versus nanotoxicity. *Chemosphere, 288*, 132533.

Bian, X. M., Liu, L., Li, H. B., Wang, C. Y., Xie, Q., Zhao, Q. L., et al. (2016). Construction of three-dimensional graphene interfaces into carbon fiber textiles for increasing deposition of nickel nanoparticles: Flexible hierarchical magnetic textile composites for strong electromagnetic shielding. *Nanotechnology, 28*(4), 045710.

Boros, B.-V., & Ostafe, V. (2020). Evaluation of ecotoxicology assessment methods of nanomaterials and their effects. *Nanomaterials, 10*, 610.

Chaud, M., Souto, E. B., Zielinska, A., Severino, P., Batain, F., Oliveira-Junior, J., & Alves, T. (2021). *Nanopesticides in agriculture: Benefits and challenge in agricultural productivity* (Vol. 9). Toxicological Risks to Human Health and Environment.

Chinnamuthu, C. R., & Boopathi, P. M. (2009). Nanotechnology and agroecosystem. *Madras Agricultural Journal, 96*(1–6), 17–31.

Costa, S., Vilas-Boas, V., Lebre, F., Granjeiro, J. M., Catarino, C. M., Teixeira, L. M., et al. (2023). Microfluidic-based skin-on-chip systems for safety assessment of nanomaterials. *Trends in Biotechnology, 41*(10), 1282–1298.

Das, H., Singh, S. K., Priyadarshini, E., Maurya, C. L., Tiwari, A. K., Gulaiya, S., et al. (2024). Nanotechnology facilitated real time soil monitoring for optimized crop production. *Asian Journal of Soil Science and Plant Nutrition, 10*(2), 582–607.
Deng, Y. Q., White, J. C., & Xing, B. S. (2014). Interactions between engineered nanomaterials and agricultural crops: Implications for food safety. *Journal of Zhejiang University Science A, 15*(8), 552–572.
Durlo, A. (2023). *1950–2022: A history of nanotechnology into physical and mathematical relationship* (Doctoral dissertation, Université de Lille).
Edwards, S. A. (2008). *The nanotech pioneers: Where are they taking us?* John Wiley & Sons.
FAO. (2017). The future of food and agriculture- Trends and challenges. Rome.
Giraldo, J. P., Landry, M. P., Faltermeier, S. M., McNicholas, T. P., Iverson, N. M., Boghossian, A. A., Reuel, N. F., Hilmer, A. J., Sen, F., Brew, J. A., & Strano, M. S. (2014). Plant nanobionics approach to augment photosynthesis and biochemical sensing. *Nature Materials, 13*(4), 400–408.
Gogos, A., Knauer, K., & Bucheli, T. D. (2012). Nanomaterials in plant protection and fertilization: Current state, foreseen applications, and research priorities. *Journal of Agricultural and Food Chemistry, 60*(39), 9781–9792.
Goswami, P., Yadav, S., & Mathur, J. (2019). *Positive and negative effects of nanoparticles on plants and their applications in agriculture* (Vol. 6, p. 232). Plant Science Today.
Hooven, L. A., Chakrabarti, P., Harper, B. J., Sagili, R. R., & Harper, S. L. (2019). Potential risk to pollinators from nanotechnology-based pesticides. *Molecules, 24*(24), 4458.
Khan, A. A., Alanazi, A. M., Alsaif, N., Al-Anazi, M., Sayed, A. Y., & Bhat, M. A. (2021). Potential cytotoxicity of silver nanoparticles: Stimulation of autophagy and mitochondrial dysfunction in cardiac cells. *Saudi Journal of Biological Sciences, 28*(5), 2762–2771.
Khot, L. R., Sankaran, S., Maja, J. M., Ehsani, R., & Schuster, E. W. (2012). Applications of nanomaterials in agricultural production and crop protection: A review. *Crop Protection, 35*, 64–70.
Kim, Y. J., Yang, S. I., & Ryu, J. C. (2010). Cytotoxicity and genotoxicity of nano-silver in mammalian cell lines. *Molecular & Cellular Toxicology, 6*, 119–125.
Kraegeloh, A., Suarez-Merino, B., Sluijters, T., & Micheletti, C. (2018). Implementation of safe-by-design for nanomaterial development and safe innovation: Why we need a comprehensive approach. *Nanomaterials, 8*, 239.
Kuhlbusch, T. A., Wijnhoven, S. W., & Haase, A. (2018). Nanomaterial exposures for worker, consumer and the general public. *NanoImpact, 10*, 11–25.
Kumari, R., Suman, K., Karmakar, S., Mishra, V., Lakra, S. G., Saurav, G. K., & Mahto, B. K. (2023). Regulation and safety measures for nanotechnology-based agri-products. *Frontiers in Genome Editing, 5*, 1200987.
Kwak, S. Y., Giraldo, J. P., & Wong, M. H. (2019). Nano-enabled plant science: Emerging opportunities for plant nanobionics. *ACS Nano, 13*(10), 10435–10443.
Lahiani, M. H., Dervishi, E., Chen, J., Nima, Z. A., Gaume, A., Biris, A. S., & Khodakovskaya, M. V. (2013). Impact of carbon nanotube exposure to seeds of valuable crops. *ACS Applied Materials & Interfaces, 5*(16), 7965–7973.
Lee, W. M., Kwak, J. I., & An, Y. J. (2012). Effect of silver nanoparticles in crop plants Phaseolus radiatus and Sorghum bicolor: Media effect on phytotoxicity. *Chemosphere, 86*, 491–499.
Liu, R., & Lal, R. (2015). Potentials of engineered nanoparticles as fertilizers for increasing agronomic productions. *Science of the Total Environment, 514*, 131–139.
Liu, R., Zhang, H., Lal, R., & Zhang, F. (2018). Nanotechnology for sustainable agriculture: Applications, impacts, and future implications. *Journal of Environmental Quality, 47*(2), 112–120.
Mahesha, K. N., Singh, N. K., Amarshettiwar, S. B., Singh, G., Gulaiya, S., Das, H., & Kumar, J. (2023). Entering a new agricultural era through the impact of nano-fertilizers on crop development: A review. *International Journal of Plant & Soil Science, 35*(20), 94–102.
Malik, R., & Patil, S. (2020). Nanotechnology: Regulatory outlook on nanomaterials and nanomedicines in United States, Europe and India. *Applied Clinical Research, Clinical Trials and Regulatory Affairs, 7*(3), 225–236.

Mastrangeli, M., Millet, S., & van den Eijnden-van Raaij, J. (2019). Organ-on-chip in development: Towards a roadmap for organs-on-chip. *ALTEX-Alternatives to Animal Experimentation, 36*(4), 650–668.

Mittal, D., Kaur, G., Singh, P., Yadav, K., & Ali, S. (2020). Nanoparticle-based sustainable agriculture and food science: Recent advances and future outlook. *Frontiers in Nanotechnology, 2*. https://doi.org/10.3389/fnano.2020.579954

Mukhopadhyay, S. S. (2014). Nanotechnology in agriculture: Prospects and constraints. In *Nanotechnology, science and applications* (pp. 63–71). https://doi.org/10.2147/NSA.S39409

Mustafa, I. F., & Hussein, M. Z. (2020). Synthesis and technology of nanoemulsion-based pesticide formulation. *Nanomaterials, 10*(8), 1608.

Mwaanga, P. (2018). *Risks, uncertainties, and ethics of nanotechnology in agriculture*. InTech. https://doi.org/10.5772/intechopen.76590

Oberbek, P., Kozikowski, P., Czarnecka, K., Sobiech, P., Jakubiak, S., & Jankowski, T. (2019). Inhalation exposure to various nanoparticles in work environment—contextual information and results of measurements. *Journal of Nanoparticle Research, 21*, 1–24.

Parisi, C. (2020). Nanobionics: A new frontier for sustainable agriculture. *Journal of Agricultural and Food Chemistry, 68*(15), 4049–4058.

Parisi, C., Vigani, M., & Rodríguez-Cerezo, E. (2015). Agricultural nanotechnologies: What are the current possibilities? *Nano Today, 10*, 124–127. https://doi.org/10.1016/j.nantod.2014.09.009

Paz-Trejo, C., Flores-Márquez, A. R., & Gómez-Arroyo, S. (2023). Nanotechnology in agriculture: A review of genotoxic studies of nanopesticides in animal cells. *Environmental Science and Pollution Research, 30*, 66473–66485.

Prasad, T. N. V. K. V., Sudhakar, P., Sreenivasulu, Y., Latha, P., Munaswamy, V., Raja Reddy, K., Sreeprasad, T. S., & Pradeep, T. (2012). Effect of nanoscale zinc oxide particles on the germination, growth and yield of peanut. *Journal of Plant Nutrition, 35*(6), 905–927.

Prokisch, J., Törős, G., Nguyen, D. H., Neji, C., Ferroudj, A., Sári, D., et al. (2024). Nano-food farming: Toward sustainable applications of proteins, mushrooms, nano-nutrients, and nanofibers. *Agronomy, 14*(3), 606.

Rai, M., & Ingle, A. P. (2012). Role of nanotechnology in agriculture with special reference to management of insect pests. *Applied Microbiology and Biotechnology, 94*(2), 287–293.

Ray, P. C., Yu, H., & Fu, P. P. (2009). Toxicity and environmental risks of nanomaterials: Challenges and future needs. *Journal of Environmental Science and Health. Part C, Environmental Carcinogenesis & Ecotoxicology Reviews, 27*(1), 1–35. https://doi.org/10.1080/10590500802708267. PMID: 19204862; PMCID: PMC2844666.

Schmutz M, Borges O, Jesus S, Borchard G, Perale G, Zinn M, Sips AAJAM, Soeteman-Hernandez LG, Wick P, Som C. A. (2020). Methodological Safe-by-Design Approach for the development of Nanomedicines. Front Bioeng Biotechnol. Apr 2;8:258. https://doi.org/10.3389/fbioe.2020.00258. PMID: 32300587; PMCID: PMC7145422.

Sekhon, B. S. (2014). Nanotechnology in agri-food production: An overview. *Nanotechnology, Science and Applications, 7*, 31–53.

Servin, A. D., & White, J. C. (2016). Nanotechnology in agriculture: Next steps for understanding engineered nanoparticle exposure and risk. *NanoImpact, 1*, 9–12.

Shang, Y., Hasan, M. K., Ahammed, G. J., Li, M., Yin, H., & Zhou, J. (2019). Applications of nanotechnology in plant growth and crop protection: A review. *Molecules, 24*(14), 2558. https://doi.org/10.3390/molecules24142558. PMID: 31337070; PMCID: PMC6680665.

Siddiqui, M. H., & Al-Whaibi, M. H. (2014). Role of nano-SiO2 in germination of tomato (Lycopersicum esculentum seeds Mill.). *Saudi Journal of Biological Sciences, 21*(1), 13–17.

Sohail, M. I., Waris, A. A., Ayub, M. A., Usman, M., ur Rehman, M. Z., Sabir, M., & Faiz, T. (2019). Environmental application of nanomaterials: A promise to sustainable future. In *Comprehensive analytical chemistry* (Vol. 87, pp. 1–54). Elsevier.

Subramanian, K. S., & Tarafdar, J. C. (2011). Prospects of nanotechnology in Indian farming. *Indian Journal of Agricultural Sciences, 81*(10), 887–893.

Tarafdar, J. C. (2015). Nanoparticle production, characterization and its application to horticultural crops. In *Winter School on "utilization of degraded land and soil through horticultural crops for agricultural productivity and environmental quality"* (pp. 222–229). NRCSS.

Ur Rahim, H., Qaswar, M., Uddin, M., Giannini, C., Herrera, M. L., & Rea, G. (2021). Nano-enable materials promoting sustainability and resilience in modern agriculture. *Nanomaterials, 11*(8), 2068.

Usman, M., Farooq, M., Wakeel, A., Nawaz, A., Cheema, S. A., ur Rehman, H., et al. (2020). Nanotechnology in agriculture: Current status, challenges and future opportunities. *Science of the Total Environment, 721*, 137778.

Wasef, L., Nassar, A. M., El-Sayed, Y. S., Samak, D., Noreldin, A., Elshony, N., et al. (2021). The potential ameliorative impacts of cerium oxide nanoparticles against fipronil-induced hepatic steatosis. *Scientific Reports, 11*(1), 1310.

White, J. M. L., Powell, A. M., Brady, K., & Russell-Jones, R. (2003). Severe generalized argyria secondary to ingestion of colloidal silver protein. *Clinical and Experimental Dermatology, 28*(3), 254–256.

Wong, M. H., Misra, S. K., Giraldo, J. P., & Landry, M. P. (2017). The role of nanotechnology in the enhancement of photosynthesis and crop production. *Nature Nanotechnology, 12*(4), 364–373.

World Health Organization. (2017). *WHO guidelines on protecting workers from potential risks of manufactured nanomaterials*. World Health Organization.

Yadav, A., Yadav, K., Ahmad, R., & Abd-Elsalam, K. A. (2023). Emerging frontiers in nanotechnology for precision agriculture: Advancements, hurdles and prospects. *Agrochemicals, 2*(2), 220–256.

Zhang, H., Huang, M., Zhang, W., Gardea- Torresday, J. L., White, J. C., Ji, R., and Zhao, L. (2020). *Environmental Science & Technology 54*(6), 3334–3342. https://doi.org/10.1021/acs.est.9b07562.

Zhou, M., Ge, X., Ke, D. M., Tang, H., Zhang, J. Z., Calvaresi, M., et al. (2019). The bioavailability, biodistribution, and toxic effects of silica-coated upconversion nanoparticles in vivo. *Frontiers in Chemistry, 7*, 218.

Zielińska, A., Costa, B., Ferreira, M. V., Miguéis, D., Louros, J. M. S., Durazzo, A., Lucarini, M., Eder, P. V., Chaud, M., Morsink, M., et al. (2020). Nanotoxicology and nanosafety: Safety-by-design and testing at a glance. *International Journal of Environmental Research and Public Health, 17*, 4657.

Chapter 13
Nano(Bio)Sensor Technologies: Fostering the Renaissance of Horticulture

Aditya Pratap Singh, Harini Palani, Amrita Kumari, Jitendra Kumar Kushwaha, and Utpal Das

Abstract In recent years, nano(bio)sensors have revolutionized horticulture by enabling real-time monitoring and precise data collection. These miniature devices incorporating nanomaterials contribute significantly to the renaissance of horticulture through several key applications. They allow for early disease detection by sensing subtle changes in plant physiology, thus enabling targeted actions to prevent crop losses. In nutrient management, nano(bio)sensors measure nutrient levels in soil and plant tissues, providing accurate data to optimize fertilization strategies. For water management, these sensors monitor soil moisture, aiming to implement efficient irrigation practices, which is crucial in addressing water scarcity. Additionally, they enhance pest surveillance by detecting infestations promptly, reducing reliance on chemical pesticides, and promoting more effective integrated pest management. During post-harvest handling, these sensors assess fruit ripeness, freshness, and quality, guiding timely harvesting and storage decisions to ensure produce quality. By providing real-time, accurate data on plant and soil health, these advanced sensors enable more informed decision-making, ultimately fostering the renaissance of horticulture through improved productivity and resource efficiency.

Keywords Nano(bio)sensors · Early disease detection · Nutrient management · Sustainable farming · Crop management

---

A. P. Singh
Department of Genetics and Plant Breeding, Bidhan Chandra Krishi Viswavidyalaya, Mohanpur, Nadia, West Bengal, India

H. Palani · U. Das (✉)
Department of Horticulture and Food Science, VIT School of Agricultural Innovations and Advanced Learning, Vellore Institute of Technology, Vellore, Tamil Nadu, India

A. Kumari
Department of Genetics and Plant Breeding, Narayan Institute of Agricultural Sciences, Gopal Narayan Singh University, Jamuhar, Sasaram, Bihar, India

J. K. Kushwaha
Subject Matter Specialist (Horticulture), Krishi Vigyan Kendra, Mau, Uttar Pradesh, India

V. D. Rajput et al. (eds.), *Contemporary Suitability of Nanobionics in Agriculture*, Nanotechnology in the Life Sciences, https://doi.org/10.1007/978-3-031-82779-2_13

13.1 Introduction

Nanotechnology is a revolutionary multidisciplinary approach that deals with particles of size 1–100 nm. By their size, the "nanoparticles," are considered able to control and manipulate matter at atomic and molecular levels (Balasooriya et al., 2017; Salvadori, 2019). Nanoparticles (NPs) synthesis and their utilization in a diverse array of scientific fields—electronics, biomedical engineering, chemical industries, agriculture, and environment—has gained momentum in recent times (Prasad et al., 2017; Rajput et al., 2018; Sreejivungsa et al., 2016).

Biosensors are sensors with a biological element as a receptor that is mounted to a converter in various manners (Chambers et al., 2008; Sode et al., 2016). Thus, the term "bionanosensors" could be attributed to nanosensors (sizing between 1 and 100 nm) that specifically use biological recognition elements (e.g., enzymes, antibodies, DNA) combined with nanomaterials to detect biological molecules or phenomena. Nano biosensors use biochemical reactions as sensing elements through electrical, thermal, or optical signals and provide quantitative measurements. They are the nanoscale instruments used to identify chemical compounds. They use a small probe and electrical, optical, or magnetic technology. The plants emit chemical signals that are collected and converted by a nano biosensor into digital data that can be measured and observed by other electronic devices (Ramesh et al., 2022). Using nanomaterials in the structure of biosensors reduces chemical intermediates of electron transfer, which is of great importance when developing third-generation biosensors.

Three components typically make up a nano biosensor:

- Bioreceptor
- Transducer
- Detector

Typically, the bio receptors are biological substances that interact with the microorganisms, nucleic acids, enzymes, and antibodies, shifting the signal to the transducer from the analyte to be detected. The transducer then functions as an interface that converts the bioreceptor signal into electrical signal data by measuring it within the detector. Then, the detector processes the signals to amplify and analyze the observed signals (Naresh & Lee, 2021).

Nano biosensors are extensively used in medical diagnostics (e.g., glucose sensors for diabetes management, detection of cancer biomarkers), environmental monitoring (detecting pathogens in water), food safety (detecting allergens or contaminants), and biotechnology research (Achi et al., 2023; Guo et al., 2014; Kosaka et al., 2014; Valenzuela-Amaro et al., 2023; Verma, 2017). Nano biosensors use biochemical reactions as sensing elements through electrical, thermal, or optical signals and provide quantitative measurements. They are the nanoscale instruments used to identify chemical compounds. They use a small probe and electrical, optical, or magnetic technology. The plants emit chemical signals that are collected and

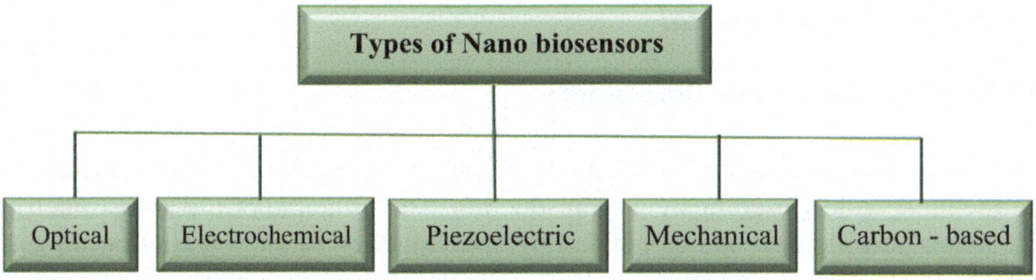

Fig. 13.1 Types of nano biosensors

converted by a nano biosensor into digital data that can be measured and observed by other electronic devices (Ramesh et al., 2022).

13.2 Types of Nano Biosensors

Various types of nano biosensors are depicted in the Fig. 13.1.

13.2.1 Optical

The most versatile class of biosensors is optical-based nano biosensors, which can be applied to a wide range of spectroscopy techniques, including dispersion spectrometry, Raman, fluorescence, phosphorescence, refraction, and absorption (Touhami, 2014).

13.2.1.1 Fluorescent Biosensors

The popular technique of optical nano biosensors is the selectivity and sensitivity of fluorescence. The fluorescence-based optical biosensors note the frequency range of the electromagnetic radiation. A fluorescence-based instrument tracks the frequency shift in the emission of electromagnetic radiation caused by radiation absorption in an excited state that lasts for an extremely short time (Suresh et al., 2014).

13.2.1.2 Plasmon Resonance Surface Biosensors

The first biosensing demonstration of a surface plasmon resonance (SPR) biosensor has undergone investigation considerably and an effective label-free instrument has been developed for examining the interactions that happen between the target and biorecognition molecules (Suresh et al., 2014).

13.2.2 *Electrochemical*

13.2.2.1 Amperometric Biosensors

Using an electrochemical method, amperometric biosensors have been reported to be the most widely used in signal transduction. The principle of the amperometric biosensor performance can be defined by applying a constant potential between a reference and a working electrode. Redox reactions from the application of potential result in a flow of net current; both anodic (oxidizing) and cathodic (reducing) reactions may be observed amperometrically, with the current magnitude being proportionate to the concentration of electroactive species in the test solution (Suresh et al., 2014).

13.2.2.2 Potentiometric Biosensors

The fundamental principle of these biosensors is to record the system potential at a working electrode concerning a specific reference electrode when there is almost no current flow (Touhami, 2014).

13.2.3 *Piezoelectric*

13.2.3.1 Microbalance-Quartz Crystal

The Quartz Crystal Microbalance (QCM) has been the most widely used acoustic device for sensor applications. A thin quartz disc that has been placed with circular electrodes on both sides builds the traditional QCM sensor. Because quartz is a piezoelectric material, mechanical oscillations of the crystal are produced by an alternating voltage between these electrodes. They are generally very stable because of the oscillations and excellent quality of the quartz (Touhami, 2014).

13.2.4 *Mechanical*

Cantilever Biosensors

Cantilever-based biosensors, which use a micromechanically produced cantilever like that of atomic force microscopy (AFM) probes, are the best example of nano transducers. This laser can detect the biosensing interaction on the surface that causes a cantilever that is several hundred nm thick to bend. One can adjust the sensitivity to analyze single-molecule interactions. Microcantilever sensors have high sensitivity, which has a potent platform for label-free, time-resolved molecular interaction detection (Touhami, 2014).

13.2.5 Carbon-Based Nano Biosensors

13.2.5.1 Nanocarbon Biosensor

Among these are the addition of various nanocarbons to the electrodes, such as graphene sheets, carbon powder, carbon nanotubes, and carbon capsules. Single-walled carbon nanotubes (SWCNT) are created by a hollow cylinder of a single carbon sheet, and concentric carbon sheets of varying diameters create multi-walled carbon nanotubes (MWCNT) bonding to translate the signals (Suresh et al., 2014).

13.3 Working Principles

Nanobiosensors employ the principles of both biotechnology and nanotechnology to identify and examine particular physical, chemical, or biological agents. Several fundamental operating principles support the operation of these sensors (Bodkhe et al., 2024).

Initially, a detector that selectively interacts with the target probe is incorporated into nano biosensors. This sensing agent could be a synthetic material with a particular binding capability or a biological molecule like an enzyme, antibody, nucleic acid, or receptor. The capacity of the sensor to discriminate between the target analyte and other variables, or its selectivity.

Secondarily, the binding reaction between the detection element and the target analyte is transformed into a measurable signal by a transducer, which is used in nano biosensors (Kakimova et al., 2023).

13.4 Integration of Nanomaterials

13.4.1 Hydrothermal Synthesis

Hydrothermal synthesis of nanomaterials has proven to be efficient in producing authoritative entities like multi-layered ceramic oxides, microporous quartzes, chemical oxides for sensing, and hypersonic electrodes. Through the use of biochemical solutions operating above the steaming point of water in a vacuum-packed temperature reaction above atmospheric pressure, hydrothermal deliberates the synthesis of nanomaterials (Kulkarni et al., 2022).

13.4.2 Sol-Gel Method

Sol-gel is one of the most popular ways to synthesize nanomaterials. This involves the solid colloidal absorption in a fluid stage, where the gel is a solid macroparticle immersed in a chemical reaction. This technique typically designates a wet-chemical process that involves a biochemical reaction that serves as a precursor to various materials into a single structure (Kulkarni et al., 2022). To develop nano biosensors, various advanced material types have been used in nanosensing applications. Nanosensors have been developed using carbon nanotube-based nanomaterials, specifically single-walled carbon nanotubes (SWCNTs) and multi-walled carbon nanotubes (MWCNTs). The carbon-based nanomaterials exhibit superior surface immobilization capabilities when it comes to biological components in biosensors. Improved conductive qualities and temperature resistance are achieved by imparting mechanical and physical properties. But in this instance, the main drawback is that these nanomaterials are less soluble in water. Creating sustainable nanostructures, producing toxicity, and producing nano biosensors are a few more difficulties. Another is the lack of selectivity. Some of the materials used in nano biosensors are quantum dots/fluorescence-based materials, carbon-based materials, metal/magnetic-based materials, and biogenic-based materials. Figure 13.2 shows the different types of nanomaterials applied in the sensors (Guruprasath et al., 2024; Thakur et al., 2022).

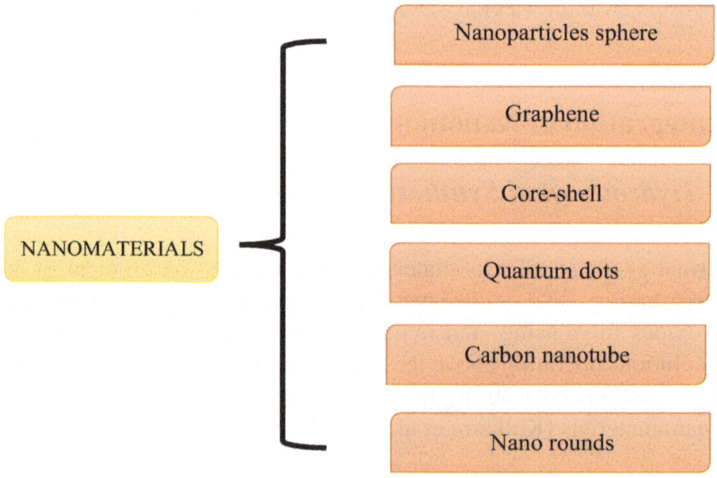

Fig. 13.2 Different types of nanomaterials applied in the sensors

13.5 Applications in Horticulture

The effective use of natural resources and nano biosensors can improve the sustainability of agriculture by increasing crop productivity. To evaluate agricultural problems, various chemo- or biosensors based on nanoparticles were developed and introduced to the market. For accurate and non-destructive monitoring of plant growth and development, nano biosensors serve as a real-time application that offers a novel platform (Kumar et al., 2024). The early detection of plant diseases, management of nutrients, water management, management of pest control, and soil control and quality management can be tracked by nano biosensors as shown in Fig. 13.3.

13.5.1 Early Disease Detection

In the early stages, the detection of specific infections in plants was supported by the time-consuming visual inspection. Novel sensing techniques are used to detect pathogens in the early stages of infection, and waste prevention and production are reduced. The specific treatment plans are aided by the accurate detection of pathogens. When compared with many conventional methods of identification, the use of

Fig. 13.3 Applications of nano biosensors in horticulture

nano biosensors could fasten the process of identification as it is used to identify which microorganisms were responsible for the contamination. To determine some of the agricultural terrorism threats, gold nanoparticle-based nano biosensors were used, and gold nanoparticle-based nano biosensor detectors were developed to identify several plant diseases (Bodkhe et al., 2024).

Novel biosensing techniques for early pathogen detection result in disease management and fewer productivity losses. Compared with traditional methods, these nano biosensing techniques are used to detect pathogens and help to develop disease-fighting treatment plans, as the detection can be done faster. Plant pathogens can be found using conventional microscopy-based and culture-dependent methods, which are labor-intensive and frequently unreliable. Additionally, the immunological and molecular techniques for pathogen detection struggle with bulky instrumentation, longer test times, and weak signal strengths. However, the use of nano biosensors in plant pathology may offer a fresh method of plant pathogen detection. The detection of plant viral disease can also be accomplished with nano biosensors. In terms of virus detection, the immune sensor exhibits excellent sensitivity and specificity (Priyanka et al., 2020).

Plant disease diagnostic systems gained a new dimension with the introduction of nanobiosensor-based technology. This technology offers a low-cost, easily operated, nondestructive, and minimally invasive system with improved sensitivity, specificity, and detection limit for the real-time identification of crop pathogens (Kashyap et al., 2019).

According to Bharti et al. (2024), nano biosensors can assist sensing systems for more precise early pathogen detection. Piezoelectric biosensors (PZ) are highly intriguing for the real-time detection of bacteria (*E. Coli*, Staphylococcal enterotoxin), and they can be used directly for label-free detection.

13.5.2 Nutrient Management

An effective analytical or sensing tool for interpreting biological signals—such as those from enzymes, antibodies, or living cells—into recognizable electrical signals is the electrochemical biosensor, which is small and portable. To monitor trace elements and make sure that plants receive the right amounts of nutrients for growth, nano biosensors are crucial because they can continuously detect and sense low concentrations of a variety of nutrients and react quickly to any excesses or deficiencies (Kumar et al., 2024; Ronkainen et al., 2010).

When carbon, nitrogen, and phosphorus are starved, lux-inserted strains of *Pseudomonas fluorescence* that express bioluminescence can be used to examine the bioavailability of nutrients in the rhizosphere and bulk soil; this will reduce their bioluminescence proportionately. During the process of organic matter decomposition, this type of biosensor is frequently used to assess the availability of nutrients in various soil forms and to look into nutrient flow in the rhizosphere (Priyanka et al., 2020).

13.5.3 *Water Management*

Bharti et al. (2024) reported that by employing the green synthesis technique of nanoengineering to create silver nanoparticles, successfully developed a nano biosensor that aids in the monitoring of water quality.

The developed flexible electronic sensing devices were used to monitor the distribution of nutrients, sap flow, and water transportation continuously. With the help of nanosensors, watermelons were analyzed, revealing that the water distribution between the fruits and leaves varied during the day and night (Kumar et al., 2024).

13.5.4 *Pest Management*

Pesticides are widely used today to improve the safety of the crops. Certain toxic residues from pesticides are released into the soil during their environmental degradation, and these residues can contaminate soil and water and make their way up the food chain. Therefore, accurate techniques for detecting pesticides are needed to assess the level of contamination in plants and soil. The conventional techniques for detecting pesticide residues are UV-V spectrometry, GC, mass spectrometry, and HPLC. These techniques are time-consuming and labor-intensive, making them unsuitable for field analysis. Therefore, more modern techniques are needed to accurately detect pesticide residue in agricultural fields (Priyanka et al., 2020).

The use of electronic sensors is the most popular technique for detecting various insecticides, herbicides, and pesticides. Gold nanoparticle-based biosensors have been successfully developed to detect carbofuran at 0.08 $\mu g/dm^3$ and chlorpyrifos at 0.06 $\mu g/dm^3$. The quantum dot-based nano biosensors were used in conjunction with immunohistochemistry and colloidal gold to detect 2,4-D levels up to 250 pg/L. By the use of biosensors, methyl parathion, carbaryl, and monocrotophos were detected (Bodkhe et al., 2024).

13.5.5 *Soil Quality Management*

One of the most crucial aspects of agriculture is soil quality, which should be assessed for pH, nutrients, and moisture content before beginning any farming operations. The relationship between the biosensor and rhizosphere microorganisms can be used to assess the quality of the soil. The microelectromechanical systems (MEMS) were developed to detect the soil quality using microelectronic circuits, while intelligent fertilizers, such as zinc fertilizers with nanoparticles, were developed to achieve the controlled release of the fertilizer to the plant roots and sense the feedback (Thakur et al., 2022).

For achieving regulated fertilizer discharge through plant origin, fertilizers like zinc fertilizer, which contains nanoparticles, were also developed. Likewise, for the determination of the condition of the soil, microelectronic (MEMS) fertilizers were developed that use microelectronic networks (Bodkhe et al., 2024).

13.6 Nano Fertilizer for Plant

A substance that provides nanoscale-level nutrients to plants is known as nano-fertilizer. Based on the function of the nanomaterials, the nano fertilizers are classified into two types based on the function of the nanomaterials:

- Macro-nutrients and micro-nutrients composed nanomaterials
- Nutrient carries in additive form and serves as nanomaterials

Some of the characteristics of nano-fertilizers include a large surface area-to-volume ratio and low toxicity, increased mobility and solubility that increase the nutrient efficiency and decrease the toxins, intentional site delivery, and limited or required release of agrochemicals to the plants. Figure 13.4 depicts the effects of smart fertilizers on the soil-plant system (Rai et al., 2022).

13.7 Post-Harvest Quality Assurance

After harvest, vegetables and fruits remain metabolically active, continuing to undergo ripening and senescence. Managing these processes is critical to extending the postharvest quality of fresh produce. Proper postharvest handling is essential to reduce losses and maintain the global supply of fresh vegetables and fruits. Poor management can lead to significant declines in nutritional value and quality, increased risk of foodborne pathogen outbreaks, and financial losses throughout the supply chain. Nanotechnology, a multidisciplinary field combining chemistry, biotechnology, and industrial processing, shows considerable promise in enhancing the preservation of fresh produce. Recently, nanomaterials have gained recognition for their ability to extend shelf life and preserve the quality of postharvest vegetables and fruits. Their unique properties, due to their extremely small size, offer significant advantages. By integrating nanomaterials with conventional preservation methods, it is possible to effectively overcome the limitations of traditional preservation techniques (Duncan, 2011).

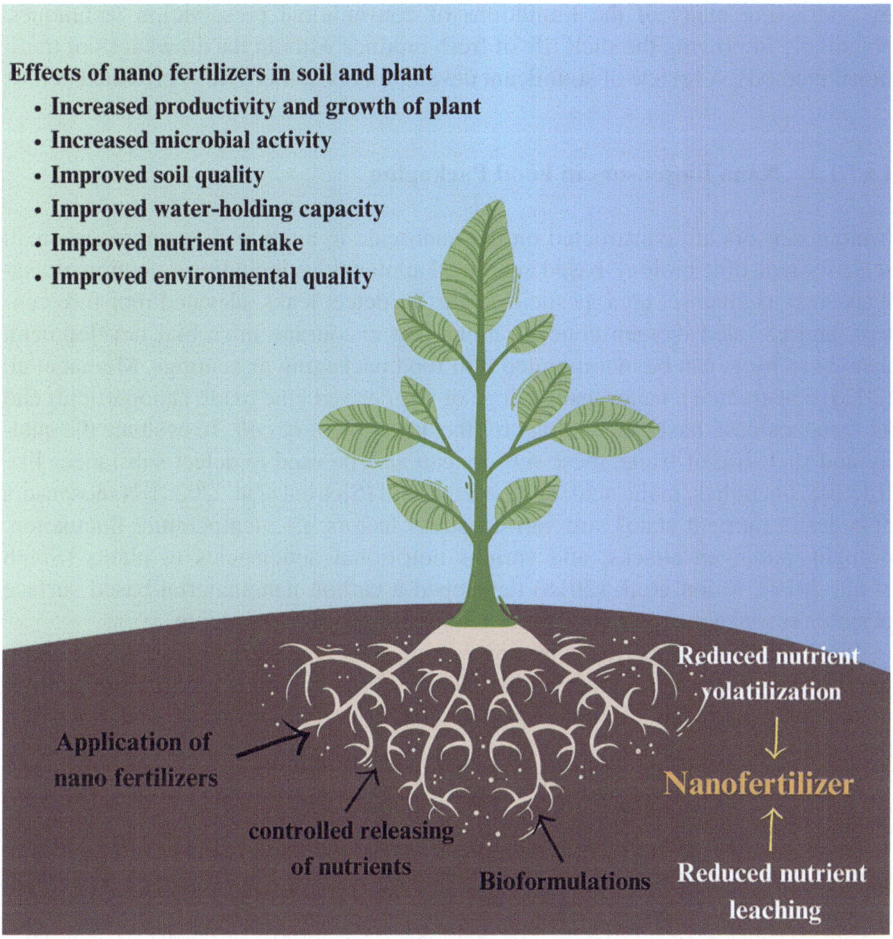

Fig. 13.4 Schematic diagram of effects of potential smart fertilizer in the soil-plant system. (Source: Rai et al. (2022), Calabi-Floody et al. (2017))

13.7.1 Ensuring the Shelf Life and Quality of Horticultural Crops

The food industry must ensure that fruit products meet quality standards in terms of biochemical composition, maturity, and ripeness before they reach consumers (Rana et al., 2010). Fruits naturally produce varying levels of organic acids throughout their growth, maturation, and shelf life. Because of their high perishability, the shelf life of most vegetables and fruits is frequently shortened when stored under standard conditions. While traditional preservation methods exist, they are typically costly and inefficient in significantly extending shelf life or leaving behind unwanted residues. Nanotechnology, with its unique properties, offers a promising alternative by addressing many of the limitations of conventional preservation techniques, potentially improving the shelf life of fresh produce without the drawbacks of traditional methods. A review of significant developments in this field is provided below:

13.7.1.1 Nano Biosensors in Food Packaging

Nanobiosensors are constructed on the nanoscale to analyze data on a very small scale for real-time biology-based analytical applications including detection of analytes such as glucose, urea, pesticides, etc. To detect leaks, elevated moisture content, and elevated oxygen concentrations that encourage microbial development, nano biosensors can be incorporated into food packaging as coatings. Meraat et al. (2020) constructed a nanosensor using sol-gel-derived zinc oxide nanoparticles and a β-galactosidase-based technology for the detection of *E. coli*. To evaluate the quality and shelf life of fruits, these sensors can also be used to detect substances like fructose, phenolics, malic acid, and benzoic acid (Sharma et al., 2020). Nanosensors also detect nutrient status and various stress factors, i.e., temperature fluctuation, drought, pathogen attacks, and various nutritional deficiencies in plants (Singh et al., 2014). Moon et al. (2018) developed a carbon nanomaterial-based surface plasmon resonance sensor to identify aflatoxins in peanuts and rice.

Incorporating nanoparticles into packaging is an emerging strategy for prolonging the shelf life of horticulture crops. These substances aid in controlling environmental elements that promote spoiling, like oxygen, moisture, and microbial development, preserving the freshness of produce for extended periods. Nanotechnology has great potential in the packaging business. It can reduce oxidation, control moisture transfer, inhibit microbiological development, regulate respiration rates, and preserve volatile flavors and fragrances (Brody et al., 2008). Researchers investigated ethylene production, sensory qualities, and nutritional content throughout storage. They concluded that this packaging effectively delayed the ripening process by scavenging ethylene. This addressed the issue of limited shelf life in freshly picked kiwifruit (Hu et al., 2011). Nano-encapsulation and processing methods employ nanomaterials to improve the overall quality and durability of food items. Metal nanoparticles, including copper, silver, magnesium oxide, and titanium dioxide, are utilized in these active packaging methods (Islam et al., 2018). Furthermore, the integration of edible antimicrobial nanocomposite coatings and gas scavengers has been implemented to enhance the preservation of food (Lee et al., 2015). Table 13.1 shows some of the achievements of nanomaterials in enhancing vegetable preservation. Table 13.2 shows some of the achievements of nanomaterials in enhancing fruit preservation. Research is also exploring the use of nanowheels, nanofibers, and nanotubes to improve the performance of food packaging, while cellulose nanofibers are being investigated as a cost-effective and environmentally friendly packaging alternative (Khalil et al., 2016).

Table 13.1 Achievements of nanomaterials in enhancing vegetable preservation

Vegetables	Nanomaterials	Effect	Reference
Carrot	A composite film consisting of chitosan and neem oil, reinforced with zinc oxide nanoparticles	Maintained quality and inhibited deterioration	Sanuja et al. (2015)
Tomato	Nanocomposite film of chitosan and titanium dioxide	Prolonged shelf life and slowed ripening	Kaewklin et al. (2018)
Okra/lady's finger	Edible coating of gelatin-chitosan/ silver and zinc oxide	Polyethylene packaging with antimicrobial properties for maintaining vegetable quality	Al-Naamani et al. (2018)
Tomato	Nano-composite particles with platinum-loaded monoclinic tungsten trioxide nanorods on zeolite (ZSM-5)	Preserved quality, diminished decay, and reduced spoilage	Kim et al. (2019)
Tomato	Packaging made with titanium oxide and Polyacrylonitrile nanomaterials	Decreases the rate of ethylene production	Yadav et al. (2021)
Carrot	A nanofilm or coating incorporating Nano zinc oxide	Mitigates microbial spoilage during storage and substantially prolongs the shelf life.	Xu et al. (2017)
Cucumber	Poly-lactic acid membranes infused with nanoparticles	Enhancements in sensory attributes are achieved by preventing microbial growth	Li et al. (2021)

13.7.1.2 Fruits and Vegetables with a Nanofilm

Metal nanoparticle-infused nanofilms serve as barrier-forming and antibacterial agents. According to Silvestre et al. (2011), preservation techniques including coating and spraying produce a micro-environment that forms a semi-permeable membrane that delays the aging and degradation of fresh fruits and vegetables. Edible nanofilms are now a very successful way to keep fresh produce fresh while it is stored. Nano-coatings have recently been used to preserve food's nutritional value while maintaining quality control (King et al., 2018). Increasing the shelf life of fresh horticulture crops while maintaining their quality is critical (Neethirajan & Jayas, 2011). A newly created edible nanocomposite film composed of glycerol, aloe vera gel, and ZnO nanoparticles has shown great potential in terms of horticultural product preservation and shelf-life extension (Jafarzadeh et al., 2021).

Table 13.2 Achievements of nanomaterials in enhancing fruit preservation

Fruits			
Apricot	Nano edible coating made from pectin and Layered Double Hydroxide (LDH) Salicylate	Lengthened shelf life	Gorrasi and Bugatti (2016)
Peach	Low-density polyethylene packaging enhanced with zinc oxide nanoparticles	Prolonged shelf life and preserved quality	Li et al. (2017)
Grapes	Nano-edible coating of chitosan nanoparticles	Delayed ripening and increased moisture	Melo et al. (2018)
Peach and nectarine	Zeolite nanoparticles coated with potassium permanganate	Eliminate ethylene produced during storage	Emadpour et al. (2015)
Banana	Nanoedible coating of chitosan/ Tripolyphosphate edible coating	Slowed the rate of ripening	Esyanti et al. (2019)
Guava fruit	Copper oxide nanomaterials based on nettle leaf extract within nano-composite packaging	Enhanced shelf life and preservation	Kalia et al. (2021)
Strawberry	Low-density polyethylene with modified titanium dioxide nanoparticles Composite film	Prolonged shelf life and reduced spoilage	Zhang et al. (2022)

13.8 Benefits and Challenges

It is unfortunate that despite the achievements in the field of nanotechnology in agriculture, which are increasingly promising, and the availability of numerous patents and commercial products (Kah et al., 2019), it is difficult to translate these results into practical applications in agriculture/horticulture (Miguel-Rojas & Pérez-de-Luque, 2023; Reddy et al., 2022). It is important to note that research in this field is highly complex and often requires the involvement of multidisciplinary teams that possess expertise in chemistry, physics, biology, and agronomy, among many other fields, as well as the use of sophisticated equipment (such as particle accelerators—synchrotrons), among others. Furthermore, nanomaterials can behave differently depending on factors such as plant species and physiology, so results obtained from one crop (especially for nanomaterials) may not be directly applicable to others (Pérez-de-Luque, 2017). It has also been noted that agriculture often operates on thin profit margins, and thus, the cost of using nanomaterials in large-scale field applications must be extremely low, or farmers are likely to not be able to adopt them. Moreover, the potential environmental impact and food safety concerns of these products must be taken into account as well, which has led to an increasing trend toward the use of natural nanomaterials (biopolymers) that are non-toxic and readily degradable.

It is also important to note that one of the major challenges to applying many of these nanodevices to crops is that it often incurs additional costs for farmers, so it is often difficult to implement them. To achieve a similar effect to conventional agrochemicals (by using machinery, irrigation, etc.), nano agrochemicals need to be able to be applied in the same way they are conventional agrochemicals.

There would, however, be a significant reduction in costs if these biosensors were to be integrated as seed coatings. Nanobiosensors could pose an even greater challenge in this regard since specific methods would need to be devised to release them on crops. In many cases, such as with non-invasive electronic systems on leaves, they can prove to be useless or too expensive unless they are used on perennial horticultural crops, which is why they are more problematic.

There is a critical need to develop research that is focused on field applications in order to address many of these challenges. It is important to note that many of the studies conducted so far have not been conducted in real-world conditions and instead have extrapolated results from laboratory experiments or greenhouse experiments. Agricultural field trials are like clinical trials in medical research in that despite a product performing well in the laboratory, it does not guarantee that it will perform well in the field, or there may be challenges involved in its manufacture or distribution as well. It is therefore only possible to verify if a nanotechnology application will have a significant impact on agriculture by conducting experiments with different crops in experimental plots, comparing them to conventional products, and following the usual practices of farmers. This is the only way to validate the effectiveness of nanotechnology applications.

13.9 Future Prospects

Nanotechnology is emerging as an advanced tool in modern agriculture, offering significant potential for promoting green agriculture. It also has a lot of potential for usage in agriculture, where several types of nanomaterials are utilized to increase productivity and quality while lowering vegetable and fruit losses after harvest. Nanotechnology increases horticultural crop productivity by leveraging nanomaterial capabilities and distribution strategies. Furthermore, it reduces the overuse of chemical fertilizers and pesticides. These nanoparticles are effective, cost-efficient, and sustainable as they can be produced quickly, with minimal effort, and without causing environmental harm. The use of nano biosensors in the horticulture industry is rapidly expanding, boosting productivity by enabling more efficient utilization of natural resources and fostering sustainability. Nanosensors in packaging films, which detect gases from food spoilage, are becoming increasingly relevant to consumers. These sensors can identify spoilage at various stages of the food supply chain, reducing overall food waste and benefiting producers, retailers, and consumers alike. Emerging concepts such as "smart packaging," the development of "antigen-specific biomarkers" and the integration of nanoparticles into nanocomposite polymer films are beginning to materialize. Further research is anticipated to drive expansion and industrial application shortly. The widespread use of nanotechnology has proved its capacity to considerably improve growth, increase yields, and reduce production costs, all while minimizing losses after harvest by preserving the quality and increasing the shelf life of newly harvested and processed vegetables and fruits. It is important to highlight that the use of nanotechnology for extending the shelf life of vegetables and fruits remains in the experimental phase, and there are numerous hurdles to address before it can be applied on an industrial scale. Nanomaterials may move into vegetables and fruits by absorption, diffusion, or disintegration during contact, although their safety consequences remain unknown. Ensuring the availability of efficient nanoparticles, as well as complete safety assessments of their field use, is critical to ensuring food and nutrition stability for the world's growing population in the face of a changing environment. The widespread application of nanomaterials will result in greater climate-resilient horticulture, lower losses after harvest, and higher overall crop quality. There is an urgent need for new methods and standardized testing procedures to assess the effects of nanoparticles on living cells and evaluate potential risks associated with human exposure. It is anticipated that, shortly, food products incorporating nanotechnology will become more widely available to consumers around the globe.

13.10 Conclusion

Nanotechnology is emerging as an advanced tool in modern agriculture, offering significant potential for promoting green agriculture. It also has a lot of potential for usage in agriculture, where several types of nanomaterials are utilized to increase productivity and quality while lowering vegetable and fruit losses after harvest. Nanotechnology increases horticultural crop productivity by leveraging nanomaterial capabilities and distribution strategies. Furthermore, it reduces the overuse of chemical fertilizers and pesticides. These nanoparticles are effective, cost-efficient, and sustainable as they can be produced quickly, with minimal effort, and without causing environmental harm.

References

Achi, F., Attar, A. M., & Lahcen, A. A. (2023). Electrochemical nanobiosensors for the detection of cancer biomarkers in real samples: Trends and challenges. *TrAC Trends Analytical Chemistry, 170*, 117423.

Al-Naamani, L., Dutta, J., & Dobretsov, S. (2018). Nanocomposite zinc oxide-chitosan coatings on polyethylene films for extending storage life of okra (Abelmoschus esculentus). *Nanomaterials, 8*(7), 479.

Balasooriya, E. R., Jayasinghe, C. D., Jayawardena, U. A., Ruwanthika, R. W. D., Mendis de Silva, R., & Udagama, P. V. (2017). Honey mediated green synthesis of nanoparticles: New era of safe nanotechnology. *Journal of Nanomaterials, 1*, 5919836.

Bharti, A., Jain, U., & Chauhan, N. (2024). From lab to field: Nano-biosensors for real-time plant nutrient tracking. *Plant Nano Biology, 9*, 100079. https://doi.org/10.1016/j.plana.2024.100079

Bodkhe, M., Chalke, T., Kulkarni, S., & Goswami, A. (2024). A review on sustainable applications of nanobiosensors in various fields and future potential. *BioNanoScience, 14*, 1940. https://doi.org/10.1007/s12668-024-01344-w

Brody, A. L., Bugusu, B., Han, J. H., Sand, C. K., & McHugh, T. H. (2008). Innovative food packaging solutions. *Journal of Food Science, 73*(8), 107–116.

Calabi-Floody, M., Medina, J., Rumpel, C., Condron, L. M., Hernandez, M., Dumont, M., & De La Luz Mora, M. (2017). Smart fertilizers as a strategy for sustainable agriculture. In *Advances in agronomy* (pp. 119–157). https://doi.org/10.1016/bs.agron.2017.10.003

Chambers, J. P., Arulanandam, B. P., Matta, L. L., Weis, A., & Valdes, J. J. (2008). Biosensor recognition elements. *Current Issues in Molecular Biology, 10*(1–2), 1–12.

Duncan, T. V. (2011). Applications of nanotechnology in food packaging and food safety: Barrier materials, antimicrobials, and sensors. *Journal of Colloid and Interface Science, 363*(1), 1–24.

Emadpour, M., Ghareyazie, B., Kalaj, Y. R., Entesari, M., & Bouzari, N. (2015). Effect of the potassium permanganate coated zeolite nanoparticles on the quality characteristic and shelf life of peach and nectarine. *Journal of Agricultural Technologies, 11*, 1263–1273.

Esyanti, R. R., Zaskia, H., & Amalia, A. (2019). Chitosan nanoparticle-based coating as postharvest technology in banana. *Journal of Physics Conference Series, 1204*(1), 012109.

Gorrasi, G., & Bugatti, V. (2016). Edible bio-nano-hybrid coatings for food protection based on pectins and LDH-salicylate: Preparation and analysis of physical properties. *LWT—Food Science and Technology, 69*, 139–145.

Guo, C., Huo, H., Han, X., Xu, C., & Li, H. (2014). Ni/CdS bifunctional Ti@ TiO2 core–shell nanowire electrode for high-performance nonenzymatic glucose sensing. *Analytical Chemistry, 86*(1), 876–883.

Guruprasath, N., Sankarganesh, P., Adeyeye, S. A. O., Babu, A. S., & Parthasarathy, V. (2024). Review on emerging applications of nanobiosensor in food safety. *Journal of Food Science, 89*(7), 3950–3972. https://doi.org/10.1111/1750-3841.17149

Hu, Q., Fang, Y., Yang, Y., Ma, N., & Zhao, L. (2011). Effect of nanocomposite-based packaging on postharvest quality of ethylene-treated kiwifruit (Actinidia deliciosa) during cold storage. *Food Research International, 44*(6), 1589–1596.

Islam, M. S., Akter, N., Rahman, M. M., Shi, C., Islam, M. T., Zeng, H., & Azam, M. S. (2018). Mussel-inspired immobilization of silver nanoparticles toward antimicrobial cellulose paper. *ACS Sustainable Chemistry & Engineering, 6*, 9178–9188.

Jafarzadeh, S., Nafchi, A. M., Salehabadi, A., Oladzad-Abbasabadi, N., & Jafari, S. M. (2021). Application of bio-nanocomposite films and edible coatings for extending the shelf life of fresh fruits and vegetables. *Advances in Colloid and Interface Science, 291*, 102405.

Kaewklin, P., Siripatrawan, U., Suwanagul, A., & Lee, Y. S. (2018). Active packaging from chitosan-titanium dioxide nanocomposite film for prolonging the storage life of tomato fruit. *International Journal of Biological Macromolecules, 112*, 523–529.

Kah, M., Tufenkji, N., & White, J. C. (2019). Nano-enabled strategies to enhance crop nutrition and protection. *Nature Nanotechnology, 14*(6), 532–540.

Kakimova, Z., Orynbekov, D., Zharykbasova, K., Kakimov, A., Zharykbasov, Y., Mirasheva, G., Toleubekova, S., Muratbayev, A., & Ntsefong, G. N. (2023). Advancements in nano bio sensors for food quality and safety assurance—A review. *Potravinarstvo Slovak Journal of Food Sciences, 17*, 728–747. https://doi.org/10.5219/1903

Kalia, A., Kaur, M., Shami, A., Jawandha, S. K., Alghuthaymi, M. A., Thakur, A., & Abd-Elsalam, K. A. (2021). Nettle-leaf extract derived ZnO/CuO nanoparticle-biopolymer-based antioxidant and antimicrobial nanocomposite packaging films and their impact on extending the postharvest shelf life of guava fruit. *Biomolecules, 11*(2), 224.

Kashyap, P. L., Kumar, S., Jasrotia, P., Singh, D. P., & Singh, G. P. (2019). *Nanosensors for plant disease diagnosis: Current understanding and future perspectives* (pp. 189–205). Springer eBooks. https://doi.org/10.1007/978-3-319-97852-9_9

Khalil, A. M., Eljamal, O., Jribi, S., & Matsunaga, N. (2016). Promoting nitrate reduction kinetics by nanoscale zero valent iron in water via copper salt addition. *Chemical Engineering Journal, 287*, 367–380.
Kim, D. S., Choi, J. H., Kim, S., & Lim, J. H. (2019). Prediction of carotenoid content in tomato fruit using a fluorescence screening method. *Postharvest Biology and Technology, 156*, 110917.
King, T., Osmond-McLeod, M. J., & Duffy, L. L. (2018). Nanotechnology in the food sector and potential applications for the poultry industry. *Trends in Food Science and Technology, 72*, 62–73.
Kosaka, P. M., Pini, V., Ruz, J. J., Da Silva, R. A., González, M. U., Ramos, D., et al. (2014). Detection of cancer biomarkers in serum using a hybrid mechanical and optoplasmonic nanosensor. *Nature Nanotechnology, 9*(12), 1047–1053.
Kulkarni, M. B., Ayachit, N. H., & Aminabhavi, T. M. (2022). Recent advancements in nano biosensors: Current trends, challenges, applications, and future scope. *Biosensors, 12*(10), 892. https://doi.org/10.3390/bios12100892
Kumar, P., Chugh, P., Ali, S. S., Chawla, W., Sushmita, S., Kumar, R., Raval, A. V., Shamim, S., Bhatia, A., & Kumar, R. (2024). Trends of nanobiosensors in modern agriculture systems. *Applied Biochemistry and Biotechnology*. https://doi.org/10.1007/s12010-024-05039-6
Lee, S. Y., Lee, S. J., Choi, D. S., & Hur, S. J. (2015). Current topics in active and intelligent food packaging for the preservation of fresh foods. *Journal of the Science of Food and Agriculture, 95*, 2799–2810.
Li, D., Li, L., Luo, Z., Lu, H., & Yue, Y. (2017). Effect of nano-ZnO-packaging on chilling tolerance and pectin metabolism of peaches during cold storage. *Scientia Horticulturae, 225*, 128–133.
Li, Z., Li, Q., Li, R., Zhou, J., & Wang, G. (2021). The distribution and impact of polystyrene nanoplastics on cucumber plants. *Environmental Science and Pollution Research, 28*, 16042–16053.
Melo, C. B. N. F., de Mendonça Soares, B. L., Diniz, K. M., Leal, C. F., Canto, D., et al. (2018). Effects of fungal chitosan nanoparticles as eco-friendly edible coatings on the quality of postharvest table grapes. *Postharvest Biology and Technology, 139*, 56–66.
Meraat, R., Issazadeh, K., Abdolahzadeh, Z. A., & Faezi Ghasemi, M. (2020). Rapid detection of *Escherichia coli* by β-galactosidase biosensor based on ZnO NPs and MWCNTs: A comparative study. *Current Microbiology, 77*, 2633–2641.
Miguel-Rojas, C., & Pérez-de-Luque, A. (2023). Nanobiosensors and nanoformulations in agriculture: New advances and challenges for sustainable agriculture. *Emerging Topics in Life Sciences, 7*(2), 229–238.
Moon, J., Byun, J., Kim, H., Lim, E. K., Jeong, J., Jung, J., & Kang, T. (2018). On-site detection of aflatoxin B1 in grains by a palm-sized surface plasmon resonance sensor. *Sensors, 18*(2), 598.
Naresh, V., & Lee, N. (2021). A review on biosensors and recent development of nanostructured materials-enabled biosensors. *Sensors, 21*(4), 1109. https://doi.org/10.3390/s21041109
Neethirajan, S., & Jayas, D. S. (2011). Nanotechnology for the food and bioprocessing industries. *Food and Bioprocess Technology, 4*(1), 39–47.
Pérez-de-Luque, A. (2017). Interaction of nanomaterials with plants: What do we need for real applications in agriculture. *Frontiers in Environmental Science, 5*, 12.
Prasad, R., Kumar, M., & Nanotechnology, V. K. (2017). *An agricultural paradigm* (pp. 1–372). Springer.
Priyanka, P., Kumar, D., Yadav, A., & Yadav, K. (2020). *Nanobiotechnology and its application in agriculture and food production* (pp. 105–134). Nanotechnology in the Life Sciences. https://doi.org/10.1007/978-3-030-31938-0_6
Rai, R., Nalini, P., & Singh, Y. P. (2022). *Nanotechnology for sustainable horticulture development: Opportunities and challenges* (pp. 191–210). Springer eBooks. https://doi.org/10.1007/978-3-030-90549-1_12
Rajput, V. D., Minkina, T., Sushkova, S., Tsitsuashvili, V., Mandzhieva, S., Gorovtsov, A., et al. (2018). Effect of nanoparticles on crops and soil microbial communities. *Journal of Soils and Sediments, 18*, 2179–2187.

Ramesh, M., Janani, R., Deepa, C., & Rajeshkumar, L. (2022). Nanotechnology-enabled biosensors: A review of fundamentals, design principles, materials, and applications. *Biosensors, 13*(1), 40. https://doi.org/10.3390/bios13010040

Rana, V. K., Pandey, A. K., Singh, R. P., Kumar, B., Mishra, S., & Ha, C. S. (2010). Enhancement of thermal stability and phase relaxation behavior of chitosan dissolved in aqueous l-lactic acid: Using 'silver nanoparticles' as nano filler. *Macromolecular Research, 18*, 713–720.

Reddy, V. S. R., Gundewadi, G., & Dasarahalli-Huligowda, L. K. (2022). Scope and applications of nanotechnology in horticulture. In *Nanotechnology applications in agricultural and bioprocess engineering* (pp. 33–77). Apple Academic Press.

Ronkainen, N. J., Halsall, H. B., & Heineman, W. R. (2010). Electrochemical biosensors. *Chemical Society Reviews, 39*(5), 1747. https://doi.org/10.1039/b714449k

Salvadori, M. R. (2019). Processing of nanoparticles by biomatrices in a green approach. In *Microbial nanobionics* (Vol. 1, pp. 1–28). State-of-the-Art.

Sanuja, S., Agalya, A., & Umapathy, M. J. (2015). Synthesis and characterization of zinc oxide-neem oil-chitosan bionanocomposite for food packaging application. *International Journal of Biological Macromolecules, 74*, 76–84.

Sharma, S., Shekhar, S., Gautam, S., Sharma, B., Kumar, A., & Jain, P. (2020). *Carbon-based nanomaterials as novel nanosensors* (pp. 323–347). In: Nanofabrication for Smart Nanosensor Applications.

Silvestre, C., Duraccio, D., & Cimmino, S. (2011). Food packaging based on polymer nanomaterials. *Progress in Polymer Science, 36*, 1766–1782.

Singh, G., & Rattanpal, H. (2014). Use of nanotechnology in horticulture: A review. *International Journal of Agricultural Sciences and Veterinary Medicine, 2*, 34–42.

Sode, K., Yamazaki, T., Lee, I., Hanashi, T., & Tsugawa, W. (2016). BioCapacitor: A novel principle for biosensors. *Biosensors & Bioelectronics, 76*, 20–28.

Sreejivungsa, K., Suchaichit, N., Moosophon, P., & Chompoosor, A. (2016). Light-regulated release of entrapped drugs from photoresponsive gold nanoparticles. *Journal of Nanomaterials, 1*, 4964693.

Suresh, S., Mathan, P., Sagadevan, S., & Periasamy, M. (2014). Recent trends in nanobiosensors and their applications—A review. *Reviews on Advanced Materials Science, 36*, 62–69.

Thakur, M., Wang, B., & Verma, M. L. (2022). Development and applications of nano biosensors for sustainable agricultural and food industries: Recent developments, challenges and perspectives. *Environmental Technology and Innovation, 26*, 102371. https://doi.org/10.1016/j.eti.2022.102371

Touhami, A. (2014). Biosensors and nanobiosensors: design and applications. In: Seifalian, A., de Mel A., & Kalaskar, D. M. (eds). *Nanomedicine (Chapter 15).* London(UK): One Central Press (OCP).

Valenzuela-Amaro, H. M., Aguayo-Acosta, A., Meléndez-Sánchez, E. R., de la Rosa, O., Vázquez-Ortega, P. G., Oyervides-Muñoz, M. A., et al. (2023). Emerging applications of nanobiosensors in pathogen detection in water and food. *Biosens, 13*(10), 922.

Verma, M. L. (2017). Nanobiotechnology advances in enzymatic biosensors for the agri-food industry. *Environmental Chemistry Letters, 15*(4), 555–560.

Xu, J., Zhang, M., Bhandari, B., & Kachele, R. (2017). ZnO nanoparticles combined radio-frequency heating: A novel method to control microorganisms and improve product quality of prepared carrots. *Innovative Food Science and Emerging Technologies, 44*, 46–53.

Yadav, S., Mehrotra, G. K., & Dutta, P. K. (2021). Chitosan-based ZnO nanoparticles loaded gallic-acid films for active food packaging. *Food Chemistry, 334*, 127605.

Zhang, W., Liu, X., Qiu, S., Chen, L., & Liu, X. (2022). LDPE/M-TiO2 composite films for modified atmosphere packaging to realize comprehensive preservation of strawberry (Fragaria × ananassa Duch.). *Journal of the Science of Food and Agriculture, 102*, 4589–4598.

Chapter 14
Interaction, Toxicity, and the Fate of Biosynthesized Nanoparticles in the Agroecosystem

Oyindamola John Samson, Racheal Oluwayemisi Fashogbon, Bukola Margaret Popoola, Theresa Abimbola Awotundun, Gideon Taiwo Otolorin, and Ismail Babatunde Onajobi

Abstract Biosynthesized nanoparticles (NPs) are emerging as promising tools in agriculture for enhancing crop productivity, nutrient uptake, and pest control. This chapter describes the interaction, toxicity, and fate of biosynthesized NPs within the agroecosystem. Various types of NPs, including metal, metal oxide, bimetallic, quantum dots, carbon-based, and organic NPs, are discussed, highlighting their unique compositions and characteristics. The complex interactions of NPs within the soil-plant system are explored, emphasizing their interactions with soil components, agricultural chemicals, and soil microbes. Factors influencing NP mobility and bioavailability, such as soil type, pH, and organic matter content, are examined. The chapter also addresses the uptake and translocation of NPs by plants and their impact on plant growth. The adsorption process within the soil matrix, which governs NP mobility and effectiveness, is thoroughly explained. As research continues, it is essential to understand these interactions thoroughly to balance the potential benefits and risks of using biosynthesized nanoparticles in agriculture, ensuring that

O. J. Samson (✉)
Department of Microbiology and Biotechnology, Faculty of Natural Sciences, Ajayi Crowther University, Oyo, Nigeria

Department of Microbiology, Faculty of Science, Olabisi Onabanjo University, Ago-Iwoye, Nigeria

R. O. Fashogbon · B. M. Popoola · T. A. Awotundun
Department of Microbiology and Biotechnology, Faculty of Natural Sciences, Ajayi Crowther University, Oyo, Nigeria

G. T. Otolorin
Department of Chemistry, Faculty of Science, University of Ibadan, Ibadan, Nigeria

I. B. Onajobi
Department of Microbiology, Faculty of Science, Olabisi Onabanjo University, Ago-Iwoye, Nigeria

V. D. Rajput et al. (eds.), *Contemporary Suitability of Nanobionics in Agriculture*, Nanotechnology in the Life Sciences, https://doi.org/10.1007/978-3-031-82779-2_14

soil health, crop productivity, and ecosystem functioning are maintained in the long term.

Keywords Biosynthesized nanoparticles · Agroecosystem · Soil-plant interactions · Crop productivity · Nutrient uptake · Pest control · Adsorption process

14.1 Introduction

Biosynthesized nanoparticles, often referred to as biogenic nanoparticles, are an innovative convergence of nanotechnology and biological systems (Devi et al., 2024). These nanoparticles, typically measuring between 1 and 100 nanometers, are synthesized through the use of living organisms or their derivatives (Bhati, 2023). Unlike chemically produced nanoparticles, biosynthesized ones utilize the natural metabolic processes of organisms such as bacteria, fungi, plants, and some animals to create nanostructures with distinctive properties (Veerasamy, 2024).

In recent years, the field of biosynthesized nanoparticles has garnered substantial attention due to its potential for producing environmentally friendly and biocompatible nanomaterials (Devi et al., 2024). These nanoparticles can be composed of various materials, including metals like gold, silver, and iron, as well as metal oxides and organic compounds. The biological synthesis process often imparts unique characteristics to the nanoparticles, such as specific shapes, sizes, and surface properties, which can be advantageous for a wide range of applications (Zafar, 2023).

The potential applications of biosynthesized nanoparticles in agriculture and environmental management are both vast and promising. In agriculture, these nanoparticles are being explored for their ability to enhance crop productivity, improve nutrient uptake, and provide protection against pests and diseases (Guleria et al., 2023). Key areas of application include the development of nanofertilizers that enable slow and controlled nutrient release, nanopesticides that offer more effective and environmentally friendly pest control (Chhipa & Joshi, 2016), nanosensors for detecting soil conditions and plant health (Johnson et al., 2021), and nanoparticle seed treatments that enhance germination rates and early plant growth (Xin et al., 2020).

Beyond agriculture, biosynthesized nanoparticles hold significant potential in environmental applications (Singh et al., 2019). These include water purification, where nanoparticles can remove contaminants like heavy metals and organic pollutants (Baby et al., 2022); soil remediation, where they can aid in breaking down pollutants or immobilizing toxic elements; air purification through nanoparticle-enhanced filters; and environmental sensing, where they can be used to develop highly sensitive and specific monitoring tools (Ojha, 2020).

The significance of biosynthesized nanoparticles lies in their ability to provide more sustainable and eco-friendly solutions compared to conventional methods. Their biological origin often results in greater biocompatibility and a reduced environmental footprint, aligning with the growing demand for green technologies in agriculture and environmental management (Santhosh et al., 2022). This chapter aims to offer an overview of the interaction, toxicity, and fate of biosynthesized nanoparticles within the agroecosystem, providing a solid foundation for understanding their role and impact.

14.2 Biosynthesis of Nanoparticles

14.2.1 Methods of Biosynthesis

The biosynthesis of nanoparticles is an innovative process that harnesses the innate abilities of living organisms to create nanoscale materials. This approach capitalizes on the natural metabolic processes of various biological entities to produce nanostructures with unique properties (Eskandar, 2023). The field has gained significant attention due to its potential for creating more environmentally friendly and biocompatible nanomaterials, with applications spanning across multiple industries (Sonkaria & Khare, 2020).

Three main approaches dominate the biosynthesis of nanoparticles: microbial synthesis, plant-mediated synthesis, and enzyme-based synthesis. Microbial synthesis utilizes microorganisms such as bacteria, fungi, and yeasts to produce nanoparticles through their metabolic pathways (Koul et al., 2021). This process typically involves culturing the microorganism, exposing it to precursor materials, allowing biosynthesis to occur under controlled conditions, and finally extracting and purifying the synthesized nanoparticles (Saravanan et al., 2021).

Plant-mediated synthesis, on the other hand, leverages the rich array of biomolecules present in plants to facilitate nanoparticle formation (Ramos & Regulacio, 2021). This method is often considered more environmentally friendly and scalable compared to microbial methods. The process involves preparing plant extracts, mixing them with precursor solutions, incubating the mixture to allow nanoparticle formation, and then separating and purifying the resulting nanoparticles (Mukherjee et al., 2024).

Enzyme-based synthesis offers a more controlled approach to nanoparticle production. This method uses isolated enzymes or enzyme systems to synthesize nanoparticles in vitro, often resulting in more uniform nanoparticles (Ovais et al., 2018). The process includes isolating or producing the desired enzymes, preparing the reaction mixture, controlling reaction conditions for optimal synthesis, and purifying the final product. Each of these biosynthesis methods presents unique advantages and challenges, contributing to the diverse landscape of nanoparticle production techniques (Savage et al., 2019).

Biosynthesized nanoparticles can be categorized based on their composition and unique properties, each offering distinct advantages for various applications. Metal nanoparticles are among the most studied in this category. Gold nanoparticles

(AuNPs) are renowned for their exceptional optical properties and biocompatibility, making them valuable in imaging and therapeutic applications (Milan et al., 2022). Silver nanoparticles (AgNPs) are highly regarded for their potent antimicrobial properties, which find utility in medical devices and coatings (Bamal et al., 2021). Platinum nanoparticles (PtNPs) are integral to catalysis and fuel cell technologies due to their catalytic efficiency, while palladium nanoparticles (PdNPs) play a crucial role in catalytic converters and chemical sensors, enhancing environmental and analytical processes (Jeyaraj et al., 2019).

Metal oxide nanoparticles, another significant class, include iron oxide nanoparticles (Fe_3O_4), which are extensively used in magnetic applications and environmental remediation efforts (Venkateshaiah et al., 2022). Zinc oxide nanoparticles (ZnONPs) are commonly employed in sunscreens and as antimicrobial agents due to their protective and antibacterial characteristics (Husen, 2019). Titanium dioxide nanoparticles (TiO_2NPs) are versatile, with applications spanning photocatalysis and solar cells, leveraging their ability to drive chemical reactions under light exposure (Thakur et al., 2024).

Bimetallic nanoparticles represent an advanced category, consisting of two different metals, such as gold-silver (Au-Ag) or palladium-platinum (Pd-Pt). These nanoparticles often exhibit synergistic effects, providing enhanced properties compared to their single-metal counterparts. The combination of metals in bimetallic nanoparticles can lead to improved catalytic activity, stability, and reactivity, making them suitable for a range of industrial and environmental applications (Li et al., 2021).

Quantum dots, another intriguing type of nanoparticle, are semiconductor particles that possess unique optical and electronic properties. Made from materials such as cadmium selenide (CdSe) or zinc sulfide (ZnS), quantum dots are used in imaging, sensing, and display technologies due to their size-tunable fluorescence (Terna et al., 2021). Carbon-based nanoparticles, including carbon nanotubes and graphene quantum dots, are produced by certain microorganisms and are known for their exceptional strength and conductivity. Finally, organic nanoparticles, which consist of materials like proteins, lipids, or polymers, are primarily used in drug delivery systems, providing targeted therapeutic delivery and reduced side effects (Alavi et al., 2021).

Figure 14.1 illustrate the biosynthesis of nanoparticles (NPs), classifying them into six categories: metal NPs, metal oxide NPs, bimetallic NPs, quantum dots, carbon-based NPs, and organic NPs. Each category has its unique composition (e.g., gold, silver, iron oxide, and carbon nanotubes) and distinct characteristics, such as antimicrobial activity, magnetic uses, optical properties, and drug delivery optimization. These nanoparticles are noted for their biocompatibility, catalysis, and applications in areas like sensors, solar cells, and fuel cells. They often offer enhanced properties compared to conventional materials.

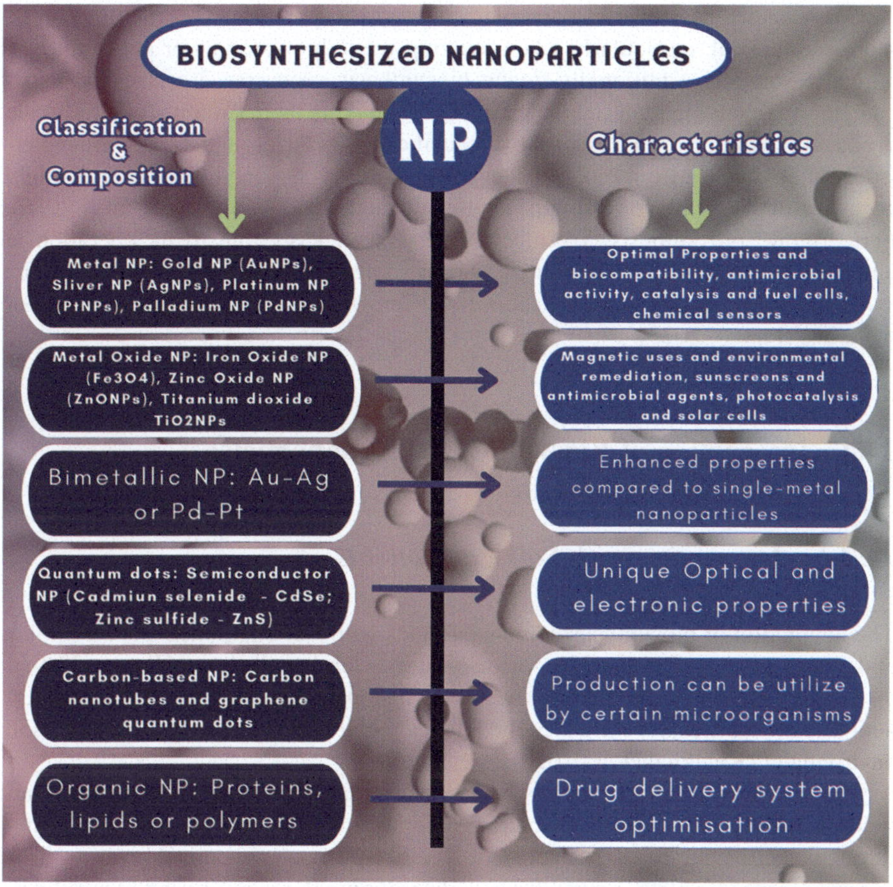

Fig. 14.1 Systematic classification and characteristics of biosynthesized of nanoparticles (NPs)

14.3 Advantages Over Chemically Synthesized Nanoparticles

Biosynthesized nanoparticles present several compelling advantages over their chemically synthesized counterparts. First, the eco-friendliness of biosynthesis is a major benefit. This process typically relies on non-toxic, renewable materials and operates under mild conditions, which minimizes environmental impact. In contrast, chemical synthesis often involves harsh reagents and energy-intensive procedures, contributing to a larger ecological footprint (Devi et al., 2024).

Another significant advantage is biocompatibility. Nanoparticles synthesized biologically are frequently coated with biomolecules from the host organism, which enhances their compatibility with biological systems and reduces potential toxicity. This natural coating can make them safer for use in medical and environmental applications compared to chemically synthesized nanoparticles, which might lack such biocompatibility (Samuel et al., 2022).

Cost-effectiveness is also a notable benefit of biosynthesis. This method can be more economical, particularly when utilizing waste biomaterials or abundant biological resources. Additionally, many biosynthesis processes are scalable, especially those involving plant-mediated synthesis, making them suitable for large-scale production. Unique properties, such as specific shapes, sizes, and surface characteristics, often emerge from biological synthesis, providing functionalities that might be difficult to achieve through chemical methods (Darroudi et al., 2020). Furthermore, biosynthesis typically requires less energy due to its operation at ambient temperatures and pressures, and in some instances, nanoparticles can be synthesized in situ, allowing for their direct integration into biological systems or materials (Pattoo, 2023).

14.4 Interaction of Biosynthesized Nanoparticles in the Agroecosystem

14.4.1 Interaction with Soil Components

The interaction of biosynthesized nanoparticles with soil components is a complex and dynamic process, shaped by various factors such as soil type, pH, organic matter content, and the inherent properties of the nanoparticles themselves. These interactions can significantly influence the fate and behavior of nanoparticles in the soil environment (Khalkhal et al., 2020).

One of the primary interactions is the adsorption of nanoparticles onto soil particles. Nanoparticles can attach to clay minerals, organic matter, and metal oxides through electrostatic forces, van der Waals interactions, and chemical bonding. This adsorption process plays a crucial role in determining the mobility and bioavailability of the nanoparticles within the soil matrix. Depending on the strength and nature of these interactions, nanoparticles may become immobilized or remain available for further reactions and interactions (Li et al., 2024).

Aggregation and dissolution are additional key processes influenced by soil conditions. In certain environments, nanoparticles may aggregate, forming larger clusters that reduce their reactivity and mobility. This aggregation can lead to decreased interaction with soil components and organisms. Conversely, some nanoparticles may undergo dissolution, releasing ions that can engage with soil constituents or be absorbed by living organisms. The balance between aggregation and dissolution is critical for understanding the long-term behavior and potential impacts of nanoparticles in soil (Shrestha et al., 2020).

Nanoparticles can also influence soil structure, potentially altering aggregation and porosity. These changes can affect soil properties such as water retention, air exchange, and nutrient dynamics, with broader implications for soil health and plant growth (Sun et al., 2020). Moreover, the interaction of nanoparticles with the soil solution, which contains dissolved organic matter, ions, and other solutes, can impact the stability and behavior of nanoparticles. Finally, nanoparticles may interact with extracellular enzymes in the soil, potentially enhancing or inhibiting their activity. Such interactions can have significant consequences for nutrient cycling and the overall functioning of the soil ecosystem (Ameen et al., 2021).

14.4.2 Interaction with Plants

The interaction between biosynthesized nanoparticles and plants is a crucial factor in understanding their potential benefits and risks within agroecosystems. These interactions can significantly influence plant growth, development, and overall health (Singh & Kumar, 2020).

One of the primary interactions is the uptake and translocation of nanoparticles by plants. Nanoparticles can be absorbed by plant roots, with smaller particles having a higher likelihood of penetrating root tissues. Once inside the plant, these nanoparticles can be translocated through the vascular system, reaching different parts of the plant, such as stems, leaves, and even fruits. This movement within the plant can affect various physiological processes and potentially alter the plant's development (Wang et al., 2023).

Nanoparticles can also impact seed germination and early seedling growth. In some cases, certain nanoparticles have been found to enhance germination rates and promote seedling vigor, possibly due to improved water uptake or their antimicrobial properties that protect seeds from pathogens (Guo et al., 2022). Additionally, nanoparticles can influence nutrient uptake in plants. For instance, iron oxide nanoparticles can serve as a direct source of iron, while other nanoparticles may alter soil chemistry, affecting the availability and absorption of essential nutrients by plants (Raiesi-Ardali et al., 2022).

Moreover, the impact of nanoparticles on photosynthesis is another critical area of concern. While some nanoparticles may enhance photosynthetic activity by improving light capture or electron transport, others might disrupt light absorption or interfere with chloroplast function, leading to reduced photosynthetic efficiency (Mony et al., 2022). The induction of oxidative stress is another potential risk associated with nanoparticle exposure. Certain nanoparticles can generate reactive oxygen species (ROS) within plant tissues, which can lead to oxidative stress, causing cellular damage and affecting overall plant health. These interactions can also result in morphological changes, such as alterations in root and shoot growth, which can have long-term effects on plant development and productivity (Sharma et al., 2019).

14.4.3 Interaction with Microorganisms

Soil microorganisms are essential for the functioning of ecosystems, playing vital roles in nutrient cycling, organic matter decomposition, and maintaining plant health. The introduction of biosynthesized nanoparticles into soil environments can have profound effects on these microbial communities, with significant implications for soil ecology. One of the most notable effects is the antimicrobial activity of certain nanoparticles, particularly those made from silver and copper (Rajput et al., 2021). These nanoparticles can disrupt the composition and diversity of soil microbial communities, potentially reducing populations of beneficial microbes and altering the balance of microbial ecosystems.

Beyond their antimicrobial properties, nanoparticles can also influence the metabolic activities of soil microorganisms. This can have far-reaching consequences for critical soil processes such as nitrogen fixation, denitrification, and the decomposition of organic matter. Changes in these metabolic activities could lead to shifts in nutrient availability, affecting plant growth and overall soil fertility (Mishra et al., 2020). Additionally, nanoparticles may impact the formation of microbial biofilms, which are crucial for soil structure and nutrient cycling. Depending on the type of nanoparticle and environmental conditions, biofilm formation could be either promoted or inhibited, with implications for soil stability and microbial interactions (Rajput et al., 2021).

Furthermore, the presence of nanoparticles in soil can affect symbiotic relationships between plants and microorganisms. Proving this, nanoparticles might interfere with the beneficial interactions between plants and mycorrhizal fungi or nitrogen-fixing bacteria, which are essential for plant nutrient uptake and soil health (Khan, 2020). Another emerging concern is the potential influence of nanoparticles on horizontal gene transfer among soil microorganisms. This process, which contributes to microbial evolution and the spread of antibiotic resistance, could be accelerated or altered by the presence of nanoparticles, leading to unexpected changes in microbial communities and their functions (Markowicz, 2023).

14.4.4 Interaction with Agricultural Chemicals

The introduction of biosynthesized nanoparticles into agricultural soils can significantly affect the behavior and effectiveness of various agrochemicals, such as pesticides and fertilizers. One of the primary ways these nanoparticles interact with agrochemicals is through adsorption. Nanoparticles can adsorb pesticides, potentially modifying their mobility, degradation rate, and bioavailability within the soil. This interaction may either limit the effectiveness of pesticides by reducing their availability to target pests or prolong their presence in the environment, raising concerns about potential residues and environmental impact (Tao et al., 2023).

Beyond adsorption, biosynthesized nanoparticles are being explored as carriers for agrochemicals like pesticides and fertilizers. By encapsulating or attaching these chemicals to nanoparticles, researchers aim to enhance their delivery to specific targets, thereby increasing efficacy while minimizing environmental exposure. This approach could lead to more efficient use of agrochemicals, reducing the amount needed and lowering the risk of contamination in non-target areas, ultimately contributing to more sustainable agricultural practices (Liu et al., 2018).

Nanoparticles may also exhibit catalytic properties, which can influence the degradation of pesticides or other organic contaminants in the soil (Tao et al., 2023). This catalytic activity could either enhance the breakdown of harmful substances, reducing their persistence in the environment, or unintentionally degrade beneficial compounds, impacting soil health. Additionally, the interaction between nanoparticles and fertilizers can affect the release and availability of nutrients, potentially improving nutrient use efficiency. However, the combination of nanoparticles with agrochemicals may also lead to synergistic or antagonistic effects on crops or target organisms, complicating predictions of their overall impact (Rasool et al., 2022).

Figure 14.2 systematically describes the complex interactions of biosynthesized nanoparticles (NPs) within the soil-plant system. It highlights how NPs interact with soil components (influenced by soil type, pH, organic matter), affecting their mobility and bioavailability. NPs also interact with agricultural chemicals, potentially altering pesticide mobility and degradation. Their impact extends to soil microbes, influencing processes like nitrogen fixation and organic matter decomposition. Additionally, plants take up and translocate NPs, impacting their overall growth. The adsorption process within the soil matrix governs the mobility and effectiveness of these NPs.

14.5 Toxicity of Biosynthesized Nanoparticles

The toxicity of biosynthesized nanoparticles arises from their ability to generate reactive oxygen species (ROS), leading to oxidative stress that damages critical biomolecules like proteins, lipids, and DNA. This stress disrupts cellular functions and compromises cell viability. Additionally, nanoparticles physically interact with and disrupt cellular membranes, creating entry points for deeper infiltration, worsening their toxic effects. The release of toxic ions from metal-based nanoparticles, such as silver and copper, further adds to their toxicity by interfering with essential metabolic processes, such as enzyme activity and nutrient transport, making them potent agents of cellular harm (Mammari et al., 2022).

Certain biosynthesized nanoparticles can benefit plant growth by enhancing nutrient uptake, promoting seed germination, and acting as protective agents against pathogens. For example, nanoparticles providing essential micronutrients like iron or zinc can be more efficiently absorbed by plants, leading to improved growth and development. However, at high concentrations or specific sizes, nanoparticles can become toxic, causing inhibited root and shoot growth, chlorosis, and reduced

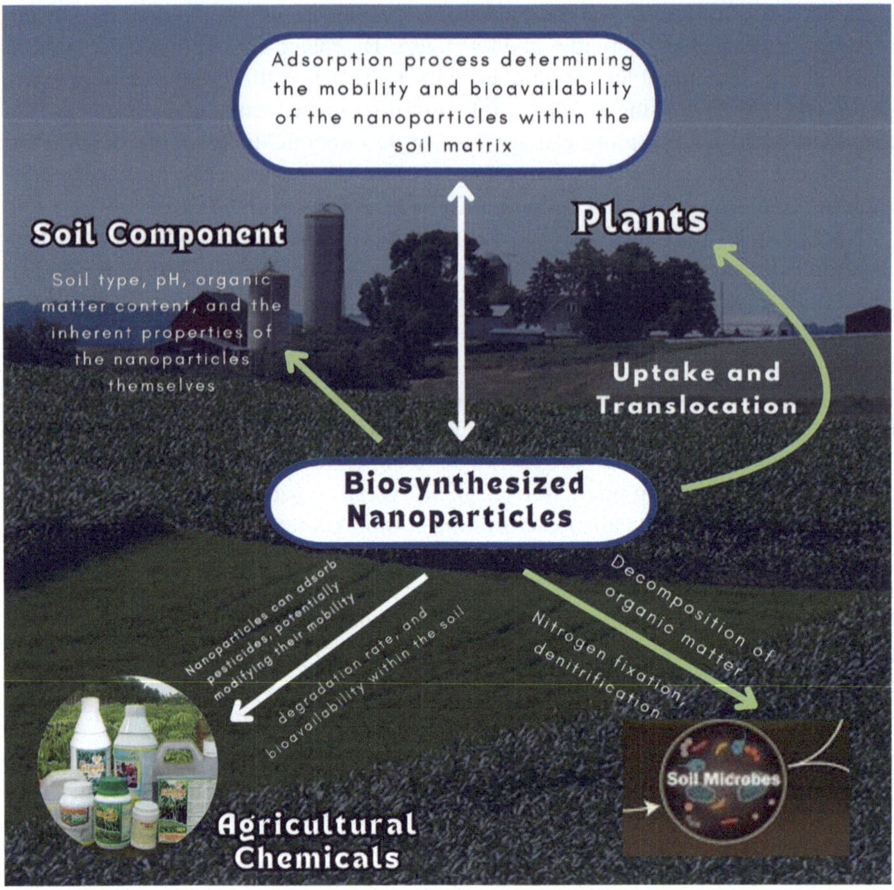

Fig. 14.2 Systematic interactions of biosynthesized nanoparticles in agricultural ecosystems

photosynthetic efficiency. This toxicity is due to oxidative stress, disrupted cellular functions, or interference with water and nutrient absorption. Factors such as nanoparticle concentration, size, surface charge, and plant species significantly influence toxicity levels (Amna Ali et al., 2022).

Soil microorganisms are vital for nutrient cycling and soil health, but biosynthesized nanoparticles, especially silver or copper-based, can disrupt microbial communities, reducing biodiversity and impairing ecosystem functions like nitrogen fixation and decomposition. Additionally, nanoparticles can harm soil fauna, such as earthworms and insects, affecting mobility, reproduction, and survival, depending on nanoparticle type, concentration, and exposure (Parada et al., 2018).

Bioaccumulation occurs when nanoparticles are absorbed faster than they are eliminated, leading to their buildup in organisms like earthworms, insects, and microbes, which are key to soil health. As predators consume these organisms, nanoparticles can biomagnify, accumulating in higher concentrations at each

Table 14.1 Mechanisms and impacts of biosynthesized nanoparticle toxicity

Mechanism categories	Action	Impact	Factors influencing toxicity	References
Mechanisms of toxicity	ROS generation, membrane disruption, release of toxic ions (e.g., silver, copper)	Oxidative stress, damage to proteins, lipids, DNA, impaired cell viability	Size, surface charge, type of metal ions, interaction with cellular components	Mammari et al. (2022)
Toxicity to plants	Enhanced growth at low concentrations (nutrient uptake, seed germination); toxicity at high concentrations	Inhibited root/shoot growth, chlorosis, reduced photosynthetic efficiency	Concentration, size, surface charge, plant species	Amna Ali et al. (2022)
Toxicity to soil microorganisms	Antimicrobial effects on microbial communities, reduced nitrogen fixation, organic matter decomposition	Reduced biodiversity, impaired nutrient cycling, negative effects on soil fauna (earthworms, nematodes, insects)	Nanoparticle type (silver, copper), concentration, exposure duration	Parada et al. (2018)
Bioaccumulation and biomagnification	Nanoparticles accumulate in organisms, potentially magnifying up the food chain	Bioaccumulation in earthworms, insects, microbes, leading to toxic effects in predators (fish, birds, mammals)	Nanoparticle size, type, surface chemistry, ability to penetrate membranes	Singh and Gurjar (2022)

trophic level and potentially disrupting ecosystems. Factors influencing bioaccumulation and biomagnification include nanoparticle size, surface chemistry, and metal content (e.g., silver, copper). Surface coatings can also affect how nanoparticles interact with organisms. In agricultural settings, the long-term ecological impact of nanoparticle use requires careful consideration to prevent harm to the food web (Singh & Gurjar, 2022). Further details are discussed in Table 14.1.

14.6 Fate of Biosynthesized Nanoparticles in the Agroecosystem

Biosynthesized nanoparticles move through soil via leaching, diffusion, and adsorption. Leaching carries nanoparticles with water, while diffusion moves them from higher to lower concentrations. Adsorption binds nanoparticles to soil particles, impacting mobility based on factors like soil texture, pH, and organic matter. Sandy soils allow greater movement, while clayey soils restrict it. Smaller nanoparticles are more mobile, raising concerns about groundwater contamination, particularly in agriculture, as they can carry contaminants deeper into the soil, posing environmental and water safety risks (Khalkhal et al., 2020).

The above outlines the mechanisms of toxicity and impacts of biosynthesized nanoparticles across various systems. Key actions include ROS generation, membrane disruption, and toxic ion release, leading to oxidative stress in cells. In plants, nanoparticles enhance growth at low concentrations but cause toxicity at high levels. For soil microorganisms, nanoparticle exposure reduces biodiversity and impairs nutrient cycling. Additionally, bioaccumulation and biomagnification present risks for organisms higher up the food chain.

Once biosynthesized nanoparticles enter the soil, they undergo transformations such as oxidation, dissolution, and aggregation. Oxidation can alter their surface chemistry, affecting reactivity and interactions with soil components. Dissolution, especially in metal-based nanoparticles like silver or zinc, may release toxic ions harmful to soil organisms and plants. Aggregation reduces their surface area and reactivity but may also form larger particles that impact mobility and biota interactions differently. These processes, influenced by pH, moisture, and organic matter, make predicting the environmental impact of nanoparticles complex (Abbas et al., 2020).

The persistence of biosynthesized nanoparticles in the environment is largely due to their resistance to natural degradation processes like chemical breakdown, microbial degradation, and physical weathering. Nanoparticles that resist these processes can accumulate in the soil, leading to prolonged exposure to plants, microorganisms, and soil organisms. This long-term presence raises concerns about chronic effects, as even low levels of exposure can disrupt ecological balance, impair soil processes such as nutrient cycling and water retention, and potentially affect plant growth and human health through the food chain.

External environmental factors, such as soil pH, organic matter content, and the presence of competing ions, further influence the bioavailability of nanoparticles. For instance, a low soil pH can increase the solubility of certain metal-based nanoparticles, making them more bioavailable and thus more likely to be taken up by plants and microorganisms. Conversely, high levels of organic matter can bind to nanoparticles, reducing their bioavailability by preventing them from interacting directly with soil organisms. Similarly, the presence of competing ions in the soil can alter the nanoparticles' charge or binding affinity, affecting their mobility and uptake (Abbas et al., 2020).

14.7 Environmental and Agricultural Implications

Impact on Soil Health and Fertility

Some nanoparticles may enhance soil fertility by increasing the availability of essential nutrients, thereby supporting plant growth and improving overall soil quality. For instance, nanoparticles that facilitate the release of nutrients or improve their uptake by plants can contribute to more fertile and productive soils. Additionally, certain nanoparticles can stimulate beneficial microbial activity, leading to enhanced decomposition of organic matter and improved nutrient cycling.

Conversely, the presence of nanoparticles in the soil can also have detrimental effects, such as disrupting microbial communities or contributing to soil contamination. Nanoparticles with antimicrobial properties may inhibit the growth of beneficial soil microorganisms, leading to imbalances in microbial populations and impaired soil functions. Furthermore, nanoparticles that persist in the soil may accumulate and potentially cause long-term contamination, impacting soil health and ecosystem stability. Achieving a balance between the positive and negative effects of nanoparticles is essential for promoting sustainable agricultural practices and ensuring that their application does not compromise soil health (Ameen et al., 2021).

The interaction of biosynthesized nanoparticles with crops can lead to both positive and negative outcomes. While certain nanoparticles may promote plant growth and enhance crop yields, others may inhibit root development, reduce photosynthetic efficiency, or induce oxidative stress. The impact on crop quality, including nutrient content and safety, must also be considered (Landa, 2021).

The potential for biosynthesized nanoparticles to be absorbed by crops and enter the food chain raises significant food safety concerns. If nanoparticles accumulate in edible plant parts, they may pose risks to human health, especially if they exhibit toxic properties or accumulate over time. Due to their novel and varied chemical compositions and sizes, nanoparticles' interactions with biological systems can be unpredictable. Therefore, rigorous monitoring and regulation of nanoparticle levels in agricultural products are crucial to ensure food safety. Additionally, the ecological implications are significant, as nanoparticles may disrupt food webs, affect ecosystem functions, and influence species interactions (Chen et al., 2023). Nanoparticle risk assessment indicates potential hazards of nanoparticles to human health and the environment, focusing on toxicity, exposure pathways, bioaccumulation, and interactions with biological systems, aiding safe design, regulation, and sustainable application of nanotechnology. This includes variability in properties as shown in Table 14.2, lack of long-term data, and limited testing methods. Solutions proposed include standardized characterization methods, extended monitoring programs, and development of new analytical tools.

Table 14.2 Addressing challenges in nanoparticle risk assessment

Challenges in risk assessment	Description	Potential solution	References
Variability in nanoparticle properties	Inconsistent behavior in the environment	Standardized characterization methods	Sheteiwy et al. (2021)
Lack of long-term data	Uncertain long-term impacts	Extended monitoring programs	Chen et al. (2023)
Limited testing methods	Insufficient risk prediction	Development of new analytical tools	Sheteiwy et al. (2021)

14.8 Conclusion

Biosynthesized nanoparticles hold considerable promise for advancing agricultural practices, offering benefits such as increased crop productivity, enhanced nutrient delivery, and minimized environmental impact. Nevertheless, their use brings substantial concerns related to toxicity, persistence, and potential effects on soil health and ecosystems. This chapter has discussed the intricate interactions between these nanoparticles and soil components, plants, and microorganisms. A thorough knowledge of these interactions is vital for assessing the complete advantages and risks of using biosynthesized nanoparticles in agriculture. Ensuring their safe and sustainable application will require robust regulation, thorough risk assessments, and the advancement of analytical techniques. Looking forward, the successful integration of biosynthesized nanoparticles into agriculture hinges on ongoing research and innovation, particularly in understanding their long-term environmental effects. As our insights into these materials expand more and more, it will be crucial to refine regulatory frameworks and establish best practices for their use. Biosynthesized nanoparticles can revolutionize agriculture, emphasizing the need for a balanced approach that maximizes technological benefits while addressing ecological and health risks for sustainable and responsible application. Effective collaboration among scientists, policymakers, and industry stakeholders will be essential for overcoming these challenges and leveraging nanotechnology to foster a sustainable agricultural future.

References

Abbas, Q., Yousaf, B., Ali, M. U., Munir, M. A. M., El-Naggar, A., Rinklebe, J., & Naushad, M. (2020). Transformation pathways and fate of engineered nanoparticles (ENPs) in distinct interactive environmental compartments: A review. *Environment International, 138*, 105646. https://doi.org/10.1016/j.envint.2020.105646

Alavi, M., Jabari, E., & Jabbari, E. (2021). Functionalized carbon-based nanomaterials and quantum dots with antibacterial activity: A review. *Expert Review of Anti-Infective Therapy, 19*(1), 35–44. https://doi.org/10.1080/14787210.2020.1810569

Ameen, F., Alsamhary, K., Alabdullatif, J. A., & Al Nadhari, S. (2021). A review on metal-based nanoparticles and their toxicity to beneficial soil bacteria and fungi. *Ecotoxicology and Environmental Safety, 213*, 112027.

Amna Ali, B., Azeem, M. A., Qayyum, A., Mustafa, G., Ahmad, M. A., Javed, M. T., & Chaudhary, H. J. (2022). Bio-fabricated silver nanoparticles: A sustainable approach for augmentation of plant growth and pathogen control. In *Sustainable agriculture reviews 53: Nanoparticles: A new tool to enhance stress tolerance* (pp. 345–371). Springer International Publishing. https://doi.org/10.1007/978-3-030-86876-5_14

Baby, R., Hussein, M. Z., Abdullah, A. H., & Zainal, Z. (2022). Nanomaterials for the treatment of heavy metal contaminated water. *Polymers, 14*(3), 583. https://doi.org/10.3390/polym14030583

Bamal, D., Singh, A., Chaudhary, G., Kumar, M., Singh, M., Rani, N., Mundlia, P., & Sehrawat, A. R. (2021). Silver nanoparticles biosynthesis, characterization, antimicrobial activities, applications, cytotoxicity and safety issues: An updated review. *Nanomaterials, 11*(8), 2086. https://doi.org/10.3390/nano11082086

Bhati, M. (2023). Biogenic synthesis of metallic nanoparticles: Principles and applications. *Materials Today Proceedings, 81*, 882–887. https://doi.org/10.1016/j.matpr.2021.04.272

Chen, J., Guo, Y., Zhang, X., Liu, J., Gong, P., Su, Z., Fan, L., & Li, G. (2023). Emerging nanoparticles in food: Sources, application, and safety. *Journal of Agricultural and Food Chemistry, 71*(8), 3564–3582. https://doi.org/10.1021/acs.jafc.2c06740

Chhipa, H., & Joshi, P. (2016). Nanofertilisers, nanopesticides and nanosensors in agriculture. In *Nanoscience in food and agriculture 1* (pp. 247–282). https://doi.org/10.1007/978-3-319-39303-2_9

Darroudi, M., Yazdi, M. E. T., & Amiri, M. S. (2020). Plant-mediated biosynthesis of nanoparticles. In *21st century nanoscience–A handbook* (pp. 1–1). CRC Press.

Devi, L., Kushwaha, P., Ansari, T. M., Kumar, A., & Rao, A. (2024). Recent trends in biologically synthesized metal nanoparticles and their biomedical applications: A review. *Biological Trace Element Research, 202*(7), 3383–3399. https://doi.org/10.1007/s12011-023-03920-9

Eskandar, K. (2023). Revolutionizing biotechnology and bioengineering: Unleashing the power of innovation. *Journal of Applied Biology & Biotechnology, 10*(3), 81–88.

Guleria, G., Thakur, S., Shandilya, M., Sharma, S., Thakur, S., & Kalia, S. (2023). Nanotechnology for sustainable agro-food systems: The need and role of nanoparticles in protecting plants and improving crop productivity. *Plant Physiology and Biochemistry, 194*, 533–549. https://doi.org/10.1016/j.plaphy.2022.12.004

Guo, H., Liu, Y., Chen, J., Zhu, Y., & Zhang, Z. (2022). The effects of several metal nanoparticles on seed germination and seedling growth: A meta-analysis. *Coatings, 12*(2), 183. https://doi.org/10.3390/coatings12020183

Husen, A. (2019). Natural product-based fabrication of zinc-oxide nanoparticles and their applications. In *Nanomaterials and plant potential* (pp. 193–219). https://doi.org/10.1007/978-3-030-05569-1_7

Jeyaraj, M., Gurunathan, S., Qasim, M., Kang, M. H., & Kim, J. H. (2019). A comprehensive review on the synthesis, characterization, and biomedical application of platinum nanoparticles. *Nanomaterials, 9*(12), 1719. https://doi.org/10.3390/nano9121719

Johnson, M. S., Sajeev, S., & Nair, R. S. (2021). Role of nanosensors in agriculture. In *2021 international conference on computational intelligence and knowledge economy (ICCIKE)* (pp. 58–63). IEEE. https://doi.org/10.1109/ICCIKE51210.2021.9410709

Khalkhal, K., Asgari Lajayer, B., & Ghorbanpour, M. (2020). An overview on the effect of soil physicochemical properties on the immobilization of biogenic nanoparticles. In *Biogenic nanoparticles and their use in agro-ecosystems* (pp. 133–160). https://doi.org/10.1007/978-981-15-2985-6_8

Khan, S. T. (2020). Interaction of engineered nanomaterials with soil microbiome and plants: Their impact on plant and soil health. In *Sustainable agriculture reviews 41: Nanotechnology for plant growth and development* (pp. 181–199). https://doi.org/10.1007/978-3-030-33996-8_10

Koul, B., Poonia, A. K., Yadav, D., & Jin, J. O. (2021). Microbe-mediated biosynthesis of nanoparticles: Applications and future prospects. *Biomolecules, 11*(6), 886. https://doi.org/10.3390/biom11060886

Landa, P. (2021). Positive effects of metallic nanoparticles on plants: Overview of involved mechanisms. *Plant Physiology and Biochemistry, 161*, 12–24. https://doi.org/10.1016/j.plaphy.2021.01.039

Li, G., Zhang, W., Luo, N., Xue, Z., Hu, Q., Zeng, W., & Xu, J. (2021). Bimetallic nanocrystals: Structure, controllable synthesis and applications in catalysis, energy and sensing. *Nanomaterials, 11*(8), 1926. https://doi.org/10.3390/nano11081926

Li, X., Li, J., Wang, Z., Bol, R., & Zou, H. (2024). Heteroaggregation of carbon nanomaterials with mineral-based nanomaterials: A review. *Journal of Environmental Chemical Engineering*, 113594. https://doi.org/10.1016/j.jece.2024.113594

Liu, Y., Lonappan, L., Brar, S. K., & Yang, S. (2018). Impact of biochar amendment in agricultural soils on the sorption, desorption, and degradation of pesticides: A review. *Science of the Total Environment, 645*, 60–70. https://doi.org/10.1016/j.scitotenv.2018.07.099

Mammari, N., Lamouroux, E., Boudier, A., & Duval, R. E. (2022). Current knowledge on the oxidative-stress-mediated antimicrobial properties of metal-based nanoparticles. *Microorganisms, 10*(2), 437. https://doi.org/10.3390/microorganisms10020437

Markowicz, A. (2023). The significance of metallic nanoparticles in the emerging, development and spread of antibiotic resistance. *Science of the Total Environment, 871*, 162029. https://doi.org/10.1016/j.scitotenv.2023.162029

Milan, J., Niemczyk, K., & Kus-Liśkiewicz, M. (2022). Treasure on the Earth—gold nanoparticles and their biomedical applications. *Materials, 15*(9), 3355. https://doi.org/10.3390/ma15093355

Mishra, S., Yang, X., & Singh, H. B. (2020). Evidence for positive response of soil bacterial community structure and functions to biosynthesized silver nanoparticles: An approach to conquer nanotoxicity? *Journal of Environmental Management, 253*, 109584. https://doi.org/10.1016/j.jenvman.2019.109584

Mony, C., Kaur, P., Rookes, J. E., Callahan, D. L., Eswaran, S. V., Yang, W., & Manna, P. K. (2022). Nanomaterials for enhancing photosynthesis: Interaction with plant photosystems and scope of nanobionics in agriculture. *Environmental Science: Nano, 9*(10), 3659–3683. https://doi.org/10.1039/D2EN00451H

Mukherjee, S., Verma, A., Kong, L., Rengan, A. K., & Cahill, D. M. (2024). Advancements in green nanoparticle technology: Focusing on the treatment of clinical Phytopathogens. *Biomolecules, 14*(9), 1082. https://doi.org/10.3390/biom14091082

Ojha, A. (2020). Nanomaterials for removal of waterborne pathogens: Opportunities and challenges. In *Waterborne pathogens* (pp. 385–432). https://doi.org/10.1016/B978-0-12-818783-8.00019-0

Ovais, M., Khalil, A. T., Ayaz, M., Ahmad, I., Nethi, S. K., & Mukherjee, S. (2018). Biosynthesis of metal nanoparticles via microbial enzymes: A mechanistic approach. *International Journal of Molecular Sciences, 19*(12), 4100.

Parada, J., Rubilar, O., Fernández-Baldo, M. A., Bertolino, F. A., Durán, N., Seabra, A. B., & Tortella, G. R. (2018). The nanotechnology among US: Are metal and metal oxides nanoparticles a nano or mega risk for soil microbial communities? *Critical Reviews in Biotechnology, 39*(2), 157–172. https://doi.org/10.1080/07388551.2018.1523865

Pattoo, T. A. (2023). Flora to nano: Sustainable synthesis of nanoparticles via plant-mediated green chemistry. *Plant Science Archives*, 12–17. https://doi.org/10.5147/PSA.2023.8.1.12

Raiesi-Ardali, T., Ma'mani, L., Chorom, M., & Moezzi, A. (2022). Improved iron use efficiency in tomato using organically coated iron oxide nanoparticles as efficient bioavailable Fe sources. *Chemical and Biological Technologies in Agriculture, 9*(1), 59. https://doi.org/10.1186/s40538-022-00318-y

Rajput, V. D., Singh, A., Singh, V. K., Minkina, T. M., & Sushkova, S. (2021). Impact of nanoparticles on soil resource. In *Nanomaterials for soil remediation* (pp. 65–85). Elsevier. https://doi.org/10.1016/B978-0-12-822891-3.00004-9

Ramos, R. M. C. R., & Regulacio, M. D. (2021). Controllable synthesis of bimetallic nanostructures using biogenic reagents: A green perspective. *ACS Omega, 6*(11), 7212–7228. https://doi.org/10.1021/acsomega.1c00692

Rasool, S., Rasool, T., & Gani, K. M. (2022). A review of interactions of pesticides within various interfaces of intrinsic and organic residue amended soil environment. *Chemical Engineering Journal Advances, 11*, 100301.

Samuel, M. S., Ravikumar, M., John, J. A., Selvarajan, E., Patel, H., Chander, P. S., Soundarya, J., Vuppala, S., Balaji, R., & Chandrasekar, N. (2022). A review on green synthesis of nanoparticles and their diverse biomedical and environmental applications. *Catalysts, 12*(5), 459. https://doi.org/10.3390/catal12050459

Santhosh, P. B., Genova, J., & Chamati, H. (2022). Green synthesis of gold nanoparticles: An eco-friendly approach. *Chemistry, 4*(2), 345–369. https://doi.org/10.3390/chemistry4020026

Saravanan, A., Kumar, P. S., Karishma, S., Vo, D. V. N., Jeevanantham, S., Yaashikaa, P. R., & George, C. S. (2021). A review on biosynthesis of metal nanoparticles and its environmental applications. *Chemosphere, 264*, 128580.

Savage, D. T., Hilt, J. Z., & Dziubla, T. D. (2019). In vitro methods for assessing nanoparticle toxicity. In Q. Zhang (Ed.), *Nanotoxicity. Methods in molecular biology* (Vol. 1894, p. 1). Humana Press. https://doi.org/10.1007/978-1-4939-8916-4_1

Sharma, S., Singh, V. K., Kumar, A., & Mallubhotla, S. (2019). Effect of nanoparticles on oxidative damage and antioxidant defense system in plants. In *Molecular plant abiotic stress: Biology and biotechnology* (pp. 315–333).

Sheteiwy, M. S., Shaghaleh, H., Hamoud, Y. A., Holford, P., Shao, H., Qi, W., Hashmi, M. Z., & Wu, T. (2021). Zinc oxide nanoparticles: Potential effects on soil properties, crop production, food processing, and food quality. *Environmental Science and Pollution Research, 28*(28), 36942–36966. https://doi.org/10.1007/s11356-021-14542-w

Shrestha, S., Wang, B., & Dutta, P. (2020). Nanoparticle processing: Understanding and controlling aggregation. *Advances in Colloid and Interface Science, 279*, 102162. https://doi.org/10.1016/j.cis.2020.102162

Singh, D., & Gurjar, B. R. (2022). Nanotechnology for agricultural applications: Facts, issues, knowledge gaps, and challenges in environmental risk assessment. *Journal of Environmental Management, 322*, 116033. https://doi.org/10.1016/j.jenvman.2022.116033

Singh, D., & Kumar, A. (2020). Understanding the effect of the interaction of nanoparticles with roots on the uptake in plants. *Environmental Nanotechnology, 3*, 277–304. https://doi.org/10.1007/978-3-030-26672-1_9

Singh, T., Singh, A., Wang, W., Yadav, D., Kumar, A., & Singh, P. K. (2019). Biosynthesized nanoparticles and its implications in agriculture. In *Biological synthesis of nanoparticles and their applications* (pp. 257–274). CRC Press.

Sonkaria, S., & Khare, V. (2020). Exploring the landscape between synthetic and biosynthetic materials discovery: Important considerations via systems connectivity, cooperation and scale-driven convergence in biomanufacturing. *Biomanufacturing Reviews, 5*, 1–23. https://doi.org/10.1007/s40898-020-0007-7

Sun, W., Dou, F., Li, C., Ma, X., & Ma, L. Q. (2020). Impacts of metallic nanoparticles and transformed products on soil health. *Critical Reviews in Environmental Science and Technology, 51*(10), 973–1002. https://doi.org/10.1080/10643389.2020.1740546

Tao, Z., Zhou, Q., Zheng, T., Mo, F., & Ouyang, S. (2023). Iron oxide nanoparticles in the soil environment: Adsorption, transformation, and environmental risk. *Journal of Hazardous Materials*, 132107. https://doi.org/10.1016/j.jhazmat.2023.132107

Terna, A. D., Elemike, E. E., Mbonu, J. I., Osafile, O. E., & Ezeani, R. O. (2021). The future of semiconductors nanoparticles: Synthesis, properties and applications. *Materials Science and Engineering B, 272*, 115363. https://doi.org/10.1016/j.mseb.2021.115363

Thakur, N., Thakur, N., Kumar, A., Thakur, V. K., Kalia, S., Arya, V., Kumar, A., Kumar, S., & Kyzas, G. Z. (2024). A critical review on the recent trends of photocatalytic, antibacterial, antioxidant and nanohybrid applications of anatase and rutile TiO_2 nanoparticles. *Science of the Total Environment*, 169815. https://doi.org/10.1016/j.scitotenv.2023.169815

Veerasamy, R. (2024). Development of nanoparticles: Recent developments and future prospects. In S. Bose, A. C. Shukla, M. R. Baig, & S. Banerjee (Eds.), *Concepts in pharmaceutical biotechnology and drug development. Interdisciplinary biotechnological advances*. Springer. https://doi.org/10.1007/978-981-97-1148-2_5

Venkateshaiah, A., Černík, M., & Padil, V. V. (2022). Metal oxide nanoparticles for environmental remediation. In *Nanotechnology for environmental remediation* (pp. 183–213). https://doi.org/10.1002/9783527834143.ch11

Wang, X., Xie, H., Wang, P., & Yin, H. (2023). Nanoparticles in plants: Uptake, transport and physiological activity in leaf and root. *Materials, 16*(8), 3097. https://doi.org/10.3390/ma16083097

Xin, X., Zhao, F., Rho, J. Y., Goodrich, S. L., Sumerlin, B. S., & He, Z. (2020). Use of polymeric nanoparticles to improve seed germination and plant growth under copper stress. *Science of the Total Environment, 745*, 141055. https://doi.org/10.1016/j.scitotenv.2020.141055

Zafar, S. (2023). Synthesis, properties, and characterization of biological nanomaterial. In *Nanomaterials: The building blocks of modern technology: Synthesis, properties and applications* (pp. 91–114). Springer Nature Singapore. https://doi.org/10.1007/978-981-99-4149-0_5

Chapter 15
Harnessing Nanomaterials for Sustainable Agriculture: Unveiling Plant-Nanoparticle Interactions and Crop Improvement

Rahul Gogoi, Panthor Debbarma, Fung Swrangshee Daimari, Indrani Debasmita Borah, Amarjeet Singh Bhogal, Madhurjya Protim Borah, and Sudipta Sankar Bora

Abstract Nanomaterials bear promises for a range of agricultural applications, such as environmental remediation, improving stress tolerance, and delivering nutrients to crop plants. Although they enhance the effectiveness of agricultural chemical, their addition could pose risks to food supply, environment, and human health. Therefore, monitoring nanoparticles' uptake, transport, conversion, interaction with higher plants, and potential hazards in agriculture is crucial. To maximize the positive effects of nanomaterials while minimizing their ecological consequences, it is imperative to comprehend the complex methods by which plants interact with them. This book chapter gives a thorough review of the interactions that occur between plants and nanomaterials, with special focus on the absorption, trans-

R. Gogoi
Department of Agricultural Biotechnology, Assam Agricultural University, Jorhat, Assam, India

P. Debbarma
Department of Horticulture, Aromatic and Medicinal Plants, Mizoram University, Aizawl, Mizoram, India

F. S. Daimari
Department of Botany, Bodoland University, Deborgaon, Assam, India

I. D. Borah
Department of Agronomy, Assam Agricultural University, Jorhat, Assam, India

A. S. Bhogal
Department of Plant Breeding and Genetics, Assam Agricultural University, Jorhat, Assam, India

M. P. Borah
Department of Biosciences and Bioengineering, Indian Institute of Technology Jammu, Jagti, Jammu and Kashmir, India

S. S. Bora (✉)
Multi-disciplinary Research Unit, Jorhat Medical College and Hospital, Jorhat, Assam, India

V. D. Rajput et al. (eds.), *Contemporary Suitability of Nanobionics in Agriculture*, Nanotechnology in the Life Sciences, https://doi.org/10.1007/978-3-031-82779-2_15

port, and accumulation processes. It explores the various routes that plants use to absorb nanomaterials, including seed priming, foliar application, and root uptake. Moreover, the chapter provides insight into the mechanisms of nanomaterial translocation and transport throughout plant tissues. The consequences of nanomaterial accumulation in edible plant parts are also covered in this chapter, emphasizing the significance of determining the safety of nanomaterials and their bioavailability for human ingestion in nutrient-loaded nanomaterials. Furthermore, it offers insights into the intricate interactions between nanomaterials and plants by reviewing the present knowledge and research gaps. This lays the groundwork for future investigations into the topic of nanotechnology for environmental stewardship and sustainable agriculture.

Keywords Agricultural · Food supply · Remediation · Stress · Nutrients · Bioavailability

15.1 Introduction

Current developments in nanotechnology in agriculture offer transformational improvements to farming. Nanomaterials (NMs) enhance crop yield by distributing nutrients and insect repellents more precisely, boosting resource utilization and reducing their ecological effects. Nanotechnology and nanomaterials play a multifaceted role in promoting agricultural productivity and environmental sustainability (Fig. 15.1). Nanoparticles are used for precise delivery of agrochemicals, reducing the overuse of fertilizers, pesticides, and herbicides, thereby minimizing harmful runoff into surrounding ecosystems (Duhan et al., 2017). Nanosensors provide real-time monitoring of soil health status including moisture and nutrient availability, enabling precision agriculture to optimize crop yields and resource allocation. In soil health, nanomaterials improve soil structure, increase water retention capacity, and facilitate the absorption of essential nutrients by plants. Furthermore, nanotechnology plays a critical role in environmental protection by ameliorating contaminated soils and immobilizing or degrading pollutants such as heavy metals and organic toxins (Sharma & Kumar, 2024). This approach contributes to more sustainable agricultural practices, enhancing both food production and ecosystem preservation.

In parallel, continuous research and collaboration are necessary to address issues such as security concerns and regulatory frameworks. Despite these barriers, there is enormous potential for using nanotechnology to boost agriculture in a sustainable way and successfully address global issues related to agriculture (Wu & Li, 2022). Plants can absorb nanoparticles (NPs) through their leaves or roots, and once they enter, they travel through the phloem and xylem. External variables such as pH and soil composition, plant age and species, and NP characteristics (size, shape, and surface chemistry) all affect uptake. Spectrophotometry, microscopy, spectroscopic imaging, electrochemical approaches, and bio-sensing techniques are examples of

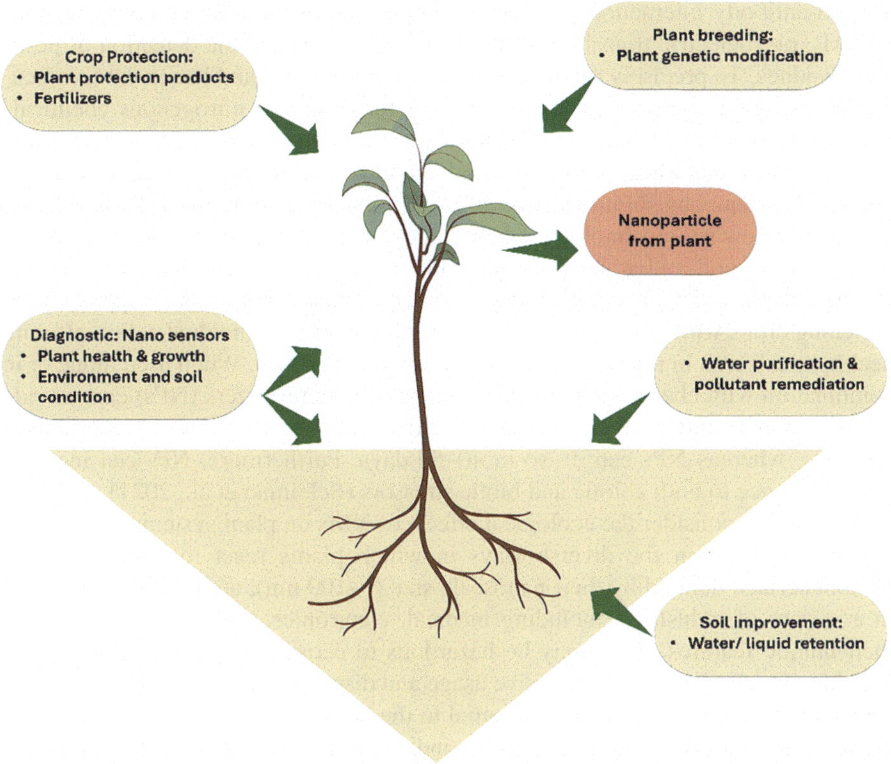

Fig. 15.1 Applications of nanoparticles in plant growth, protection, and environmental management

detection techniques allowing to detect and visualize NPs in plant tissues precisely (Sembada & Lenggoro, 2024). Nanopesticides, composed of NPs intended for effective pesticide transportation, offer increased efficacy and lower ecological effects via specific administration and controlled release. Similarly, nanofertilizers can offer advantages such as increased absorption of nutrients effectiveness, decreased leaching of nutrients, and regulated nutrient release corresponded to crop requirements (Kah et al., 2018). With their compact size, increased sensitivity, quick response times, and long operating lives, nanobiosensors are a big improvement over traditional biosensors. Analytes can be detected at extremely low concentrations, frequently in parts per trillion (ppt). With dangerous metal ions and other pollutants that pose major dangers to human health and the environment, this capability is especially helpful for early pollutant identification in soil and water. Numerous agricultural domains, such as the identification of toxins from pesticides and fertilizers as well as soil pollution, employ nano biosensors. For example, optical sensors that employ reduced metal oxides or biological macromolecules can detect metal ions through modifications in the characteristics of electromagnetic radiation, providing accurate measuring capabilities. Comparably, using antigen-antibody interactions, enzymatic inhibition, or the binding characteristics of NMs, piezoelectric transducers in nanobiosensors enable the detection of pesticide residues. To precisely regulate nutrients in agriculture and lessen environmental effects, nanobiosensors are also essential for detecting nitrogenous chemicals obtained from urea fertilizers. Novel NMs are still being developed to improve sensor properties including sensitivity and stability, while most present uses are restricted to aqueous solutions and lab conditions. Electronic noses (e-noses) are example of how nanotechnology is being integrated into agriculture to improve the monitoring of soil and water pollutants, plant diseases, and production processes (Usman et al., 2020). Nano-biofortification is another revolutionary approach for enriching crops with essential elements to supplement an individual's diet, offering healthy meals rich in micronutrients to prevent malnutrition. When used alone or in combination with chemical or organic fertilizers, nanofertilizers (NFs) can provide nutrients slowly and steadily. Chemical fertilizers require 4–10 days to release their nutrients, whereas NPs can do so in 40–50 days. Furthermore, NFs can increase plant tolerance to both abiotic and biotic stressors (Seleiman et al., 2021).

It is vital to consider the ecological effects of NMs on plant systems, with a particular emphasis on the diverse ways in which plants react to nano-pollution. Nanomaterials, defined by their nanoscale size (1–100 nm), are rapidly being used in a variety of industries, including medical, electronics, and agriculture, due to their unique features. They may be hazardous to ecosystems and have a lengthy half-life, which makes their extensive usage and disposal problematic from an environmental standpoint. Plants are essential to the health of ecosystems because they are primary producers and important bio indicators of environmental contaminants (Gayathiri et al., 2024). Bio-based NMs are used in environmental remediation to successfully remove organic contaminants and heavy metals from polluted sources while also purifying water and soil. They are also essential for improving soil quality and reducing the effects of pollution. These NPs are used in agriculture to administer fertilizers, herbicides, and insecticides in a tailored manner that minimizes environmental contamination and maximizes efficiency through accurate application and absorption. They also serve as regulators and growth stimulants, enhancing crop productivity and resistance to environmental stressors (Saraswat et al., 2023). This chapter explores the overall idea of plant-nanomaterial interactions in agricultural systems.

15.2 Uptake and Translocation of Nanomaterials by Plants

Nanotechnology is making waves in agriculture, offering exciting opportunities to boost plant growth, productivity, and resilience. NMs have shown great potential in enhancing agrochemical efficiency and promoting plant growth (Rani et al., 2020). The uptake of NMs by plants occurs through various routes, including foliar and root exposure, with subsequent translocation via xylem and phloem pathways (Rai et al., 2022). At the cellular level, NMs modify cell wall composition and structure, facilitating internalization and interaction with plant cells, influencing germination, seedling physiology, and plant metabolism (Sharma et al., 2023). NMs, such as nano-iron, can lead to oxidative stress at higher concentrations but promote plant growth at lower concentrations, affecting morpho-physiological, biochemical, and molecular traits of plants (Oliveira et al., 2023). Understanding the mechanisms of NM uptake and translocation within plants is crucial for harnessing their potential in agriculture sustainably and effectively. To harness these benefits, it is crucial to understand how plants take up and move NMs within their systems (Fig. 15.2).

15.2.1 Mechanisms of Nanomaterial Uptake by Plants

NMs can enter plants through their roots and leaves, using the plants natural methods for absorbing nutrients and water. NM uptake by plants involves complex mechanisms influenced by various factors such as NM size, shape, surface chemistry, and plant species. Understanding these mechanisms is crucial for assessing the potential

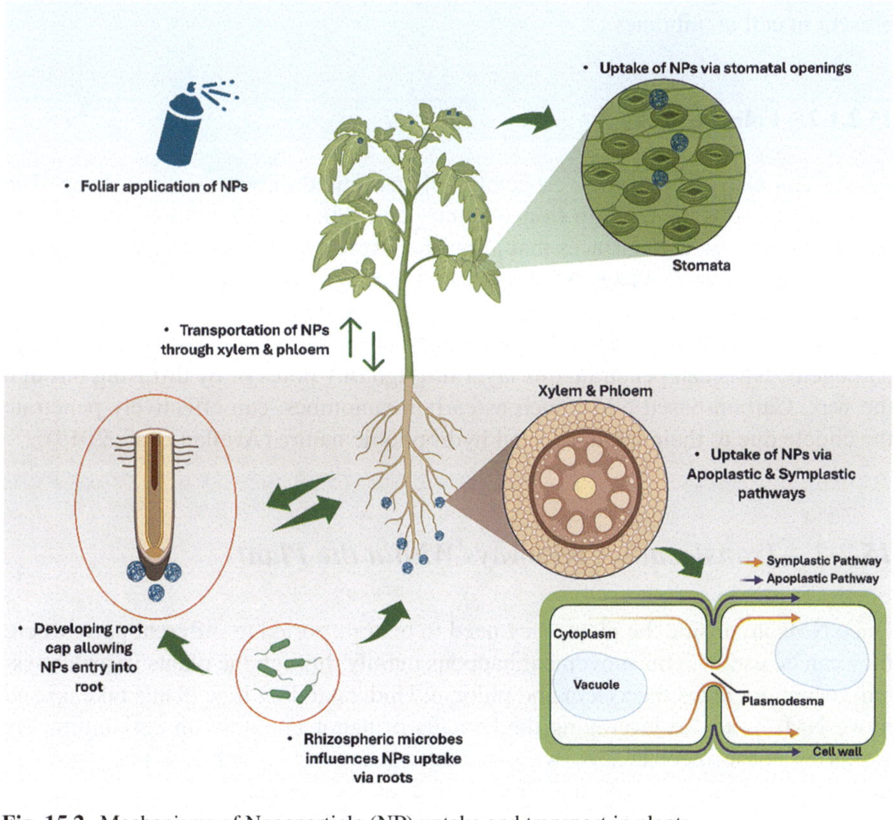

Fig. 15.2 Mechanisms of Nanoparticle (NP) uptake and transport in plants

environmental impact of NMs and designing efficient NM-based agricultural applications.

15.2.1.1 Root Uptake

Roots are the main gateway for nutrients and water from the soil, and they play a vital role in the uptake of NMs. The uptake takes place through different pathways, viz., apoplastic pathway, symplastic pathway and transmembrane transport. In apoplastic pathway, NMs move through the spaces between and along the cell walls, without crossing into the cells. This method is effective for NMs like cerium oxide (CeO_2) and zinc oxide (ZnO), which can travel through these spaces due to their tiny size and surface characteristics (Zhao et al., 2015). In symplastic pathway, NM crosses the cell membrane to move through the cytoplasm and from cell to cell via plasmodesmata. For example, gold NMs (Au NPs) have been seen entering root cells in Arabidopsis thaliana through tiny vesicles, suggesting they use this route (Schwab et al., 2016). However, some NMs can get into cells by mimicking or utilizing existing nutrient transport mechanisms via transmembrane transportation. Silicon dioxide (SiO_2) NMs are taken up by plant cells using silicon transporters present in cell membranes.

15.2.1.2 Foliar Uptake

Leaves can also absorb NMs, especially when applied directly to the foliage. The main mechanisms are through stomatal entry and cuticular penetration. Stomata are small openings on leaf surfaces that allow gas exchange. NMs like silver (Ag NPs) and titanium dioxide (TiO_2 NPs) can enter through these openings, particularly when stomata are open during photosynthesis (Eichert et al., 2008). The leaf surface is covered with a waxy layer called the cuticle, which protects against water loss and pathogens. NMs can penetrate this layer through tiny pores or by diffusing through the wax. Carbon-based NMs, such as carbon nanotubes, can effectively penetrate the cuticle due to their small size and hydrophobic nature (Avellan et al., 2019).

15.2.2 Translocation Pathways Within the Plant

Once NMs are inside the plant, they need to be transported to different parts where they can be useful. This movement happens mainly through the plants vascular system, which includes the xylem and phloem. Understanding how plants take up and move NMs is key to leveraging the benefits of nanotechnology in agriculture. By exploring these mechanisms, we can better use NMs to enhance plant growth, increase resistance to stress, and improve overall crop productivity.

15.2.2.1 Xylem Transport

The xylem is responsible for moving water and nutrients from the roots to the above-ground parts of the plant. NMs taken up by the roots are transported upwards through the xylem. The crucial factors influencing this transport include size and surface properties of NMs and transpiration rate. Smaller NMs and those with hydrophilic (water-attracting) surfaces move more easily through the xylem. Iron oxide NMs (Fe_3O_4 NPs) are effectively transported through the xylem due to their small size and appropriate surface properties (Zhu et al., 2008). The rate of water movement driven by transpiration also impacts NM transport. Higher transpiration rates enhance the upward movement of NMs like copper oxide (CuO) and zinc oxide (ZnO) (Wang et al., 2016).

15.2.2.2 Phloem Transport

The phlóem distributes sugars and other organic molecules from the leaves to other parts of the plant. NMs can enter the phloem and be transported throughout the plant. NMs taken up by leaf cells can enter the phloem through plasmodesmata or via endocytosis. Quantum dots or semiconductor nanocrystals have been seen entering the phloem in rice plants, allowing them to move to various tissues (Rico et al., 2011). The phloem can transport substances in both directions, enabling NMs to reach different parts of the plant. This bidirectional flow is crucial for the distribution of NMs like multi-walled carbon nanotubes (MWCNTs), which have been shown to move from leaves to roots and other parts of the plant (Lahiani et al., 2013).

15.3 Effects of Nanomaterials on Plant Physiology and Metabolism

NMs with their unique physicochemical properties hold great promise for revolutionizing agriculture by improving crop yield, quality, and sustainability. NMs have been extensively studied for their effects on plant physiology and metabolism, showing both positive and negative impacts. NMs can modulate secondary metabolites in plants by influencing reactive nitrogen/oxygen species, gene expression, and signaling pathways, ultimately affecting plant growth and development (Selvakesavan et al., 2023). These interactions can lead to oxidative stress, metabolic disorders, and cell damage, with NMs affecting processes like protein folding, calcium homeostasis, and photosynthesis in plants (Kang et al., 2023). Engineered nanomaterials (ENPs) and nanoscale formulations of agrochemicals offer targeted delivery systems and enhanced efficacy compared to conventional counterparts. Furthermore, NMs can act as elicitors, stimulating the production of secondary metabolites in response to stress conditions, thus enhancing plant tolerance to various environmental challenges (Gonçalves et al., 2023). The complex relationship between plants and NMs extends from cellular interactions to ecosystem-level effects, impacting plant growth, symbiotic relationships, and even pollinators, highlighting the importance of understanding the bidirectional effects of NMs on plants.

15.3.1 Impact on Plant Growth and Development

NMs have shown significant potential in enhancing plant growth and development by improving physiological processes and metabolic pathways, especially under environmental stresses. Various NMs like zinc, iron, silicon, cerium, silver, titanium dioxide, and carbon have been reported to positively impact plant physiology and antioxidant enzyme activity, leading to improved growth and development (Roy, 2023). However, it is crucial to note that the effects of metal and metal oxide NMs on plants can be dose-dependent and vary among different plant species, with some studies reporting phytotoxicity issues (Nawaz et al., 2023). Additionally, the synergistic interactions between NMs and plant growth promoting rhizobacteria (PGPR) have emerged as a novel approach to enhance plant health, productivity, and stress tolerance, activating plant defense systems and increasing antioxidant enzyme activity (Venkateswaran et al., 2023).

Numerous studies have reported positive effects of NMs on plant growth parameters such as shoot and root biomass, leaf area, and germination rates. Engineered NMs, such as carbon-based NMs and metal oxide NMs, can enhance nutrient availability, hormone signaling and photosynthetic efficiency, promoting plant growth and development (Khan et al., 2019). Conversely, some NMs have been shown to exert inhibitory effects on plant growth under certain conditions. High concentrations or prolonged exposure to NMs may disrupt cellular processes, impairing root elongation, leaf expansion, and reproductive development (Thakur et al., 2017). Moreover, NM-induced oxidative stress and genotoxicity can impede plant growth and productivity (Rico et al., 2011). Table 15.1 summarizes the various positive impacts of nanomaterials on plant growth and productivity, as well as their role in mitigating different stress conditions.

15.3.2 Influence on Nutrient Uptake and Utilization

Nutrient uptake and utilization in plants are influenced by various factors such as intercropping systems, physiological adaptations, and genetic traits. Research has shown that maize-soybean relay intercropping enhances nitrogen (N), phosphorous (P), and potassium (K) uptake efficiency, leading to a higher land equivalent ratio (LER) and yield advantage compared to sole cropping. Additionally, plants have evolved adaptive strategies to cope with nutrient limitations, emphasizing the importance of nutrient use efficiency (NUE) in sustainable agriculture development

Table 15.1 Beneficial effects of nanomaterials on plant growth, productivity, and stress tolerance

Sl. no.	Type of nanomaterial	Positive effects on plants	References
1	Silver Nanoparticles	Improved root growth, enhanced disease resistance	Khan et al. (2023)
2	Zinc Oxide Nanoparticles	Enhanced growth under drought conditions	El-Zohri et al. (2021)
3	Titanium Dioxide Nanoparticles	Enhanced seed germination and stress tolerance	Basah,(2021)
4	Silicon Nanoparticles	Increased drought tolerance and disease resistance	Nazim et al. (2024)
5	Carbon Nanotubes	Stimulates root development and nutrient absorption	Hao et al. (2023)
6	Iron Nanoparticles	Enhanced chlorophyll content, improved photosynthesis	Rezayian et al. (2023)
7	Graphene Oxide	Enhanced nutrient uptake and plant growth	Zhao et al. (2023)
8	Cerium Oxide Nanoparticles	Reduces oxidative stress, improves plant resilience	Preetha et al. (2023)
9	Aluminum Oxide Nanoparticles	Improved root morphology and nutrient uptake	Pokharel et al. (2024)
10	Magnesium Oxide Nanoparticles	Promotes chlorophyll synthesis and growth	Salas-Leiva et al. (2021)
11	Calcium Carbonate Nanoparticles	Enhances soil quality and root development	Chen et al. (2023)
12	Chitosan Nanoparticles	Improves nutrient absorption and stress tolerance	Attaran Dowom et al. (2022)
13	Boron Nanoparticles	Enhances seed germination and plant growth	Sutulienė et al. (2022)
14	Selenium Nanoparticles	Enhances antioxidant activity and stress tolerance	Liu et al. (2022)
15	Graphene Quantum Dots	Promotes photosynthesis and plant growth	Xu et al. (2020)
16	Noble Metal Nanoparticles (e.g., Gold)	Enhances growth and stress tolerance	Ferrari et al. (2021)

(Aslam et al., 2022; Dalir, 2023). Physiological mechanisms, including root structure modifications and genetic factors, play crucial roles in improving nutrient absorption efficiency and overall plant growth (Dalir, 2023; Jose, 2023). Understanding these processes is essential for maximizing crop productivity, reducing fertilizer application, and promoting genetic modifications for enhanced plant nutrition in biofortification programs.

NMs can improve nutrient uptake by plants through various mechanisms, including increased root surface area, enhanced root exudation, and modulation of nutrient transporters (Hao et al., 2018). Nano-enabled fertilizers and nutrient delivery systems offer targeted release and efficient utilization of essential elements, optimizing nutrient uptake and crop yield. However, excessive accumulation of certain

NMs in plant tissues may disrupt nutrient homeostasis and metabolism, leading to nutrient imbalances and physiological disorders (González-Melendi et al., 2008). Metal-based NMs, in particular, can interfere with ion uptake and translocation, impairing nutrient assimilation and partitioning in plants (Li et al., 2016).

15.4 Role of Nanomaterials in Improving Plant Tolerance to Abiotic Stresses

15.4.1 Drought Stress

Efforts to reduce the detrimental effects on crop plants of drought stress or inadequate water availability have involved the use of NMs. Farmers may be able to recognize the effects of stress on plants early on thanks to nanosensors in conjunction with global positioning systems that provide satellite photographs of fields (Khot et al., 2012). NMs like nano-silica or nano-hydrogels can improve water retention in soil and enhance water use efficiency in plants. They help maintain optimal hydration levels, reduce transpiration rates, and ensure sustained growth even under water-deficient conditions. Nano zinc oxide helps speed up the process of using seed reservoirs for seedling growth, which can improve seed germination in soybeans grown in water stress (Sedghi et al., 2013). Under drought stress, chitosan nanoparticles (NPs) have been demonstrated to raise the relative water content, photosynthetic rate, yield, biomass, and the activity of the enzymes catalase (CAT) and superoxide dismutase (SOD).

15.4.2 Salinity Stress

High salt levels in the soil cause plants to absorb excessive sodium (Na^+) and chloride (Cl^-) ions, disrupting their internal balance and leading to toxicity within the cells. NPs such as nano-clays or nano-iron oxides can mitigate the harmful effects of salt stress by reducing sodium uptake in plants and enhancing the absorption of essential nutrients like potassium. They also improve soil structure, reducing soil salinity and increasing plant productivity in saline environments. Recent research on the use of multi-walled carbon nanotubes (MWCNTs) in broccoli (Martínez-Fernández et al., 2015), chitosan NPs in maize (Oliveira et al., 2016) and tomato (Hernández-Hernández et al., 2018), and silver NPs in wheat (Ismail & Abou-Zeid, 2018; Mohamed, 2017) has demonstrated their significant alleviating effects on salinity stress. Research has shown that NPs of silica (SiO_2) and titanium dioxide (TiO_2) enhance plant growth and yield under stress conditions. This increased tolerance is likely due to the absorption of silicon NPs by the roots that create a thin layer within the cell wall. This layer supports the plant's ability to withstand diverse stressors and maintain yield (Abdel Latef et al., 2018; DeRosa et al., 2010).

15.4.3 Temperature Extremes (Heat and Cold)

NMs like nano-calcium carbonate or nano-ceramics can regulate plant temperature by reflecting excess heat or insulating against cold temperatures. They help plants maintain optimal physiological processes and prevent damage due to extreme temperature fluctuations. Heat stress disrupts photosystem II, leading to disturbances in electron flow and reduction in carbon fixation. This stress also induces chlorophyll degradation, thereby disrupting the photosynthesis process and ultimately reducing plant growth. It leads to the overproduction of reactive oxygen species (ROS), resulting in oxidative stress. Membrane lipids fall apart, electrolyte leaks, and biomolecules become denaturated as a result of this oxidative stress (Aref et al., 2016; Karuppanapandian et al., 2011; Savicka & Škute, 2010). Additionally, heat stress reduces chlorophyll content and the rate of photosynthesis (Prasad et al., 2011). The application of low concentrations of selenium (Se) NPs has been shown to increase chlorophyll content, hydration levels, and growth in tomato and sorghum plants due to their antioxidative properties (Djanaguiraman et al., 2018). The application of biologically synthesized selenium nanoparticles (Se NPs) at a concentration of 100 μg/ml improved wheat growth by enhancing heat stress tolerance in plants (El-Saadony et al., 2021).

Cold stress (0–15 °C) occurs at temperatures reduced to the point where plant tissues are harmed without developing ice crystals. On the other hand, below-freezing temperatures (<0 °C) cause ice crystals to develop inside plant tissues, which can cause frost damage and possibly even crop mortality (Hasanuzzaman et al., 2013). By controlling temperature drop-induced damage and modifying metabolism to promote plant growth, TiO_2 nanoparticles (NPs) applied to chickpea plants improve their resistance to cold stress (Hasanpour et al., 2015). Furthermore, applying TiO_2 NPs to liquorice plants increases their glycyrrhizin content and reduces the negative effects of cold stress (Kardavan Ghabel & Karamian, 2020).

15.4.4 Heavy Metal Toxicity

Under heavy metal (HM) stress conditions, applying NPs to plants decreases HM concentrations in the soil. Additionally, it enhances physiological processes, controls the expression of HMs transfer genes in plants, boosts the activity of antioxidant systems in plants, and encourages the synthesis of protective compounds such as organic acids, phytochelatins, and root secretions. NPs such as nano-zeolites or carbon nanotubes can immobilize heavy metals in soil, reducing their uptake by plants and mitigating toxicity. They also facilitate the detoxification and sequestration of heavy metals, thereby improving plant health and productivity in contaminated soils. Hydroxyapatite nanoparticles (NPs) have been shown to mitigate the toxic effects of metals in soil by releasing phosphate ions and helping to maintain soil pH (Cui et al., 2018). Additionally, NPs induce the formation of apoplast barriers in plants, which can reduce the uptake of heavy metals into roots. Furthermore, specific NPs regulate metal transporter genes in plants, thereby intercepting heavy metals and forming complexes that inhibit their translocation (Wang et al., 2021).

15.4.5 Flooding Stress

Excessive water from river overflows, rising groundwater tables, and over-irrigation can lead to flooding, potentially damaging standing crops. These conditions can induce either partial anaerobiosis (hypoxia) or complete anaerobic conditions (anoxia) in the soil, displacing air and affecting soil health (Vartapetian et al., 2014). NPs have shown promise in alleviating flooding stress in plants. For instance, silver NPs were found to mitigate stress in soybean plants by regulating processes such as amino acid synthesis, protein expression, glycolysis, and wax formation, ultimately promoting soybean growth under stress conditions. In another study involving soybeans under flooding stress, the application of Al_2O_3 NPs helped alleviate growth impairments. These NPs increased root length, including the hypocotyl, suppressed glycolysis-related proteins, mediated ROS scavenging through upregulation of the ascorbate/glutathione pathway (AsA/GSH), and enhanced ribosomal protein production (Mustafa et al., 2016).

15.4.6 Nutrient Deficiency

Using fertilizers is a standard agricultural method to enhance crop productivity. In addition to increasing agricultural yield and productivity, researchers are trying to find NMs that minimize fertilizer-related nutrient loss. This approach helps prevent nutrient deficiencies and increases the effective availability of nutrients to plants. Nanofertilizers or nanocarriers loaded with essential nutrients (e.g., nitrogen, phosphorus, potassium) can enhance nutrient availability and uptake efficiency in plants. They deliver nutrients directly to roots or foliage, ensuring optimal plant nutrition and growth even in nutrient-deficient soils. The application of nano$CaCO_3$ combined with humic acid and organic manures significantly improved the growth and development of peanut plants. This treatment resulted in significant increases in the amounts of protein and soluble sugar in the stems and leaves. It also improved the absorption of nutrients like calcium, nitrogen, phosphorus, and potassium (Xiumei et al., 2005). A study on two Iranian rice cultivars found that the use of nitrogen and nanopotassium fertilizer improved their yields (Lemraski et al., 2017).

15.4.7 UV Radiation

Exposure of plants to ultraviolet-B (UV-B) radiation, which falls within the wavelength range of 280–315 nm and is non-ionizing and photosynthetically inactive, results in elevated levels of reactive oxygen species (ROS) within cells. These ROS have the potential to harm DNA, disturb chloroplast structure, and hinder several cellular functions, including photosynthesis (Hideg et al., 2013; Wang et al., 2012). NPs such as zinc oxide or titanium dioxide NPs can act as UV filters, absorbing or scattering harmful UV radiation. They protect plant tissues from UV-induced damage, maintain photosynthetic efficiency, and ensure sustained growth in environments with high UV exposure. NMs have been shown to shield photosynthetic systems from UV-B stress by increasing chlorophyll content, enhancing Rubisco activity, and reducing oxidative stress (Sicard et al., 2011).

15.5 Mitigation of Biotic Stresses Through Nanotechnology

Various types of NMs are utilized in agriculture, containing metal or metal oxide nanoparticles (NPs), such as micelles or chitosan nanoparticles (CS NPs), as well as organic-based NPs, such as silver (AgNPs) and copper oxide (CuO_2) NPs (Pascoli et al., 2018). For biotic stress applications, the following comparison of these two primary types of NPs is made:

(a) Inorganic Nanoparticles (e.g., AgNPs, CuO_2 NPs):Inorganic NPs typically exhibit inherent activity against pathogens. For instance, silver nanoparticles (AgNPs) have direct antifungal and antibacterial effects against plant-infecting nematodes, bacteria, and fungi (Khan et al., 2021).
(b) Polymeric Nanoparticles (e.g., micelles, CS NPs): Polymeric NPs are primarily used as nanocarriers for the efficient delivery of active agents. They facilitate the controlled release of active compounds, protecting them against adverse environmental conditions (Shakiba et al., 2020).

These NPs are valued for their stability and ability to release active ingredients precisely where needed in plant tissues. Polymeric NPs are also characterized by biodegradability, biocompatibility, and low toxicity, making them safe for agricultural applications. Among different polymeric nanostructures, nanomicelles, nanocapsules, and nanospheres are synthesized and utilized across various fields. These nanostructures are designed to encapsulate and deliver active ingredients efficiently, offering controlled release and protection against environmental degradation. Their diverse applications span industries such as pharmaceuticals, agriculture, and materials science, where their specific properties and structures cater to targeted delivery and enhanced efficacy of bioactive compounds.

15.6 Environmental Implications of Nanomaterial Use in Agriculture

Nanotechnology resources could be useful for forestry, agriculture, and environmental issues like urbanization, energy scarcity, and sustainable resource use. Engineered nanomaterials (ENMs) in agro-food technologies and unintentional emissions from other nanotechnologies interact with agricultural food crop plants. The food chain, human and environmental health, and agricultural yield and quality are all impacted by these forms of exposure (Kranjc & Drobne, 2019). Nevertheless, the use of nanotechnology may pose new risks to human health and the environment. This is concerning for agricultural laborers, as their work duties may expose them to these xenobiotics most of the time (Iavicoli et al., 2017). Due to the population's continuous growth, there is a demand for more productive ways to maximize agricultural practices and increase agricultural yields. Nanopesticides and other nanotechnologies have potential uses in the agro-food industry. Improve the quality of the soil by applying hydrogels, nanofertilizers, nanozeolites (SiO_2, TiO_2, and carbon), and nanotubes (CNTs) to encourage the growth of plants and intelligent observation using nanosensors connected to wireless communication devices. Moreover, pesticide degradation, pesticide residue detection, plant germination and growth, water purification, and crop disease prevention could all be achieved with engineered NPs or NMs. However, concerns have been raised about the potential short- and long-term effects of NMs on plants, including food crop productivity, trophic transfer, and ultimately environmental and human health, because of their increasing unintentional emissions from consumer products (Schwab et al., 2016).

In this case, it is anticipated that routes that expose soil and crops to NM residues will expose people to more of these residues in the environment, with a chance that they will bioaccumulate in the ecosystem and food chain. Consequently, applying NMs in agriculture may have advantages and disadvantages. The use of NMs in agriculture could endanger both humans and the environment if improper supervision is provided. Moreover, any potential hazards connected to NM exposure should be assessed using a suitably tailored life-cycle perspective. This means considering all potential uses for nano-solutions, such as field use, potential integration with food supplies, disposal, or repurposing, in addition to any potential impacts of agro-system conditions that might affect the hazardous properties and risk characterization of NMs. The Working Group for the Safety of Novel Foods and Feeds (WG-SNFF) acknowledged at its 25th meeting in June 2018 that because nanotechnology is still a relatively new field of application and research, the possible risks and benefits of products based on this technology are evaluated on a case-by-case basis. It was acknowledged, nevertheless, that to keep up with the advancements in this field, new strategies might be required in the future.

Concern over the potential toxicological and environmental effects of using and manufacturing more ENMs has sparked interest in the possible uptake, accumulation, and transformation of ENMs by plants (cells). The interactions between plants and ENM have forced plant biologists to re-evaluate the role of NPs as sorbents and/

or suppliers of heavy metals, insecticides, and essential nutrients for plants and phytoplankton (Van Moorleghem et al., 2013).

The last 10 years have seen a rise in the consideration of the potential effects on the environment that plant-enmity bioaccumulation and transformation may have on ENMs. With the growing production of NMs of all kinds, there is a growing concern about the possibility of their release into the environment and the ensuing effects on ecosystem health. Regulatory agencies need to take this into consideration (Klaine et al., 2008). Both deliberate and accidental releases, such as atmospheric emissions and solid or liquid waste streams from manufacturing facilities, are ways that manufactured NMs end up in the environment. One application of NMs that is deliberate is the remediation of contaminated soils. One example of this is the remediation of groundwater with iron NPs. NMs will ultimately deposit on agricultural land and water bodies and contaminate soils and water bodies. NPs in surface water bodies may sit down to the groundwater and contaminate it. The largest hazards for environmental releases, with respect to fugitive releases from the manufacturing process, are spills related to manufactured particulate matter being transported from production facilities to various other manufacturing locations, releases intended for use in the environment, and diffuse releases from general use related to wear and erosion. As a result, CNTs, metal oxides, and the NMs with the highest production have received most of the attention in research so far. When applied as biosolids, agricultural nanotechnologies (such as growth regulators, insecticides, and fertilizers) and atmospheric deposition (particularly in urban and industrial areas) are likely to expose plants to NMs in agricultural settings (Gardea-Torresdey et al., 2014). When absorbed by or internalized by edible plants, the NM load can be passed to higher echelons of the food chain as well as into human consumption. Entrapment on the outermost layer of cuticular wax and internalization through stomata, or openings that control the balance of gas and water, are two possible ways that leaves retain material (Kranjc & Drobne, 2019).

15.7 Safety Considerations for Plant-Nanomaterial Interactions

Safety considerations while using NMs in agriculture are crucial. NPs, when used in agricultural lands, get incorporated into crops through the root system, and NPs accumulate in edible plant parts, raising food chain concerns. Enzyme assays may not accurately assess NP-toxicity in plants. Evaluating the potential health risks associated with agricultural chemicals is a challenging task due to the wide range of substances used, combinations used in the field, and variations in exposure dosage, as well as the agricultural region's geographic and meteorological features in regions that use agricultural chemicals (Bolognesi, 2003). Nanopesticide formulations that show improved stability, better efficacy, high dispersion, and regulated release to target species in comparison to traditional pesticides can be developed with nanocarriers or nanosized active components. This may reduce the amount of insecticide used on crops worldwide. Better dispersion of the active components, however, can result in interactions between non-target organisms and nanopesticides; also, the higher chemical stability of the active ingredients suggests longevity. Therefore, non-target creatures may be adversely affected by nanopesticides, a possibility that is frequently disregarded by the individuals who create these kinds of substances. Risk assessment specific to nanotechnology is a difficult problem because of the use of tests and assumptions to evaluate the hazards associated with conventional chemicals. Strategies and modeling paradigms for environmental behavior and potential human absorption might not be suitable for nano-enabled goods. If the nanocomponent serves only to shield the active ingredient from deterioration, its behavior and fate might be identical to those of a conventional pesticide. When using pristine, inorganic, engineered nanoparticles (NPs) directly as biologically effective fertilizers, pesticides, or for soil or water remediation, it is important to carefully consider the hazardous behavior of these particles from a lifecycle perspective. This begins with the distribution of the particles through application fields and ends alongside the disposal of working residues (Shatkin & Kim, 2015).

The deliberate and increased introduction of NMs into agricultural ecosystems presents many concerns about these materials' transportation and environmental fate into the surroundings that remain unanswered. With so many nanoformulations potentially used in agricultural practices and the unknowns surrounding potential interactions with changing environmental factors, this seems like an even more pressing issue to address. The impact of naturally occurring NPs on the fate of nanoagrochemicals is still unknown. Additional uncertainties stem from soil, water features, aging, and different work procedures. They also arise from the difficulties of quantifying all these factors in an appropriate risk assessment process. These factors should be carefully considered, though, as they might have an impact on the physicochemical characterization of NMs, which could alter their toxicological profile and consequently the risks associated with occupational exposure.

Since exposure assessments have historically relied on estimating the soluble portion of the contaminant, traditional methods used in aquatic ecological risk assessments may be less applicable to NMs. Furthermore, it has been assumed that the soluble form of the total contaminant constitutes most of the bioavailable portion (Di Toro et al., 1991). Careful consideration and modification of these assumptions and approaches are necessary to account for the topics of toxicity, bioavailability, and particle fate and behavior that are essential to the quantitative environmental threat evaluation of NMs.

Unplanned NP usage has a detrimental impact on the soil, the environment, and human health. Risk evaluation requires a thorough understanding of the interaction between NMs and plants. Plant interaction with metal-based nanoparticles (MNPs) influences nutrient uptake and growth. For safety, standardization of cultivation conditions and methods is essential (Skiba et al., 2021). These nanoparticles' durability and presence in the soil can modify a variety of biological processes, encouraging modifications in plant growth and transforming plant diseases into toxicological impacts that reverberate throughout the food chain (Feregrino Pérez et al., 2023).

15.8 Innovative Strategies for Sustainable Agriculture

15.8.1 Targeted Delivery of Nutrients Using Nanomaterials

Agriculture has shown great promise for the application of nanotechnology. The agricultural industry has undergone a radical transformation thanks to recent developments in the creation of agro-products like nanopesticides, nanocomposites, nanosensors, nanofertilizers, and nanoformulations (Khan et al., 2024). Numerous NMs have been researched for use in agriculture to reduce nutrient losses through effective resource management and to replace the overuse of agrichemicals by using smart delivery platforms. Through improved control over monitoring, assessing, and managing appropriate agricultural practices, agro-nanotechnology can help develop sustainable agricultural systems by reducing the effects of environmental changes, improving yield, guaranteeing sustainable production, protecting crops, and managing diseases. In addition to reducing chemical dosages to boost crop yield, nanotechnology offers new agrochemicals and distribution methods. The current generation of agrochemicals efficiently reduces pests and diseases while increasing crop yield. However, it necessitates a high level of affectivity and indiscriminate application (Khan et al., 2019). The goal of applying nanotechnology is to lower the dosage of agrochemicals by encapsulating, coating, and other methods (Ayyaril et al., 2023). NMs facilitate the targeted delivery of nutrients for sustainable agricultural systems. Intelligent delivery systems increase agricultural output by reducing nutrient losses (Khan et al., 2024). NMs enable targeted nutrient delivery for environmentally friendly farming practices. Plant nutrition is monitored and balanced with the help of NMs equipped with smart sensors (Malini et al., 2022). Nanosensors monitor plant illnesses, pesticide residues, pH, and soil fertility (Kaur et al., 2022). For delivering specific nutrients, NMs allow the controlled release of bioactive compounds. Nanocarriers transfer DNA fragments for improved plant productivity (Rani et al., 2020). The field of agrochemicals has given a lot of attention to NMs because of their unique characteristics, which include small particle size, surface structure, solubility, and chemical composition. The shortcomings of conventional agrochemicals, such as low bioavailability, easy photolysis, organic solvent pollution, etc., are significantly overcome by the application of NMs and nanotechnology in agrochemicals. Several facets of agrochemicals, including pesticide formulations and pesticide residue analysis have been the subject of subsequent research on the application of N as plant growth regulators (PGRs), fertilizers, and the creation of bio-pesticides, which include pheromones and nucleic acids (de Oliveira et al., 2014).

Nutraceuticals have the following drawbacks: poor solubility, chemical instability, bitter taste, and unpleasant odor. Examples of these include curcumin, carotenoids, anthocyanins, omega-3 fatty acids, vitamins C and B12, and quercetin. Furthermore, some nutraceuticals degrade due to pH variations, the presence of gastrointestinal (GIT) membrane barriers, and interactions with GIT enzymes. Using nanotechnology-based nutrient delivery systems can increase oral bioavailability by decreasing first-pass metabolism in the gut and liver, increasing nutraceutical solubility in intestinal fluids, and improving nutraceutical stability in meals and the gastrointestinal tract. For the efficient delivery of bioactive molecules, consider using nanocarriers like polymeric NPs, micelles, liposomes, noisome, solid lipid nanocarriers, nanostructured lipid carriers, microemulsions, dendrimers in organic NPs, and nanocomposites (Kaur et al., 2022).

The current plant delivery methods for NMs are not precise enough to target specific subcellular compartments, making it more challenging to manipulate plant function. A study demonstrated a nanoscale platform that uses a guiding peptide recognition motif to target and deliver NMs with biochemicals to plant photosynthetic organelles (chloroplasts) (Santana et al., 2020). Quantum dot (QD) fluorescence emission in a low background window allows for the quantitative detection of plant cells and organelles using elemental analysis and the imaging of these structures using confocal microscopy. The loading and delivery of various chemicals is made possible by QD functionalization with β-cyclodextrin molecular baskets, and the delivery of these chemicals to chloroplasts is targeted by coating NPs with a conserved and well-designed guiding peptide. Apart from QD with chemical cargoes having a high in vivo effectiveness in delivery and specificity ($74.6 \pm 10.8\%$) to chloroplasts within plant cells, peptide biorecognition provides more precise and adjustable alterations of chloroplast redox activity than chemicals alone. Biorecognition patterns can be used to direct the targeted release of NMs bearing chemical payloads, opening new avenues for nanotechnology applications including plant biology and bioengineering, NP-plant interactions, and nano-enabled agriculture. Among the ten chemical technologies that the IUPAC identified in 2019 as those that are most likely to have an impact on people and society in the future, nanopesticides ranked top due to their potential low impact on the environment and human health (Gomollón-Bel, 2019). All agrochemicals with nanotechnology applications in agriculture will benefit from enhanced penetration, coverage, and target absorption of the active components (AIs) in the form of NMs (García-Gómez et al., 2018). Food security and agricultural system sustainability can be maintained and enhanced by utilizing nano-agrochemical delivery systems via NMs. These systems have the potential to reduce environmental pollution, increase labor savings, and boost agricultural input efficacy. NMs will have a prospective future in agriculture (Pulizzi, 2019). While NMs have shown promise in the delivery of agrochemicals, scholars have also expressed some concern, primarily regarding the unknown health risks to humans and the regulatory challenges posed by NMs' small size effect.

15.8.2 Nanotechnology-Based Pest and Disease Management

Concerning agriculture and food, pests are among the most serious biotic issues. Researchers are continuously looking for novel methods of control. Aside from the negative effects chemical pesticide application has on humans and animals, it also reduces soil fertility, increases pest resistance, eliminates natural enemies, pollutes the environment, reduces biodiversity, and poses health risks to humans. Traditional agricultural strategies, like integrated pest management, are insufficient. Using nanotechnology as an alternative to standard chemical pesticides is one of the most promising ways to overcome their disadvantages in pest management (Yousef et al., 2023). Nano pesticides commenced marching into agriculture and food production. The use rate of conventional pesticides for the targeted crops is less than 30%, and they have coarse carrier particles, poor stability, poor dispersion, and poor biological activity (Zhou et al., 2023). Furthermore, negligent pesticide use increases pesticide bioaccumulation, decreases nitrogen fixation, increases resistance to pesticides, and thus increases the intensity of diseases and pests (Tilman et al., 2002). These are significant issues that require careful consideration if we are to reduce costs for stakeholders, increase productivity, increase efficiency, and decrease our impact on the environment. By increasing pesticide efficacy, lowering dosage requirements, and improving payload stability delivered from the environment (i.e., UV degradation), the use of innovative, nanotechnology-based pesticide ideas—sometimes referred to as "nanopesticides"—is expected to alleviate these problems; consequently, run-off will be minimized, and environmental concerns will be reduced.

Compared to their traditional analogs, nanocides may be more effective, flexible, sustainable, and environmentally safe (Wang et al., 2022). Through improvements in efficiency against target species in field testing of 31.5% and 18.9%, he has proven to have advantages over conventional insecticides. Nanocides cause insufficient collateral environmental damage, as evidenced by their low toxicity of less than 43.1% to non-target organisms. The pesticide's effectiveness and efficiency against action targets like insects, crops, and pathogens are increased by the control-targeted delivery system, which allows the effective ingredients to be released sustainably at the targeted distinction for a longer duration (Hayles et al., 2017). Using currently available pesticides to formulate the active ingredients into nanoemulsions and dispersions, create new nanopesticide formulations by employing NMs (silica, titania, silver, etc.) as active pesticide agents or using these NMs as nanocarriers to deliver the pesticides (polymer nanoparticles). The main shortcomings of current pest management strategies are expected to be addressed by nanopesticides, which will provide new, innovative, nanobased formulations that are stable and active in spray environments (sun, heat, rain), penetrate organisms (fungi, insects), can be delivered to targets, withstand defenses from pests or pathogens, safe for plants and mammals, cost-effective, deterministic to create and produce, and ideally have an innovative mode of action (Smith et al., 2008). In addition to improving the stability and dispersion of pesticides, pesticide nano-carriers increase the bioavailability of the active components by facilitating their delivery to the intended organisms (Sharma et al., 2017). Nano pesticide formulations containing nanocarriers or nano-sized active ingredients exhibit several advantages over conventional pesticides, including enhanced stability, superior efficacy, good dispersion, and controlled release to target organisms. This could lower the global amount of pesticide applied to crops. Conversely, better dispersion of active ingredients may lead to interactions between nanopesticides and non-target organisms, and improved chemical stability of active ingredients implies that they will endure. Therefore, non-target organisms may be negatively impacted by nanopesticides, a possibility that is frequently ignored by the companies that produce these products. The development of a pesticide colloid serves multiple purposes in preventing the pesticide from being demised with water. First off, creaming and sedimentation can be effectively avoided by creating a colloid in which the great majority of particles have a diameter of less than 200 nm. This is due to Brownian motion's opposition to gravity and buoyancy, respectively (McClements, 2012). Second, by producing a strong repulsive force between the droplets/particles (e.g., by steric hindrance or electrostatic repulsion), surfactants can aid in the prevention of by constructing a substantial energy barrier that needs to be overcome in order for flocculation and coalescence to take place (McClements, 2012). Liquid artificial intelligence, or nano-emulsions are made of colloids that have been dissolved in a tiny quantity of organic solvent. Solid particles are called nanodispersions when they are either made up of liquid pesticides or solid pesticides ensconced in an inert, nonpolar matrix (Green & Beestman, 2007). Pesticides are dispersed as nanosized droplets in water in nanoemulsions, also known as oil-in-water (O/W) emulsions, having molecules of surfactant localized at the pesticide-water contact. Nanoemulsions can be categorized as thermodynamically or kinetically stable, depending on the kind and concentration of surfactant employed. For as long as the initial conditions are maintained, these nanoemulsions will remain stable indefinitely. This suggests that when the ingredients of the surfactant, pesticide, and water are mixed, these emulsions also form on their own. However, in practical applications, the system might need to be heated or mechanically agitated to prepare the nanoemulsion due to kinetic energy barriers and/or slow mass transport processes (Green & Beestman, 2007). Given that the swollen micelle state has the least free energy, this kind of the nanoemulsion appears thermodynamically stable (within a specific range of concentrations and temperatures). Many commercial formulations of these nanoemulsions are currently on the market due to their great stability and ease of manufacturing. As a result, they are particularly useful for pesticide formulations.

Examples of pesticide delivery systems that use NPs include Mesoporous Silicon-Based Nano-Carriers (MPSNPs), Trimethylammonium-functionalized MSN, Nano-organized biogenic silica particles, Solid Lipid Nanoparticles (SLN), Nano-Capsules, Nanogels, Nano-emulsions, Micelles Triblock copolymer Shell, etc. With regulated release to the intended pathogen, nanopesticides can increase the effectiveness of conventional pesticides. Nevertheless, it has been noted that the targeted pathogen receives only 0.1% of the nanopesticide. However, the remaining 99.9% is directed toward the pathogen's surrounding environment, which results in a decrease in biodiversity, contamination of the water and soil, and an increase in the body's defense against infections and pests (de Albuquerque et al., 2020). According to recent research, weed species and insect pests can be effectively managed and controlled with the help of ENM platforms.

15.9 Interdisciplinary Research Collaborations and Regulatory Frameworks

15.9.1 Importance of Researchers in Different Fields

Interdisciplinary research collaborations and regulatory frameworks are essential components of advancing nanotechnology applications, particularly in fields like agriculture, medicine, and environmental science. These collaborations bring together experts from diverse disciplines such as biology, chemistry, engineering, and policy-making to address complex challenges that cannot be effectively tackled within the scope of a single discipline. This approach fosters innovation by combining diverse perspectives, methodologies, and techniques and maximizes the benefits of nanotechnology while minimizing risks. Interdisciplinary research collaborations benefit from clear understanding and compliance with regulatory frameworks across disciplines.

Some key points to consider are:

(a) *Regulatory Compliance*: Different disciplines may have specific regulations and ethical guidelines that researchers need to adhere to. For example, biomedical research involving human subjects must comply with institutional review boards (IRBs) and national regulations like the Common Rule in the United States.
(b) *Data Management and Sharing*: Interdisciplinary research often involves diverse data sources and types. Regulatory frameworks such as data protection laws (e.g., GDPR in Europe) govern how data can be collected, stored, and shared across disciplines and borders. Researchers must navigate these regulations to ensure compliance and ethical use of data.
(c) *Intellectual Property (IP) Issues*: Collaborations across disciplines may lead to innovations that are subject to intellectual property rights. Researchers must understand IP laws and regulations to protect their inventions, discoveries, or creative works and navigate ownership and commercialization rights.
(d) *Funding and Grant Requirements*: Funding agencies often have specific regulations regarding interdisciplinary research collaborations. Researchers must ensure that their projects meet the funding agency's criteria and comply with financial regulations governing grant expenditures.
(e) *Ethical Considerations*: Interdisciplinary research can raise unique ethical challenges that may not be fully addressed by traditional disciplinary guidelines. Regulatory frameworks help ensure that research involving complex

ethical issues (e.g., dual-use technologies and environmental impacts) is conducted responsibly.

(f) *Communication and Collaboration Challenges*: Different disciplines may have varying terminology, methodologies, and priorities. Regulatory frameworks can provide a common ground for interdisciplinary teams to communicate effectively, set expectations, and align their research goals. Regulatory frameworks play a crucial role in translating research findings into policies and practices that benefit society while mitigating potential risks.

15.9.2 Need for Stringent Regulations to Ensure Safe Applications of Nanotechnology

The need for stringent regulations to ensure the safe application of nanotechnology is paramount due to the unique properties and potential risks associated with NMs. Nanotechnology operates at the scale of nanometers, where materials exhibit novel physical, chemical, and biological characteristics that differ from their bulk counterparts. These properties can lead to innovative applications across various sectors such as medicine, electronics, and environmental remediation. However, the same properties that make NMs beneficial also raise concerns regarding their safety for human health and the environment. Without robust regulations, there is a risk of unintended exposure to NMs during manufacturing, use, and disposal, which could have unforeseen health consequences. For instance, NPs can penetrate biological barriers more easily than larger particles, potentially causing cellular damage or toxicity. Moreover, their small size and large surface area increase the likelihood of interactions with biological systems, raising concerns about long-term impacts and bioaccumulation. Stringent regulations are essential to address these risks by establishing clear guidelines for the safe handling, production, and disposal of NMs. These regulations may include requirements for risk assessment, toxicity testing, exposure limits, labeling requirements, and environmental impact assessments. By imposing such regulations, governments and regulatory bodies can ensure that nanotechnology developments proceed responsibly, minimizing risks while maximizing benefits. This approach not only protects human health and the environment but also fosters public confidence in the safe and ethical application of nanotechnology across diverse industries.

International regulatory bodies have put policies and procedures in place to guarantee the secure use of nanotechnology in agricultural goods. Here are some of the key regulatory aspects concerning nanotechnology-based agro-products:

(a) *Regulatory Oversight*: Each country has its own regulatory framework governing the use of nanotechnology in agriculture. For example, in the United States, agricultural biotechnology products are managed by the U.S. Department of Agriculture (USDA), while nanotechnology in pesticides is regulated by the Environmental Protection Agency (EPA). In the European Union, the European

Chemicals Agency (ECHA) and the European Food Safety Authority (EFSA) regulate nanotechnology in foods and pesticides.

(b) *Risk Assessment*: Regulatory bodies mandate rigorous risk assessments prior to approving any agricultural product based on nanotechnology. As part of this procedure, the toxicity of NMs is assessed, assessing the potential for environmental release, and examining impacts on human health.

(c) *Labeling*: Agriculture items that include NMs must be labelled in accordance with regulations. With the help of this labeling, customer may choose the things they buy and use with knowledge.

(d) *International Standards*: The safety and caliber of agro-products based on nanotechnology are greatly enhanced by international standards. Standards like ISO/TS 80004-1, which offer vocabulary and definitions for NMs, have been produced by the International Organization for Standardisation (ISO).

(e) *Research and Development*: Agro-products based on nanotechnology are actively being researched and developed by regulatory organizations to guarantee their sustainability and safety for the environment. In an attempt to monitor and evaluate the hazards connected to these novel goods, researchers, producers, regulatory agencies, and consumers are cooperating. In India, the regulation of nanotechnology-based agro-products involves multiple agencies and laws aimed at ensuring safety, environmental protection, and regulatory compliance. The relevant laws, safety measures, and regulations are:

   (i) *Environment (Protection) Act, 1986*: Administered by the Ministry of Environment, Forest and Climate Change (MoEFCC), this law regulates hazardous substances, including NMs, to protect the environment and human health.

   (ii) *Regulations for the Management, Handling, and Transboundary Movement of Hazardous Waste, 2016*: In order to ensure the handling and disposal of hazardous wastes, including NMs, these regulations require facilities handling them to register and receive authorization.

   (iii) *Food Safety and Standards Act, 2006*: Governed by the Food Safety and Standards Authority of India (FSSAI), this act ensures the food goods' quality and safety, including those using nanotechnology in their production.

   (iv) *Insecticides Act, 1968*: Regulates the registration, sale, distribution, and use of insecticides, including those incorporating nanotechnology for enhanced efficacy and safety.

   (v) *Seeds Act, 1966*: Governs the quality of seeds used in agriculture, including genetically modified seeds that may employ nanotechnology, ensuring their safety and effectiveness.

   (vi) *DBT Guidelines for Evaluating the Safety of Foods Produced from Microorganisms and Genetically Modified Plants (2017)*: Provides a framework for assessing the safety of food products developed from genetically altered organisms, particularly those containing nanotechnology.

(vii) *MoEFCC Notification on Hazardous Chemicals Rules (1989): Manufacture, Storage, and Import*: Mandates that producer and importer of dangerous chemicals, including NMs, adhere to safety and environmental laws.

(viii) *FSSAI Regulations on Food Additives (2011)*: Specifies conditions for the use of food additives, including those derived from nanotechnology, in food products to ensure consumer safety.

(ix) *ICAR Guidelines for Nanotechnology Research in Agriculture (2010)*: Provides principles for the ethical and secure application of nanotechnology within agricultural research and development.

(x) *Indian Pharmacopoeia Commission (IPC) Guidelines on Nanoparticle Characterization (2019)*: Provides standards for the characterization of NPs used in various applications, including agro-products, ensuring consistency and safety.

15.10 Conclusion

The strategic use of nanotechnology can transform the agricultural industry by optimizing the effectiveness and efficiency of agrochemicals and stimulating vigorous plant development. NMs improve physiological and metabolic activities in plants by being transported through the xylem and phloem through sophisticated strategies during nutrient and water intake. In addition to boosting plants' resistance to environmental stresses such as drought, salinity, and high temperatures, these compounds enhance the uptake and utilization of nutrients. Ensuring targeted dispersion reduces the need for high chemical dosages and minimizes environmental impact by combining nanoscale agrochemical formulations with engineered NPs. Plant growth-promoting rhizobacteria (PGPR) and natural killers (NMs) work in concert to improve plant health, yield, and stress tolerance. To ensure safety, strict regulatory procedures, and thorough risk assessments are necessary, as the potential hazards of nanotechnology in agriculture, such as oxidative stress, nutritional imbalances, and ecological bioaccumulation, balance with its advantages. Because of their enhanced stability and regulated release mechanisms, nanopesticides hold the potential to be a more environmentally friendly and successful method of controlling pests, lowering non-target toxicity and contamination. Biofortification initiatives and nutrient delivery systems based on nanotechnology provide creative ways to raise crop nutrition and ensure food security. Nanotechnology's ability to improve crop resilience, reduce the need for chemical inputs, and revolutionize agricultural methods highlights its critical role in development of sustainable agriculture, even in the face of obstacles. As regulatory frameworks change to address safety concerns, the ethical deployment of nanotechnology will be essential for optimizing its benefits and minimizing its hazards in a healthy and productive agricultural environment.

References

Abdel Latef, A. A. H., Srivastava, A. K., El-sadek, M. S. A., et al. (2018). Titanium dioxide nanoparticles improve growth and enhance tolerance of broad bean plants under saline soil conditions. *Land Degradation and Development, 29*, 1065–1073. https://doi.org/10.1002/ldr.2780

Aref, I. M., Khan, P. R., Khan, S., et al. (2016). Modulation of antioxidant enzymes in Juniperus procera needles in relation to habitat environment and dieback incidence. *Trees, 30*, 1669–1681. https://doi.org/10.1007/s00468-016-1399-0

Aslam, M. M., Waseem, M., Felix, A., & Shaheen, I. (2022). Nitrogen uptake and utilization toward sustainable agriculture development. In *Nitric oxide in plants* (pp. 78–94). Wiley.

Attaran Dowom, S., Karimian, Z., Mostafaei Dehnavi, M., & Samiei, L. (2022). Chitosan nanoparticles improve physiological and biochemical responses of Salvia abrotanoides (Kar.) under drought stress. *BMC Plant Biology, 22*, 364. https://doi.org/10.1186/s12870-022-03689-4

Avellan, A., Yun, J., Zhang, Y., et al. (2019). Nanoparticle size and coating chemistry control foliar uptake pathways, translocation, and Leaf-to-Rhizosphere transport in wheat. *ACS Nano, 13*, 5291–5305. https://doi.org/10.1021/acsnano.8b09781

Ayyaril, S. S., Shanableh, A., Bhattacharjee, S., et al. (2023). Recent progress in micro and nano-encapsulation techniques for environmental applications: A review. *Results in Engineering, 18*, 101094. https://doi.org/10.1016/j.rineng.2023.101094

Basahi, M. (2021). Seed germination with titanium dioxide nanoparticles enhances water supply, reserve mobilization, oxidative stress and antioxidant enzyme activities in pea. *Saudi Journal of Biological Sciences, 28*, 6500–6507. https://doi.org/10.1016/j.sjbs.2021.07.023

Bolognesi, C. (2003). Genotoxicity of pesticides: A review of human biomonitoring studies. *Mutation Research, 543*, 251–272. https://doi.org/10.1016/S1383-5742(03)00015-2

Chen, S., Dong, C., Gao, Y., et al. (2023). Effects of nanocarbon and nano-calcium carbonate on soil enzyme activities and soil microbial community in wheat (Triticum aestivum L.) rhizosphere soil. *Journal of Plant Nutrition and Soil Science*. https://doi.org/10.1002/jpln.202300146

Cui, H., Shi, Y., Zhou, J., et al. (2018). Effect of different grain sizes of hydroxyapatite on soil heavy metal bioavailability and microbial community composition. *Agriculture, Ecosystems and Environment, 267*, 165–173. https://doi.org/10.1016/j.agee.2018.08.017

Dalir, N. (2023). Nutrients use efficiency in Plants. In *Plant ionomics* (pp. 171–179). Wiley.

de Albuquerque, F. P., Preisler, A. C., Fraceto, L. F., et al. (2020). Overview of nanopesticide environmental safety aspects and regulatory issues: The case of nanoatrazine. In *Nanopesticides* (pp. 281–298). Springer International Publishing.

de Oliveira, J. L., Campos, E. V. R., Bakshi, M., et al. (2014). Application of nanotechnology for the encapsulation of botanical insecticides for sustainable agriculture: Prospects and promises. *Biotechnology Advances, 32*, 1550–1561. https://doi.org/10.1016/j.biotechadv.2014.10.010

DeRosa, M. C., Monreal, C., Schnitzer, M., et al. (2010). Nanotechnology in fertilizers. *Nature Nanotechnology, 5*, 91–91. https://doi.org/10.1038/nnano.2010.2

Di Toro, D. M., Zarba, C. S., Hansen, D. J., et al. (1991). Technical basis for establishing sediment quality criteria for nonionic organic chemicals using equilibrium partitioning. *Environmental Toxicology and Chemistry, 10*, 1541–1583. https://doi.org/10.1002/etc.5620101203

Djanaguiraman, M., Belliraj, N., Bossmann, S. H., & Prasad, P. V. V. (2018). High-temperature stress alleviation by selenium nanoparticle treatment in grain sorghum. *ACS Omega, 3*, 2479–2491. https://doi.org/10.1021/acsomega.7b01934

Duhan, J. S., Kumar, R., Kumar, N., et al. (2017). Nanotechnology: The new perspective in precision agriculture. *Biotechnology Reports, 15*, 11–23. https://doi.org/10.1016/j.btre.2017.03.002

Eichert, T., Kurtz, A., Steiner, U., & Goldbach, H. E. (2008). Size exclusion limits and lateral heterogeneity of the stomatal foliar uptake pathway for aqueous solutes and water-suspended nanoparticles. *Physiologia Plantarum, 134*, 151–160. https://doi.org/10.1111/j.1399-3054.2008.01135.x

El-Saadony, M. T., Saad, A. M., Najjar, A. A., et al. (2021). The use of biological selenium nanoparticles to suppress Triticum aestivum L. crown and root rot diseases induced by Fusarium species and improve yield under drought and heat stress. *Saudi Journal of Biological Sciences, 28*, 4461–4471. https://doi.org/10.1016/j.sjbs.2021.04.043

El-Zohri, M., Al-Wadaani, N. A., & Bafeel, S. O. (2021). Foliar sprayed green zinc oxide nanoparticles mitigate drought-induced oxidative stress in tomato. *Plants, 10*, 2400. https://doi.org/10.3390/plants10112400

Feregrino Pérez, A. A., Páramo Serrano, L. A., Hernández Reséndiz, J. R., et al. (2023). Interactions between nanomaterials and plant–microbe partnership. In *Agricultural and environmental nanotechnology* (pp. 353–392).

Ferrari, E., Barbero, F., Busquets-Fité, M., et al. (2021). Growth-promoting gold nanoparticles decrease stress responses in arabidopsis seedlings. *Nanomaterials, 11*, 3161. https://doi.org/10.3390/nano11123161

García-Gómez, C., Obrador, A., González, D., et al. (2018). Comparative study of the phytotoxicity of ZnO nanoparticles and Zn accumulation in nine crops grown in a calcareous soil and an acidic soil. *Science of the Total Environment, 644*, 770–780. https://doi.org/10.1016/j.scitotenv.2018.06.356

Gardea-Torresdey, J. L., Rico, C. M., & White, J. C. (2014). Trophic transfer, transformation, and impact of engineered nanomaterials in terrestrial environments. *Environmental Science & Technology, 48*, 2526–2540. https://doi.org/10.1021/es4050665

Gayathiri, E., Prakash, P., Pandiaraj, S., et al. (2024). Investigating the ecological implications of nanomaterials: Unveiling plants' notable responses to nano-pollution. *Plant Physiology and Biochemistry, 206*, 108261. https://doi.org/10.1016/j.plaphy.2023.108261

Gomollón-Bel, F. (2019). Ten chemical innovations that will change our world: IUPAC identifies emerging technologies in chemistry with potential to make our planet more sustainable. *Chemistry International, 41*, 12–17. https://doi.org/10.1515/ci-2019-0203

Gonçalves, S., Mansinhos, I., & Romano, A. (2023). Role of nanoparticles on modulation of plant secondary metabolism. In *Engineered nanomaterials for sustainable agricultural production, soil improvement and stress management* (pp. 447–473). Elsevier.

González-Melendi, P., Fernández-Pacheco, R., Coronado, M. J., et al. (2008). Nanoparticles as smart treatment-delivery systems in plants: assessment of different techniques of microscopy for their visualization in plant tissues. *Annals of Botany, 101*, 187–195. https://doi.org/10.1093/aob/mcm283

Green, J. M., & Beestman, G. B. (2007). Recently patented and commercialized formulation and adjuvant technology. *Crop Protection, 26*, 320–327. https://doi.org/10.1016/j.cropro.2005.04.018

Hao, Y., Ma, C., Zhang, Z., et al. (2018). Carbon nanomaterials alter plant physiology and soil bacterial community composition in a rice-soil-bacterial ecosystem. *Environmental Pollution, 232*, 123–136. https://doi.org/10.1016/j.envpol.2017.09.024

Hao, Y., Yu, Y., Sun, G., et al. (2023). Effects of multi-walled carbon nanotubes and nano-silica on root development, leaf photosynthesis, active oxygen and nitrogen metabolism in maize. *Plants, 12*, 1604. https://doi.org/10.3390/plants12081604

Hasanpour, H., Maali-Amir, R., & Zeinali, H. (2015). Effect of TiO_2 nanoparticles on metabolic limitations to photosynthesis under cold in chickpea. *Russian Journal of Plant Physiology, 62*, 779–787. https://doi.org/10.1134/S1021443715060096

Hasanuzzaman, M., Nahar, K., & Fujit, M. (2013). Extreme temperature responses, oxidative stress and antioxidant defense in plants. In *Abiotic stress – Plant responses and applications in agriculture*. InTech.

Hayles, J., Johnson, L., Worthley, C., & Losic, D. (2017). Nanopesticides: a review of current research and perspectives. In *New pesticides and soil sensors* (pp. 193–225). Elsevier.

Hernández-Hernández, H., González-Morales, S., Benavides-Mendoza, A., et al. (2018). Effects of Chitosan–PVA and Cu nanoparticles on the growth and antioxidant capacity of tomato under saline stress. *Molecules, 23*, 178. https://doi.org/10.3390/molecules23010178

Hideg, É., Jansen, M. A. K., & Strid, Å. (2013). UV-B exposure, ROS, and stress: Inseparable companions or loosely linked associates? *Trends in Plant Science, 18*, 107–115. https://doi.org/10.1016/j.tplants.2012.09.003

Iavicoli, I., Leso, V., Beezhold, D. H., & Shvedova, A. A. (2017). Nanotechnology in agriculture: Opportunities, toxicological implications, and occupational risks. *Toxicology and Applied Pharmacology, 329*, 96–111. https://doi.org/10.1016/j.taap.2017.05.025

Ismail, G., & Abou-Zeid, H. (2018). The role of priming with biosynthesized silver nanoparticles in the response of Triticum aestivum L. to salt stress. *Egyptian Journal of Botany*. https://doi.org/10.21608/ejbo.2017.1873.1128

Jose, J. V. (2023). Physiological and molecular aspects of macronutrient uptake by higher plants. In *Sustainable plant nutrition* (pp. 1–21). Elsevier.

Kah, M., Kookana, R. S., Gogos, A., & Bucheli, T. D. (2018). A critical evaluation of nanopesticides and nanofertilizers against their conventional analogues. *Nature Nanotechnology, 13*, 677–684. https://doi.org/10.1038/s41565-018-0131-1

Kang, M., Weng, Y., Liu, Y., et al. (2023). A review on the toxicity mechanisms and potential risks of engineered nanoparticles to plants. *Reviews of Environmental Contamination and Toxicology, 261*, 5. https://doi.org/10.1007/s44169-023-00029-x

Kardavan Ghabel, V., & Karamian, R. (2020). Effects of TiO_2 nanoparticles and spermine on antioxidant responses of Glycyrrhiza glabra L. to cold stress. *Acta Botanica Croatica, 79*, 137–147. https://doi.org/10.37427/botcro-2020-025

Karuppanapandian, T., Wang, H. W., Prabakaran, N., et al. (2011). 2,4-dichlorophenoxyacetic acid-induced leaf senescence in mung bean (Vigna radiata L. Wilczek) and senescence inhibition by co-treatment with silver nanoparticles. *Plant Physiology and Biochemistry, 49*, 168–177. https://doi.org/10.1016/j.plaphy.2010.11.007

Kaur, S., Sharma, K., Singh, R., & Kumar, N. (2022). Advancement in crops and agriculture by nanomaterials. In *Synthesis and applications of nanoparticles* (pp. 319–335). Springer Nature Singapore.

Khan, I., Saeed, K., & Khan, I. (2019). Nanoparticles: Properties, applications and toxicities. *Arabian Journal of Chemistry, 12*, 908–931. https://doi.org/10.1016/j.arabjc.2017.05.011

Khan, M., Khan, A. U., Bogdanchikova, N., & Garibo, D. (2021). Antibacterial and antifungal studies of biosynthesized silver nanoparticles against plant parasitic nematode meloidogyne incognita, plant pathogens ralstonia solanacearum and fusarium oxysporum. *Molecules, 26*, 2462. https://doi.org/10.3390/molecules26092462

Khan, S., Zahoor, M., Sher Khan, R., et al. (2023). The impact of silver nanoparticles on the growth of plants: The agriculture applications. *Heliyon, 9*, e16928. https://doi.org/10.1016/j.heliyon.2023.e16928

Khan, M. A., Sarfraz, W., & Ditta, A. (2024). Chapter 13: Applications of nano-based fertilizers, pesticides, and biosensors in sustainable agriculture and food security. In H. Tombuloglu, G. Tombuloglu, E. Al-Suhaimi, et al. (Eds.), *Nanomaterial-plant interactions* (pp. 277–303). Academic.

Khot, L. R., Sankaran, S., Maja, J. M., et al. (2012). Applications of nanomaterials in agricultural production and crop protection: A review. *Crop Protection, 35*, 64–70. https://doi.org/10.1016/j.cropro.2012.01.007

Klaine, S. J., Alvarez, P. J. J., Batley, G. E., et al. (2008). Nanomaterials in the environment: Behavior, fate, bioavailability, and effects. *Environmental Toxicology and Chemistry, 27*, 1825–1851. https://doi.org/10.1897/08-090.1

Kranjc, E., & Drobne, D. (2019). Nanomaterials in plants: A review of hazard and applications in the agri-food sector. *Nanomaterials, 9*, 1094. https://doi.org/10.3390/nano9081094

Lahiani, M. H., Dervishi, E., Chen, J., et al. (2013). Impact of carbon nanotube exposure to seeds of valuable crops. *ACS Applied Materials & Interfaces, 5*, 7965–7973. https://doi.org/10.1021/am402052x

Lemraski, M. G., Normohamadi, G., Madani, H., et al. (2017). Two Iranian rice cultivars' response to nitrogen and nano-fertilizer. *Open Journal of Ecology, 07*, 591–603. https://doi.org/10.4236/oje.2017.710040

Li, J., Hu, J., Ma, C., et al. (2016). Uptake, translocation and physiological effects of magnetic iron oxide (γ-Fe2O3) nanoparticles in corn (Zea mays L.). *Chemosphere, 159*, 326–334. https://doi.org/10.1016/j.chemosphere.2016.05.083

Liu, H., Xiao, C., Qiu, T., et al. (2022). Selenium regulates antioxidant, photosynthesis, and cell permeability in plants under various abiotic stresses: A review. *Plants, 12*, 44. https://doi.org/10.3390/plants12010044

Malini, S., Raj, K., Madhumathy, S., et al. (2022). Bioinspired advances in nanomaterials for sustainable agriculture. *Journal of Nanomaterials, 2022*, 1–11. https://doi.org/10.1155/2022/8926133

Martínez-Fernández, D., Vítková, M., Bernal, M. P., & Komárek, M. (2015). Effects of nano-maghemite on trace element accumulation and drought response of Helianthus annuus L. in a contaminated mine soil. *Water, Air, & Soil Pollution, 226*, 101. https://doi.org/10.1007/s11270-015-2365-y

McClements, D. J. (2012). Nanoemulsions versus microemulsions: terminology, differences, and similarities. *Soft Matter, 8*, 1719–1729. https://doi.org/10.1039/C2SM06903B

Mohamed, E. F. (2017). Nanotechnology: Future of environmental air pollution control. *Environmental Management and Sustainable Development, 6*, 429. https://doi.org/10.5296/emsd.v6i2.12047

Mustafa, G., Sakata, K., & Komatsu, S. (2016). Proteomic analysis of soybean root exposed to varying sizes of silver nanoparticles under flooding stress. *Journal of Proteomics, 148*, 113–125. https://doi.org/10.1016/j.jprot.2016.07.027

Nawaz, S., Maqsood, I., Batool, F., et al. (2023). Improvement of abiotic stress tolerance in plants with the application of nanoparticles. In *Abiotic stress in plants – Adaptations to climate change*. IntechOpen.

Nazim, M., Li, X., Anjum, S., et al. (2024). Silicon nanoparticles: A novel approach in plant physiology to combat drought stress in arid environment. *Biocatalysis and Agricultural Biotechnology, 58*, 103190. https://doi.org/10.1016/j.bcab.2024.103190

Oliveira, H. C., Gomes, B. C. R., Pelegrino, M. T., & Seabra, A. B. (2016). Nitric oxide-releasing chitosan nanoparticles alleviate the effects of salt stress in maize plants. *Nitric Oxide, 61*, 10–19. https://doi.org/10.1016/j.niox.2016.09.010

Oliveira, H. C., Seabra, A. B., Kondak, S., et al. (2023). Multilevel approach to plant–nanomaterial relationships: from cells to living ecosystems. *Journal of Experimental Botany, 74*, 3406–3424. https://doi.org/10.1093/jxb/erad107

Pascoli, M., Lopes-Oliveira, P. J., Fraceto, L. F., et al. (2018). State of the art of polymeric nanoparticles as carrier systems with agricultural applications: A minireview. *Energy, Ecology and Environment, 3*, 137–148. https://doi.org/10.1007/s40974-018-0090-2

Pokharel, B. R., Sheri, V., Kumar, M., et al. (2024). The update and transport of aluminum nanoparticles in plants and their biochemical and molecular phototoxicity on plant growth and development: A systematic review. *Environmental Pollution, 340*, 122875. https://doi.org/10.1016/j.envpol.2023.122875

Prasad, P. V. V., Pisipati, S. R., Momčilović, I., & Ristic, Z. (2011). Independent and combined effects of high temperature and drought stress during grain filling on plant yield and chloroplast EF-Tu expression in spring wheat. *Journal of Agronomy and Crop Science, 197*, 430–441. https://doi.org/10.1111/j.1439-037X.2011.00477.x

Preetha, J. S. Y., Sriram, D., Premasudha, P., et al. (2023). Cerium oxide as a nanozyme for plant abiotic stress tolerance: An overview of the mechanisms. *Plant Nano Biology, 6*, 100049. https://doi.org/10.1016/j.plana.2023.100049

Pulizzi, F. (2019). Nano in the future of crops. *Nature Nanotechnology, 14*, 507–507. https://doi.org/10.1038/s41565-019-0475-1

Rai, P., Sharma, S., Tripathi, S., et al. (2022). Nanoiron: Uptake, translocation and accumulation in plant systems. *Plant Nano Biology, 2*, 100017. https://doi.org/10.1016/j.plana.2022.100017

Rani, A., Rani, K., Tokas, J., et al. (2020). Nanomaterials for agriculture input use efficiency. In *Resources use efficiency in agriculture* (pp. 137–175). Springer Singapore.

Rezayian, M., Niknam, V., & Arabloo, M. (2023). Iron nanoparticle regulate succinate dehydrogenase activity in canola plants under drought stress. *Scientific Reports, 13*, 9628. https://doi.org/10.1038/s41598-023-36105-4

Rico, C. M., Majumdar, S., Duarte-Gardea, M., et al. (2011). Interaction of nanoparticles with edible plants and their possible implications in the food chain. *Journal of Agricultural and Food Chemistry, 59*, 3485–3498. https://doi.org/10.1021/jf104517j

Roy, I. (2023). Interactions of nanoparticles with plants. In *Bioinspired and green synthesis of nanostructures* (pp. 157–188). Wiley.

Salas-Leiva, J. S., Luna-Velasco, A., & Salas-Leiva, D. E. (2021). Use of magnesium nanomaterials in plants and crop pathogens. *Journal of Nanoparticle Research, 23*, 267. https://doi.org/10.1007/s11051-021-05337-8

Santana, I., Wu, H., Hu, P., & Giraldo, J. P. (2020). Targeted delivery of nanomaterials with chemical cargoes in plants enabled by a biorecognition motif. *Nature Communications, 11*, 2045. https://doi.org/10.1038/s41467-020-15731-w

Saraswat, P., Singh, S., Prasad, M., et al. (2023). Applications of bio-based nanomaterials in environment and agriculture: A review on recent progresses. *Hybrid Advances, 4*, 100097. https://doi.org/10.1016/j.hybadv.2023.100097

Savicka, M., & Škute, N. (2010). Effects of high temperature on malondialdehyde content, superoxide production and growth changes in wheat seedlings (Triticum aestivum L.). *Ekologija, 56*, 26–33. https://doi.org/10.2478/v10055-010-0004-x

Schwab, F., Zhai, G., Kern, M., et al. (2016). Barriers, pathways and processes for uptake, translocation and accumulation of nanomaterials in plants – Critical review. *Nanotoxicology, 10*, 257–278. https://doi.org/10.3109/17435390.2015.1048326

Sedghi, M., Hadi, M., & Toluie, S. G. (2013). Effect of nano zinc oxide on the germination parameters of soybean seeds under drought stress. *Annales of West University of Timisoara Series of Biology, 16*(2), 73.

Seleiman, M. F., Almutairi, K. F., Alotaibi, M., et al. (2021). Nano-fertilization as an emerging fertilization technique: Why can modern agriculture benefit from its use? *Plants, 10*, 2.

Selvakesavan, R. K., Kruszka, D., Shakya, P., et al. (2023). Impact of nanomaterials on plant secondary metabolism. In *Nanomaterial interactions with plant cellular mechanisms and macromolecules and agricultural implications* (pp. 133–170). Springer International Publishing.

Sembada, A. A., & Lenggoro, I. W. (2024). Transport of nanoparticles into plants and their detection methods. *Nanomaterials, 14*, 131. https://doi.org/10.3390/nano14020131

Shakiba, S., Astete, C. E., Paudel, S., et al. (2020). Emerging investigator series: polymeric nanocarriers for agricultural applications: Synthesis, characterization, and environmental and biological interactions. *Environmental Science. Nano, 7*, 37–67. https://doi.org/10.1039/C9EN01127G

Sharma, R., & Kumar, V. (2024). Nano enabled agriculture for sustainable soil. *Waste Management Bulletin, 2*, 152–161. https://doi.org/10.1016/j.wmb.2024.01.002

Sharma, S., Singh, S., Ganguli, A. K., & Shanmugam, V. (2017). Anti-drift nano-stickers made of graphene oxide for targeted pesticide delivery and crop pest control. *Carbon, 115*, 781–790. https://doi.org/10.1016/j.carbon.2017.01.075

Sharma, B., Chettri, D., Verma, A. K., & Soni, U. (2023). Nanomaterial based delivery of genetic material to plant systems. In *The impact of nanoparticles on agriculture and soil* (pp. 41–56). Elsevier.

Shatkin, J. A., & Kim, B. (2015). Cellulose nanomaterials: life cycle risk assessment, and environmental health and safety roadmap. *Environmental Science. Nano, 2*, 477–499. https://doi.org/10.1039/C5EN00059A

Sicard, C., Perullini, M., Spedalieri, C., et al. (2011). CeO_2 nanoparticles for the protection of photosynthetic organisms immobilized in silica gels. *Chemistry of Materials, 23*, 1374–1378. https://doi.org/10.1021/cm103253w

Skiba, E., Adamczyk-Szabela, D., & Wolf, W. M. (2021). Metal-based nanoparticles' interactions with plants. In V. P. Singh, S. Singh, D. K. Tripathi, et al. (Eds.), *Plant responses to nanomaterials: Recent interventions, and physiological and biochemical responses* (pp. 145–169). Springer International Publishing.

Smith, K., Evans, D. A., & El-Hiti, G. A. (2008). Role of modern chemistry in sustainable arable crop protection. *Philosophical Transactions of the Royal Society of London. Series B, Biological Sciences, 363*, 623–637. https://doi.org/10.1098/rstb.2007.2174

Sutulienė, R., Ragelienė, L., Duchovskis, P., & Miliauskienė, J. (2022). The effects of nano-copper, -molybdenum, -boron, and -silica on Pea (Pisum sativum L.) growth, antioxidant properties, and mineral uptake. *Journal of Soil Science and Plant Nutrition, 22*, 801–814. https://doi.org/10.1007/s42729-021-00692-w

Thakur, M., Kumawat, M. K., & Srivastava, R. (2017). Multifunctional graphene quantum dots for combined photothermal and photodynamic therapy coupled with cancer cell tracking applications. *RSC Advances, 7*, 5251–5261. https://doi.org/10.1039/C6RA25976F

Tilman, D., Cassman, K. G., Matson, P. A., et al. (2002). Agricultural sustainability and intensive production practices. *Nature, 418*, 671–677. https://doi.org/10.1038/nature01014

Usman, M., Farooq, M., Wakeel, A., et al. (2020). Nanotechnology in agriculture: Current status, challenges and future opportunities. *Science of the Total Environment, 721*, 137778. https://doi.org/10.1016/j.scitotenv.2020.137778

Van Moorleghem, C., De Schutter, N., Smolders, E., & Merckx, R. (2013). The bioavailability of colloidal and dissolved organic phosphorus to the alga Pseudokirchneriella subcapitata in relation to analytical phosphorus measurements. *Hydrobiologia, 709*, 41–53. https://doi.org/10.1007/s10750-013-1442-8

Vartapetian, B. B., Dolgikh, Y. I., Polyakova, L. I., et al. (2014). Biotechnological approaches to creation of hypoxia and anoxia tolerant plants. *Acta Naturae, 6*, 19–30. https://doi.org/10.32607/20758251-2014-6-2-19-30

Venkateswaran, S. P., Palanisamy, V. K., Divya, P., et al. (2023). Impact of metal oxide nanomaterials on seed germination, plant growth and development. In *Nanometal oxides in horticulture and agronomy* (pp. 81–99). Elsevier.

Wang, X., Han, H., Liu, X., et al. (2012). Multi-walled carbon nanotubes can enhance root elongation of wheat (Triticum aestivum) plants. *Journal of Nanoparticle Research, 14*, 841. https://doi.org/10.1007/s11051-012-0841-5

Wang, P., Lombi, E., Zhao, F.-J., & Kopittke, P. M. (2016). Nanotechnology: A new opportunity in plant sciences. *Trends in Plant Science, 21*, 699–712. https://doi.org/10.1016/j.tplants.2016.04.005

Wang, K., Wang, Y., Wan, Y., et al. (2021). The fate of arsenic in rice plants (Oryza sativa L.): Influence of different forms of selenium. *Chemosphere, 264*, 128417. https://doi.org/10.1016/j.chemosphere.2020.128417

Wang, D., Saleh, N. B., Byro, A., et al. (2022). Nano-enabled pesticides for sustainable agriculture and global food security. *Nature Nanotechnology, 17*, 347–360. https://doi.org/10.1038/s41565-022-01082-8

Wu, H., & Li, Z. (2022). Recent advances in nano-enabled agriculture for improving plant performance. *Crop Journal, 10*, 1–12. https://doi.org/10.1016/j.cj.2021.06.002

Xiumei, L., Fudao, Z., Shuqing, Z., et al. (2005). Responses of peanut to Nano-calcium carbonate. *Plant Nutrition and Fertilizer Science, 11*, 385–389.

Xu, Y., Lu, Y., Li, J., et al. (2020). Effect of graphene quantum dot size on plant growth. *Nanoscale, 12*, 15045–15049. https://doi.org/10.1039/D0NR01913E

Yousef, H. A., Fahmy, H. M., Arafa, F. N., et al. (2023). Nanotechnology in pest management: advantages, applications, and challenges. *International Journal of Tropical Insect Science, 43*, 1387–1399. https://doi.org/10.1007/s42690-023-01053-z

Zhao, L., Sun, Y., Hernandez-Viezcas, J. A., et al. (2015). Monitoring the environmental effects of CeO_2 and ZnO nanoparticles through the life cycle of corn (Zea mays) plants and in situ μ-XRF mapping of nutrients in kernels. *Environmental Science & Technology, 49*, 2921–2928. https://doi.org/10.1021/es5060226

Zhao, S., Wang, W., Chen, X., et al. (2023). Graphene oxide affected root growth, anatomy, and nutrient uptake in alfalfa. *Ecotoxicology and Environmental Safety, 250*, 114483. https://doi.org/10.1016/j.ecoenv.2022.114483

Zhou, J., Liu, G., Guo, Z., et al. (2023). Stimuli-responsive pesticide carriers based on porous nanomaterials: A review. *Chemical Engineering Journal, 455*, 140167. https://doi.org/10.1016/j.cej.2022.140167

Zhu, H., Han, J., Xiao, J. Q., & Jin, Y. (2008). Uptake, translocation, and accumulation of manufactured iron oxide nanoparticles by pumpkin plants. *Journal of Environmental Monitoring, 10*, 713. https://doi.org/10.1039/b805998e

Chapter 16
Advancing Food Science from Field to Fork: The Role of Nanobionics in Enhancing Safety, Quality, and Sustainability

Akshay Ramani, Pavel Rout, and Subhradip Bhattacharjee

Abstract The modern food system faces pressing challenges such as population growth, environmental degradation, and food security concerns, necessitating innovative solutions for sustainable and safe food production. Nanobionics, integrating nanotechnology with biological systems, emerges as a transformative approach to enhance food safety, quality, and sustainability. This chapter explores the role of nanobionics in revolutionizing food science from field to fork. In agriculture, nanobionics improves crop production and quality through precision farming techniques and smart fertilizers, optimizing resource use while reducing environmental impact. Post-harvest, nanotechnology advances food preservation with intelligent packaging and nanopreservation techniques that extend shelf life and ensure food safety. Furthermore, nanobionics enhances food safety by enabling rapid detection of contaminants and pathogens, while also addressing consumer concerns through rigorous risk management and transparency. Sustainable practices, supported by nanobionics, offer solutions to reduce food waste and resource use, although challenges remain regarding environmental impacts and regulatory frameworks. Future trends highlight promising innovations poised to address global food security challenges.

Keywords Nanoencapsulation · Nanobiosensors · Food Safety Technologies · Functional Foods Development · Antimicrobial agent

---

A. Ramani · P. Rout
Dairy Chemistry Division, ICAR- National Dairy Research Institute, Karnal, Haryana, India

S. Bhattacharjee (✉)
Agronomy Section, ICAR- National Dairy Research Institute, Karnal, Haryana, India

ICAR- Central Coastal Agricultural Research Institute, Old Goa, Goa, India

V. D. Rajput et al. (eds.), *Contemporary Suitability of Nanobionics in Agriculture*, Nanotechnology in the Life Sciences, https://doi.org/10.1007/978-3-031-82779-2_16

16.1 Introduction

The modern world faces an intricate set of challenges regarding food systems that are influenced by several factors, including population growth, environmental degradation, food security concerns, and consumer demand for safe, nutritious, and sustainable food. Food science, as a multidisciplinary field, aims to address these issues through innovations in agriculture, food production, processing, preservation, and distribution. However, current food systems remain inadequate to meet these growing demands in a sustainable manner. In this context, nanobionics—a novel approach combining nanotechnology with biological systems—offers promising solutions that can revolutionize how food is grown, processed, and consumed. This chapter will provide an overview of the primary challenges in food science and introduce the concept of nanobionics, focusing on its role in enhancing food safety, quality, and sustainability from field to fork. One of the key challenges in food science is the growing global population, expected to exceed 9.7 billion by 2050. This surge will intensify the strain on food systems, which are already struggling to feed over 820 million people suffering from chronic hunger, according to the FAO. Additionally, two billion people face "hidden hunger," lacking essential micronutrients. Addressing this issue requires more than increasing food production—it demands the creation of systems that provide safe, nutrient-rich, and accessible food.

As income rises, especially in developing countries, dietary preferences are shifting toward protein- and fat-rich foods, further straining agricultural systems. Producing animal-based products, such as beef, consumes significantly more resources than plant-based alternatives. Furthermore, the globalization of food supply chains increases the risks of contamination and foodborne illnesses. WHO reports over 600 million cases of foodborne diseases annually, emphasizing the need for innovative solutions to ensure food safety from farm to fork.

Despite advancements in agricultural technologies, food production still faces limitations that hinder sustainable growth. One major issue is the degradation of natural resources due to intensive farming practices that deplete soil nutrients, water, and biodiversity. These practices also drive environmental problems such as greenhouse gas emissions, deforestation, and water pollution. The FAO estimates that agriculture contributes roughly 30% of global greenhouse gas emissions and accounts for 70% of global freshwater withdrawals. As water scarcity intensifies, the need to produce more food without further environmental harm becomes a critical challenge. Another limitation is inefficiency in food systems. The FAO reports that approximately one-third of global food production—about 1.3 billion tons—is lost or wasted annually. In low-income countries, this loss occurs mainly during production and post-harvest stages due to poor infrastructure and inadequate storage. In high-income countries, however, food waste occurs at the retail and consumer levels. Reducing these inefficiencies is crucial to ensuring food reaches those in need while minimizing the environmental footprint. Nanobionics presents a promising solution to these challenges by offering innovations that enhance food safety, quality, and sustainability. In light of these challenges, nanobionics offers a transformative approach to enhancing food safety, quality, and sustainability. Nanobionics is the integration of nanotechnology and biological systems to create new functionalities at the molecular level. In the context of food science, nanobionics can be applied across various stages of the food supply chain, from agricultural production to food preservation and distribution. The ability to manipulate materials at the nanoscale provides unprecedented control over biological processes, enabling the development of advanced solutions for improving crop yields, extending food shelf life, detecting contaminants, and reducing food waste.

Nanobionics has the potential to revolutionize agricultural practices by enhancing crop production while minimizing environmental impacts. For instance, nanomaterials can be used to create smart fertilizers that release nutrients in a controlled manner, ensuring that plants receive the right amount of nutrients at the right time. This reduces the overuse of chemical fertilizers, which can lead to soil and water contamination. Nanoparticles can also be engineered to improve plant resistance to pests, diseases, and environmental stresses, such as drought and extreme temperatures. This is especially important in the face of climate change, which is expected to increase the frequency and severity of extreme weather events, putting additional strain on global food production.

In addition to improving crop production, nanobionics can play a significant role in food preservation and extending shelf life. Nanomaterials can be incorporated into food packaging to create smart, active packaging systems that not only protect food from environmental factors such as moisture and oxygen but also monitor food quality. For example, nanosensors embedded in packaging materials can detect changes in temperature, humidity, and the presence of pathogens, providing real-time information about the freshness and safety of the food inside. This can help prevent foodborne illnesses and reduce food waste by ensuring that food is consumed before it spoils. Nanobionics can also be used to create edible coatings for fruits and vegetables that prevent spoilage and extend shelf life by creating a barrier against microorganisms and oxidation.

Furthermore, nanobionics offers novel solutions for enhancing food safety through the development of nanosensors that can rapidly detect contaminants and pathogens in food products. Traditional methods of detecting foodborne pathogens are often slow and require specialized laboratory equipment, making them inadequate for preventing contaminated food from reaching consumers. Nanosensors, on the other hand, can provide real-time detection of harmful bacteria, viruses, and toxins, allowing for immediate action to be taken to prevent contamination. This technology can be applied both during food processing and in retail settings, ensuring that food remains safe throughout the supply chain.

The integration of nanobionics into the food supply chain also has significant implications for sustainability. By reducing the need for chemical inputs such as fertilizers and pesticides, improving resource-use efficiency, and extending food shelf life, nanobionics can help minimize the environmental impact of food production. Additionally, the use of biodegradable nanomaterials in packaging can reduce the amount of plastic waste generated by the food industry, which is a major contributor to global pollution.

16.2 Nanobionics in Agriculture: Enhancing Crop Production and Quality

Nanobionics is revolutionizing agriculture by significantly improving crop production and quality. Advanced nanosensors enable precise monitoring of soil conditions, water levels, and nutrient needs, allowing for targeted interventions that optimize resource use and reduce waste (Reddy & Chhabra, 2022; Saranya et al., 2019). Controlled-release fertilizers and pesticides, developed using nanotechnology, enhance the efficiency and effectiveness of these inputs while minimizing environmental impact (Zhao et al., 2019). Additionally, nanobionics contributes to plant growth enhancement by improving photosynthesis and stress resistance through the application of nanoparticles (Gao et al., 2020). These innovations not only increase crop yield and quality but also promote sustainability by reducing water, fertilizer, and pesticide usage, thereby lowering the overall environmental footprint of agricultural practices (Fig. 16.1) (Khan et al., 2021).

16.2.1 Nanobionics for Precision Agriculture

Precision agriculture represents a transformative approach to farming that leverages technology to optimize crop production and resource use. Nanobionics, integrating nanotechnology and biotechnology, is at the forefront of this innovation. Smart sensors are a key component of nanobionic precision agriculture. These nanoscale sensors can monitor soil conditions, water levels, and nutrient needs with unprecedented accuracy. For instance, nanosensors embedded in the soil can provide real-time data on soil moisture and nutrient levels, allowing for more precise irrigation and fertilization practices (Reddy & Chhabra, 2022). This real-time monitoring helps farmers make data-driven decisions, improving resource efficiency and crop health.

Controlled-release fertilizers and pesticides are another significant advancement enabled by nanobionics. Traditional fertilizers and pesticides often suffer from inefficiencies related to rapid degradation or leaching, which can lead to environmental pollution and reduced efficacy. Nanotechnology allows for the development of controlled-release formulations where nanoparticles encapsulate nutrients or pesticides, releasing them in a targeted manner over time (Zhao et al., 2019). This approach not only enhances the effectiveness of these inputs but also minimizes their environmental impact by reducing runoff and lowering the overall quantities required (Khan et al., 2021). By improving the precision of input applications, nanobionics contributes to more sustainable and productive agricultural practices.

Fig. 16.1 Applications of nanobionic in various sectors of agriculture

16.2.2 Plant Nanobionics for Growth Enhancement

Nanobionics also plays a crucial role in enhancing plant growth and metabolism. Nanoparticles can influence various physiological processes in plants, leading to improved growth and productivity. For example, certain nanoparticles, such as zinc oxide and titanium dioxide, have been shown to enhance photosynthesis, a critical process for plant growth (Adhikari et al., 2020; Bormashenko et al., 2015). By increasing the efficiency of photosynthesis, these nanoparticles help plants convert sunlight into energy more effectively, promoting faster and healthier growth. Table 16.1 depicts the effects of various nanomaterials on agricultural growth.

In addition to enhancing photosynthesis, nanomaterials can improve crop yield and stress resistance. Nanobionics can be used to develop nanofertilizers that provide essential nutrients to plants in a more bioavailable form, thereby enhancing their growth and yield (Gao et al., 2020). Furthermore, nanomaterials can help plants cope with environmental stressors such as drought and high salinity. For instance, nanoparticles can enhance plant resistance to these stresses by improving water and nutrient uptake or by activating stress-response pathways (Reddy & Chhabra, 2022). This increased resilience can lead to more stable crop yields even under adverse conditions, contributing to food security and agricultural sustainability.

Table 16.1 Effect of nanomaterials on plant growth and development

Plant	Plant parts	Nanomaterial	Effect	References
Trigonella foenum	Seeds	Ag NPs	Enhanced growth, diosgenin content, and secondary metabolite production	Jasim et al. (2016)
Pearl millet	Whole plant	Zn NPs	Enhanced shoot length, chlorophyll, and yield	Tarafdar et al. (2014)
Petroselinum crispum	Seeds	Nanoanatase	Enhanced shoot length, chlorophyll, and yield	Dehkourdi and Mosavi (2013)
Foeniculum vulgare Mill.	Seeds	TiO_2	Enhanced seed germination	Feizi et al. (2013)
Cabbage, maize	Root	Ag NPs	Root elongation	Pokhrel et al. (2013)
Arabidopsis thaliana	Seeds	Au NPs	Enhanced seed germination	Kumar et al. (2013)
Brassica juncea	Seeds	Au NPs	Increase in net productivity and yield	Arora et al. (2012)
Tomato	Roots	MWCNTS	Efficient water and nutrient uptake	Villagarcia et al. (2012)
Zea mays L.	Seeds	SiO_2	Enhanced seed germination	Suriyaprabha et al. (2012)
Boswellia ovalifoliata	Seeds	Au NPs	Enhanced seed germination	Savithramma et al. (2012)
Bacopa monnieri	Whole plant	Ag NPs	Increased carbohydrate, protein, and decreased total phenol content	Krishnaraj et al. (2012)
Tomato	Tissue	SWCNTS	DNA-dye molecule delivery	Srinivasan et al. (2010)
Tomato	Seeds	MWCNTS	Enhanced seed germination and vegetative biomass	Khodakovskaya et al. (2009)
Cucumber, onion	Root	SWCNTS	Root elongation	Canas et al. (2008)
Rape, rye grass, corn	Root	MWCNTS	Root elongation	Lin and Xing (2007)
Changbai larch	Seeds	SiO_2	Enhanced seed germination and chlorophyll biosynthesis	Bao-shan et al. (2004)

16.2.3 Sustainability in Agricultural Practices

Sustainability is a central theme in modern agriculture, and nanobionics offers several pathways to achieve more sustainable practices. One of the key benefits of nanobionics is the potential reduction in water, fertilizer, and pesticide usage. Through precise control and targeted delivery, nanobionics can significantly decrease the quantities of these resources required. For example, controlled-release fertilizers can reduce the frequency of application and lower the overall amounts needed, minimizing waste and environmental impact (Khan et al., 2021). Similarly, nanosensors can optimize irrigation by providing real-time data on soil moisture, helping to prevent overwatering and conserve water resources (Zhao et al., 2019). Nanobionics also plays a role in reducing the environmental footprint of agriculture. By enhancing the efficiency of resource use and minimizing waste, nanobionics can help reduce pollution and environmental degradation. For instance, the use of nanomaterials in soil and water treatment can help remediate contaminated environments, restoring soil fertility and water quality (Bormashenko et al., 2015). Additionally, the development of eco-friendly nanomaterials that degrade safely in the environment can further mitigate the ecological impact of agricultural practices (Gao et al., 2020). These advancements contribute to a more sustainable agricultural system that balances productivity with environmental stewardship.

16.3 Post-Harvest Nanobionics: Improving Food Preservation and Shelf Life

Post-harvest nanobionics is transforming food preservation and shelf life through advanced packaging and preservation techniques. Nanomaterials in packaging, such as nanosensors and nanocomposites, enable real-time spoilage detection and enhance barrier properties against oxygen and moisture, thereby extending food freshness (Barage et al., 2022; Gao et al., 2020). Active packaging systems incorporate antimicrobial nanoparticles to prevent microbial growth and spoilage, while intelligent packaging with freshness sensors provides consumers with real-time quality information (Khan et al., 2021). Additionally, nanopreservation techniques use nanoparticles in coatings to protect against oxidation and spoilage, helping to maintain food quality for longer periods. These innovations collectively contribute to a more efficient and sustainable food supply chain.

16.3.1 Nanomaterials for Packaging

The post-harvest stage of the food supply chain is crucial for maintaining food quality and safety. Nanobionics is making significant strides in this area through the development of advanced packaging materials. Smart packaging technologies, utilizing nanomaterials, are at the forefront of this innovation. These smart packages are designed to detect spoilage and contamination in real time, thereby extending the shelf life of food products and ensuring their safety. Nanomaterials, such as nanosensors, are integrated into packaging materials to monitor various factors that can affect food quality. For example, nanosensors can detect changes in pH, gas composition, and microbial contamination, providing early warnings about potential spoilage (Chelliah et al., 2021). These sensors can be embedded in packaging films or labels, offering a noninvasive method for continuous monitoring without

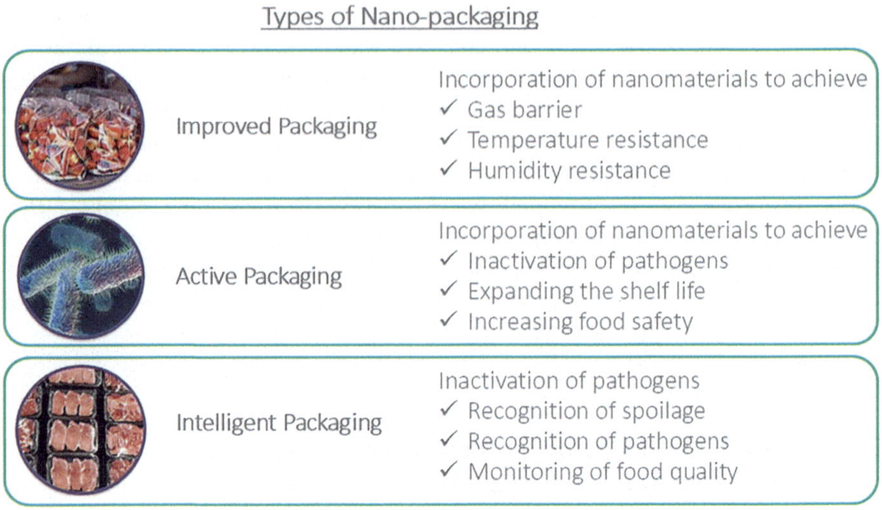

Fig. 16.2 Classification of food nanopackaging and their applications in food safety

altering the food product. Nanocomposite materials are another critical development in packaging. These materials incorporate nanoparticles, such as clay, silver, or carbon nanotubes, into packaging films to enhance their barrier properties. For instance, nanocomposite films can improve resistance to oxygen, moisture, and light, which are key factors in food degradation (Gao et al., 2020). By providing a more effective barrier against environmental factors, these materials help maintain the freshness and extend the shelf life of packaged foods. Nanopackaging can be divided into three main categories (Fig. 16.2): (i) improved packaging: contain NPs and are resistant to temperature and humidity; (ii) active packaging: contain preservatives such as inorganic NPs that can interact directly with food and provide antimicrobial properties; and (iii) intelligent/smart packaging: designed for sensing biochemical or microbial changes and specific pathogen developing in the food. The development of nanomaterials leads to manufacturing of antimicrobial active packaging that is capable of preserving foods and extending their shelf life. Since the main focus of this study is on antimicrobial nanopackages, we review the active packages in more detail.

16.3.2 Active and Intelligent Packaging

Active and intelligent packaging systems represent another significant advancement enabled by nanobionics. Active packaging involves the integration of antimicrobial agents into packaging materials to inhibit microbial growth and extend shelf life.

Nanoparticles, such as silver and zinc oxide, possess inherent antimicrobial properties that can be utilized in packaging materials to reduce spoilage and contamination (Khan et al., 2021). These active agents work by releasing antimicrobial compounds or by directly interacting with microorganisms, thereby enhancing the safety and longevity of food products. Intelligent packaging systems, on the other hand, use embedded sensors and indicators to provide real-time information about food quality. For example, freshness sensors can detect changes in the environment, such as temperature and humidity, which can impact the quality of perishable foods. Some intelligent packaging systems also feature color-changing indicators that signal the freshness or spoilage of the food. These indicators use nanoscale materials that change color in response to chemical changes in the food, allowing consumers to easily assess the quality of the product (Zhao et al., 2019).

16.3.3 Nanopreservation Techniques

Nanopreservation techniques offer innovative solutions for extending the shelf life of food products through advanced coating and preservation methods. One prominent approach is the use of nanoparticles in food coatings. These coatings can be applied to food surfaces to create a protective layer that prevents spoilage and oxidation. For example, nanoparticles of materials like silica or chitosan can form thin, transparent coatings that act as barriers to oxygen and moisture (Reddy & Chhabra, 2022; Rout et al., 2023). This protection helps to maintain the food's freshness and nutritional value for a longer period. Nanopreservation also includes techniques to prevent oxidation, which is a major cause of food spoilage. Oxidation can lead to off-flavors, loss of nutrients, and changes in texture. Nanoparticles, such as those made from metals or metal oxides, can be incorporated into food packaging or directly applied to food surfaces to inhibit oxidative reactions (Gao et al., 2020). These nanoparticles work by scavenging free radicals or by reacting with oxygen, thereby slowing down the oxidative processes that contribute to spoilage. Post-harvest nanobionics represents a transformative approach to improving food preservation and extending shelf life. Nanomaterials in packaging enhance barrier properties and enable smart monitoring of food quality. Active and intelligent packaging systems leverage antimicrobial agents and sensors to maintain freshness and provide real-time quality information. Additionally, nanopreservation techniques, including advanced coatings and oxidation prevention, offer effective solutions for prolonging the shelf life of food products. As these technologies continue to evolve, they hold the promise of significantly improving the efficiency and effectiveness of food preservation, contributing to a more sustainable and secure food supply chain.

16.4 Nanobionics for Food Safety: Detection and Prevention of Contaminants

Nanobionics is revolutionizing food safety through advanced detection and prevention methods. Nanosensors and biosensors enable rapid, real-time detection of foodborne pathogens, toxins, and contaminants, enhancing safety during both food processing and retail (Bormashenko et al., 2015). These sensors utilize nanomaterials to amplify detection signals, providing quick and accurate results. In contamination prevention, antimicrobial coatings and additives with nanoparticles, such as silver and zinc oxide, effectively inhibit pathogen growth and neutralize toxins (Gao et al., 2020). Additionally, nanoparticle-based filtration systems address toxins in food and water, further ensuring safety (Khan et al., 2021). The integration of these technologies into food quality control systems, including HACCP, ensures consistent product safety and compliance, making nanobionics a crucial component in modern food safety strategies.

16.4.1 Rapid Detection of Foodborne Pathogens

Foodborne illnesses pose a significant threat to public health globally, making rapid and accurate detection of pathogens a crucial aspect of food safety. Nanobionics offers innovative solutions for real-time detection of foodborne pathogens through advanced biosensors and nanosensors. These technologies enable the swift identification of bacteria, viruses, and toxins, improving response times and reducing the risk of widespread outbreaks. Biosensors and nanosensors leverage the unique properties of nanomaterials to detect pathogens with high sensitivity and specificity. For example, gold nanoparticles, carbon nanotubes, and quantum dots are commonly used in biosensor applications due to their ability to enhance signal detection. These sensors can be designed to target specific pathogens or toxins by employing recognition elements such as antibodies or nucleic acids that bind to the target molecules. When these recognition elements interact with the pathogens or toxins, the biosensor produces a detectable signal, often amplified by the nanomaterials, allowing for rapid identification even at low concentrations (Zhao et al., 2019). The applications of these technologies extend across various stages of the food supply chain. In food processing, nanosensors can be integrated into production lines to continuously monitor for contaminants, providing immediate feedback and enabling corrective actions to be taken swiftly. In retail environments, nanosensor-based detection systems can be employed in packaging or point-of-sale systems to ensure that food products meet safety standards before reaching consumers. This real-time monitoring helps to reduce the incidence of foodborne illnesses and enhances overall food safety.

16.4.2 Nanobionics in Contaminant Prevention

Beyond detection, nanobionics plays a vital role in the prevention of contaminants through the use of antimicrobial coatings and additives. These technologies are designed to control pathogen growth and reduce the risk of contamination in food products and packaging.

Antimicrobial coatings utilize nanoparticles with inherent antimicrobial properties to create surfaces that inhibit microbial growth. For example, silver nanoparticles are well known for their antibacterial properties and are often incorporated into coatings for food packaging and processing equipment (Gao et al., 2020). These coatings can be applied to a variety of surfaces, including food contact surfaces and packaging materials, to provide an additional layer of protection against pathogens. By continuously releasing antimicrobial agents or directly interacting with microorganisms, these coatings help to prevent contamination and extend the shelf life of food products. In addition to coatings, nanobionics also offers solutions for neutralizing toxins in food and water. Nanomaterials can be used to develop filtration systems and additives that remove or neutralize harmful substances. For example, nanoparticle-based filters can effectively capture and remove toxins from water, while additives containing nanomaterials can bind to and neutralize toxins in food (Khan et al., 2021). These technologies enhance food safety by addressing potential contamination sources before they pose a risk to consumers.

16.4.3 Nanobionics and Food Quality Control

Ensuring consistent food quality during production is essential for maintaining consumer trust and meeting regulatory standards. Nanobionics contributes to food quality control through the use of nanosensors and their integration into HACCP (Hazard Analysis and Critical Control Points) systems. Nanosensors play a crucial role in monitoring various quality parameters, such as temperature, pH, and humidity, which can impact food safety and quality. These sensors provide precise and real-time data, enabling manufacturers to track and control production conditions more effectively. For instance, temperature-sensitive nanosensors can detect deviations from optimal storage conditions, allowing for timely interventions to prevent spoilage or degradation (Bormashenko et al., 2015). Similarly, nanosensors can monitor the freshness of food products by detecting changes in chemical markers associated with spoilage. The integration of nanobionics into HACCP systems further enhances food safety management. HACCP is a systematic approach to identifying and controlling hazards throughout the food production process. By incorporating nanosensors into HACCP systems, manufacturers can improve the accuracy and efficiency of hazard detection and control measures. For example, nanosensors can provide continuous monitoring of critical control points, such as temperature or contamination levels, allowing for real-time adjustments and ensuring that food products meet safety and quality standards (Shruti et al., 2024).

Nanobionics offers transformative solutions for food safety through rapid pathogen detection, contaminant prevention, and quality control. The development of advanced biosensors and nanosensors enhances the ability to detect and respond to foodborne pathogens quickly and accurately. Antimicrobial coatings and nanoparticle-based additives provide effective means for preventing contamination and neutralizing toxins. Additionally, the integration of nanobionics into food quality control systems supports consistent product quality and regulatory compliance. As these technologies continue to advance, they hold the potential to significantly improve food safety and enhance the overall integrity of the food supply chain. Consumer perception of nanobionics in food is critical for its adoption. Public concerns often revolve around the safety of nanomaterials, with apprehensions about potential health risks and environmental impacts (Wahab et al., 2024). Addressing these concerns requires clear regulatory frameworks and transparent labeling to inform consumers about the presence and safety of nanomaterials in food. Safety assessments, including toxicological studies, are essential to evaluate the potential health effects of nanoparticles (Warheit et al., 2008). Additionally, long-term impacts on both human health and the environment must be considered. Effective risk management strategies, guided by regulatory oversight, are crucial for ensuring that nanobionics contribute positively to food science while safeguarding consumer trust (FDA, 2013; Trusted Science for Safe Food, 2020).

16.5 Public Perception of Nanotechnology in Food

The integration of nanobionics in food science presents both opportunities and challenges, particularly concerning consumer perception. Nanotechnology's potential to enhance food safety, quality, and sustainability is significant; however, public acceptance often hinges on addressing concerns about nanomaterials. Consumers frequently express apprehension regarding the safety of ingesting nanoparticles, driven by limited understanding and transparency (Siegrist, 2017). Effective communication about the benefits and risks of nanobionics is crucial for building trust. Regulatory frameworks play a pivotal role in this process. For example, the European Union mandates labeling of food products containing nanomaterials, providing consumers with essential information (Trusted Science for Safe Food, 2020). In contrast, the United States has more lenient requirements, which can lead to variability in consumer awareness and perception (Sullivan, 2019).

16.5.1 Safety and Ethical Considerations

The safety of nanomaterials used in food systems is a critical concern that encompasses both toxicological and long-term impacts. Toxicological assessments are vital to determine the potential health risks associated with nanoparticles, which can differ significantly from those of bulk materials due to their unique properties (Warheit et al., 2008). Long-term health impacts are particularly challenging to assess, as the behavior of nanomaterials in the human body and environment is still under extensive study. Environmental impacts are also a concern; nanoparticles may accumulate in ecosystems and affect non-target organisms, leading to potential ecological disruptions (Oberdörster et al., 2007). Ethical considerations, including the right to informed consent and transparency about the presence of nanomaterials in food, are crucial for maintaining public trust (Singh & Kumar, 2023). Comprehensive safety evaluations and ethical considerations are necessary to ensure that the benefits of nanobionics do not compromise consumer health or environmental integrity.

16.5.2 Risk Management Strategies

Effective risk management is essential for the safe application of nanobionics in food systems. Regulatory oversight provides a framework for ensuring that nanomaterials are rigorously tested and monitored. For instance, the US Food and Drug Administration (FDA) and the European Food Safety Authority (EFSA) have established guidelines for the evaluation of nanotechnology in food, focusing on safety and risk assessment (FDA, 2013; Trusted Science for Safe Food, 2020). Risk management strategies also involve the development of best practices for the use and disposal of nanomaterials, as well as ongoing research to address emerging risks. Collaboration between scientists, regulators, and industry stakeholders is crucial to maintaining safety and addressing any issues that arise. By implementing robust regulatory frameworks and promoting transparent communication, it is possible to mitigate risks and ensure that nanobionics contribute positively to food science advancements.

16.6 Sustainability and Environmental Impact of Nanobionics

Nanobionics significantly contributes to sustainability in food systems by reducing food waste and enhancing resource efficiency. Through advanced nanosensors and smart packaging, it helps monitor food quality and extend shelf life, thereby minimizing spoilage (Gao et al., 2020). Additionally, nanobionics supports circular economy approaches by enabling the recycling and repurposing of food by-products (Khan et al., 2021). However, environmental concerns arise from the potential accumulation of nanomaterials in ecosystems, which could impact wildlife and plants (Nel et al., 2006). Developing biodegradable and eco-friendly nanomaterials is crucial to address these risks (Khan et al., 2021). Life cycle assessment (LCA) is essential for evaluating the overall environmental impact of nanobionics, helping to ensure that their benefits for food science and sustainability outweigh any potential ecological drawbacks (Finnveden et al., 2009).

16.6.1 Nanobionics and Sustainable Food Systems

Nanobionics has the potential to revolutionize sustainability in food systems by enhancing resource efficiency and reducing food waste. Through innovations such as nanosensors and smart packaging, nanobionics can monitor the freshness and quality of food products in real time, which helps in reducing spoilage and extending shelf life (Gao et al., 2020). This efficiency not only minimizes food waste but also optimizes resource use throughout the supply chain. Moreover, nanobionics can play a role in the circular economy by facilitating the recycling and repurposing of food by-products. For example, nanomaterials can be employed to recover nutrients from food waste, supporting a closed-loop system that conserves resources and reduces environmental impact (Khan et al., 2021).

16.6.2 Environmental Concerns and Solutions

Despite its advantages, the use of nanobionics raises environmental concerns that need to be addressed. Nanomaterials can potentially accumulate in ecosystems, posing risks to wildlife and plant life due to their small size and high reactivity (Nel et al., 2006). To mitigate these risks, it is crucial to develop biodegradable and eco-friendly nanomaterials. Researchers are focusing on creating materials that break down naturally without leaving harmful residues, such as biopolymers and organic nanoparticles (Khan et al., 2021). These advancements aim to minimize the ecological footprint of nanobionics and align their use with sustainable practices.

16.6.3 Life Cycle Assessment (LCA) of Nanobionics

A comprehensive evaluation of the life cycle of nanomaterials is essential to understand their overall environmental impact. Life cycle assessment (LCA) provides a systematic approach to assessing the environmental effects associated with the production, use, and disposal of nanobionic products (Finnveden et al., 2009). LCA helps identify potential environmental hotspots and opportunities for improvement, such as enhancing manufacturing processes or increasing recyclability. Recent studies highlight the importance of incorporating LCA into the development and assessment of nanobionics to ensure that their benefits for food science and sustainability outweigh any potential drawbacks (Fava, 2025). By integrating LCA, stakeholders can make informed decisions that support both technological advancement and environmental protection.

16.7 Future Trends and Innovations in Nanobionics for Food Science

Nanobionics is set to drive significant advancements in food science through emerging technologies such as biosensors, nanodevices, and molecular engineering. Innovations include highly sensitive nanoscale biosensors for real-time contamination detection and smart packaging that adapts to environmental changes to extend shelf life (Gao et al., 2020). These technologies are expected to revolutionize food production, preservation, and quality control. Nanobionics also holds promise for addressing global food security by enhancing agricultural practices in developing regions and mitigating challenges like climate change and resource scarcity (Zhao et al., 2019). However, challenges such as regulatory hurdles, safety concerns, and technological limitations must be addressed to fully integrate these innovations into existing food systems (Oberdörster et al., 2007; Sullivan, 2019).

16.7.1 Emerging Nanobionic Technologies

Innovations in biosensors are particularly noteworthy; these devices are increasingly capable of detecting contaminants and pathogens with high precision and sensitivity. Recent advances include nanoscale biosensors that use nanomaterials to enhance signal detection, allowing for rapid and accurate monitoring of food quality and safety (Bormashenko et al., 2015). These sensors can identify spoilage indicators or contamination at early stages, significantly improving food safety.

Nanodevices are also making significant strides. Nanoscale delivery systems, such as nanocarriers, are being developed to deliver nutrients, flavors, and preservatives in a controlled and targeted manner. These systems can enhance the efficiency of nutrient delivery and preservation while minimizing waste (Reddy & Chhabra, 2022). Additionally, molecular engineering techniques are facilitating the creation of advanced materials that can respond to environmental changes, leading to innovations in smart packaging that can adapt to temperature and humidity variations to prolong shelf life.

These advancements hold promise for breakthroughs in food production, preservation, and quality control. For instance, nanotechnology could lead to the development of self-healing materials that repair minor damages in packaging, or smart systems that can alert consumers to potential food spoilage (Gao et al., 2020). As these technologies evolve, they have the potential to enhance the efficiency and safety of food systems globally.

16.7.2 Nanobionics for Global Food Security

Nanobionics offers significant potential for addressing global food security challenges, particularly in developing regions. In areas where traditional agricultural practices face limitations, nanotechnology can provide innovative solutions. For example, nanoscale fertilizers and pesticides can enhance nutrient delivery and pest control, which can lead to increased crop yields and reduced environmental impact (Zhao et al., 2019). These technologies are particularly beneficial in regions with limited access to conventional agricultural inputs, providing a means to improve food production and security.

Moreover, nanobionics can help address global challenges such as climate change and resource scarcity. Advanced nanomaterials can be used for water purification, which is crucial in areas suffering from water shortages or contamination. Nanotechnology can also play a role in soil remediation, helping to restore soil quality and fertility, thus supporting sustainable agricultural practices (Khan et al., 2021). By enhancing the resilience and efficiency of food systems, nanobionics can contribute to a more sustainable and secure global food supply.

16.7.3 Potential Challenges and Barriers

Despite its promising potential, the integration of nanobionics into food systems faces several challenges. Regulatory hurdles are a major concern, as existing regulations may not fully address the unique aspects of nanotechnology. Developing comprehensive and coherent regulatory frameworks is essential to ensure that nanobionic products are safe and effective while promoting innovation (Sullivan, 2019). These frameworks must address issues related to the labeling, safety, and environmental impact of nanomaterials.

Safety concerns also persist, particularly regarding the long-term effects of nanomaterials on human health and the environment. There is a need for extensive toxicological studies to understand the potential risks associated with nanoparticles, as well as their behavior and accumulation in ecosystems (Oberdörster et al., 2007). Addressing these safety concerns is crucial for gaining public trust and ensuring the responsible use of nanobionics.

Technological limitations, such as high costs and the complexity of scaling up from laboratory to field, also pose challenges. The high cost of advanced nanobionic technologies can be a barrier to widespread adoption, particularly in developing regions. Additionally, integrating these technologies into existing food systems requires overcoming technical and logistical challenges, ensuring that they are practical and cost-effective. Addressing these barriers will be essential for realizing the full potential of nanobionics in food science.

16.8 Conclusion

Nanobionics has already demonstrated a significant impact on food science by enhancing food safety, quality, and sustainability. Through innovations in biosensors, nanodevices, and molecular engineering, nanobionics has improved the detection of contaminants, extended food shelf life, and optimized resource use. These advancements bridge the gap from laboratory research to practical applications, contributing to safer and more sustainable food systems (Gao et al., 2020; Khan et al., 2021; Reddy & Chhabra, 2022). Looking ahead, nanobionics has the potential to revolutionize the global food system. By addressing critical challenges such as climate change, resource scarcity, and food security, nanobionics can contribute to the development of more resilient and efficient food production and distribution systems. To fully realize these potential benefits, interdisciplinary collaboration among researchers, industry professionals, and policymakers will be essential. Continued research and development, combined with effective policy frameworks, will enable the integration of nanobionics into food systems and drive advancements that ensure a more sustainable and secure global food future (Khan et al., 2021).

References

Adhikari, B., Adhikari, M., & Park, G. (2020). The effects of plasma on plant growth, development, and sustainability. *Applied Sciences, 10*(17), 6045.

Arora, S., Sharma, P., Kumar, S., Nayan, R., Khanna, P. K., & Zaidi, M. G. H. (2012). Gold-nanoparticle induced enhancement in growth and seed yield of Brassica juncea. *Plant Growth Regulations, 66*, 303–310.

Bao-shan, L., Shao-qi, D., Chun-hui, L., Li-jun, F., Shu-chun, Q., & Min, Y. (2004). Effect of TMS (nanostructured silicon dioxide) on growth of Changbai larch seedlings. *Journal Forestry Research, 15*, 138–140.

Barage, S., Lakkakula, J., Sharma, A., Roy, A., Alghamdi, S., Almehmadi, M., Hossain, M. J., Allahyani, M., & Abdulaziz, O. (2022). Nanomaterial in food packaging: A comprehensive review. *Journal of Nanomaterials, 2022*(1), 6053922. https://doi.org/10.1155/2022/6053922

Bormashenko, E., Shapira, Y., Grynyov, R., Whyman, G., Bormashenko, Y., & Drori, E. (2015). Interaction of cold radiofrequency plasma with seeds of beans (Phaseolus vulgaris). *Journal of Experimental Botany, 66*(13), 4013–4021.

Canas, J. E., Long, M., Nations, S., Vadan, R., Dai, L., Luo, M., Ambikapathi, R., Lee, E. H., & Olszyk, D. (2008). Effects of functionalized and nonfunctionalized single-walled carbon nanotubes on root elongation of select crop species. *Environmental Toxicology Chemistry, 27*(9), 1922–1931.

Chelliah, R., Wei, S., Daliri, E. B.-M., Rubab, M., Elahi, F., Yeon, S.-J., Jo, K. H., Yan, P., Liu, S., & Oh, D. H. (2021). Development of nanosensors based intelligent packaging systems: Food quality and medicine. *Nanomaterials, 11*(6), 1515.

Dehkourdi, E. H., & Mosavi, M. (2013). Effect of anatase nanoparticles (TiO2) on parsley seed germination (Petroselinm crispum) in vitro. *Biology Trace Element Research, 155*, 283–286.

Fava, J. (2025). Sustainable innovation and tools for product sustainability. In *Greener products* (pp. 128–179). CRC Press. https://www.taylorfrancis.com/chapters/edit/10.1201/9781003441939-7/sustainable-innovation-tools-product-sustainability-james-fava

Feizi, H., Kamali, M., Jafari, L., & Moghaddam, P. R. (2013). Phytotoxicity and stimulatory impacts of nanosized and bulk titanium dioxide on fennel (Foeniculum vulgare Mill). *Chemosphere, 91*(4), 506–511.

Finnveden, G., Hauschild, M. Z., Ekvall, T., Guinée, J., Heijungs, R., Hellweg, S., Koehler, A., Pennington, D., & Suh, S. (2009). Recent developments in life cycle assessment. *Journal of Environmental Management, 91*(1), 1–21.

Food And Drug Administration. (2013). *Guidance for industry oversight of clinical investigations – A risk-based approach to monitoring*. In U.S. https://www.fda.gov/media/116754/download

Gao, C., El-Sawah, A. M., Ali, D. F. I., Alhaj Hamoud, Y., Shaghaleh, H., & Sheteiwy, M. S. (2020). The integration of bio and organic fertilizers improve plant growth, grain yield, quality and metabolism of hybrid maize (Zea mays L.). *Agronomy, 10*(3), 319.

Jasim, B., Thomas, R., Mathew, J., & Radhakrishnan, E. K. (2016). Plant growth and diosgenin enhancement effect of silver nanoparticles in Fenugreek (Trigonella foenum-graecum L.). *Saudi Pharmaceutical Journal, 25*(3), 443–447.

Khan, N., Ray, R. L., Sargani, G. R., Ihtisham, M., Khayyam, M., & Ismail, S. (2021). Current progress and future prospects of agriculture technology: Gateway to sustainable agriculture. *Sustainability, 13*(9), 4883.

Khodakovskaya, M., Dervishi, E., Mahmood, M., Xu, Y., Li, Z., Watanabe, F., & Biris, A. S. (2009). Carbon nanotubes are able to penetrate plant seed coat and dramatically affect seed germination and plant growth. *ACS Nanotechnology, 3*, 3221–3227.

Krishnaraj, C., Jagan, E. G., Ramachandran, R., Abirami, S. M., Mohan, N., & Kalaichelvan, P. T. (2012). Effect of bio-logically synthesized silver nanoparticles on Bacopa monnieri (Linn.) Wettst. Plant growth metabolism. *Process Biochemistry, 47*, 651–658.

Kumar, V., Guleria, P., Kumar, V., & Yadav, S. K. (2013). Gold nanoparticle exposure induces growth and yield enhance-ment in Arabidopsis thaliana. *Science Total Environment, 461*, 462–468.

Lin, D., & Xing, B. (2007). Phytotoxicity of nanoparticles: Inhibition of seed germination and root growth. *Environmental Pollution, 150*(2), 243–250.

Nel, A., Xia, T., Mädler, L., & Li, N. (2006). Toxic potential of materials at the Nanolevel. *Science, 311*(5761), 622–627. https://doi.org/10.1126/science.1114397

Oberdörster, G., Stone, V., & Donaldson, K. (2007). Toxicology of nanoparticles: A historical perspective. *Nanotoxicology, 1*(1), 2–25. https://doi.org/10.1080/17435390701314761

Pokhrel, L. R., Dubey, B., Patel, A., Dalwadi, P., Prasad, R., Patel, P., & Selvaraj, K. (2013). Evaluation of developmen-tal responses of two crop plants exposed to silver and zinc oxide nanoparticlesNicotiana tobaccum. *Science Total Environment, 10*(41), 321–332.

Reddy, S. S., & Chhabra, V. (2022). Nanotechnology: Its scope in agriculture. *Journal of Physics: Conference Series, 2267*(1), 012112. https://iopscience.iop.org/article/10.1088/1742-6596/2267/1/012112/meta

Rout, P., Ramani, A., Bhattacharjee, S., & Rajput, V. D. (2023). Micro- and nanoparticle of chitosan for vitamin encapsulation: A nutshell overview. In V. D. Rajput, H. El-Ramady, S. K. Upadhyay, T. Minkina, B. Ahmed, & S. Mandzhieva (Eds.), *Nano-biofortification for human and environmental health* (pp. 187–210). Springer International Publishing. https://doi.org/10.1007/978-3-031-35147-1_11

Saranya, S., Aswani, R., Remakanthan, A., & Radhakrishnan, E. K. (2019). Nanotechnology in agriculture. In D. G. Panpatte & Y. K. Jhala (Eds.), *Nanotechnology for agriculture* (pp. 1–17). Springer Singapore. https://doi.org/10.1007/978-981-32-9370-0_1

Savithramma, N., Ankanna, S., Bhumi, G., Kale, A., Tolcheva, E., Levchenko, T. S., & Torchilin, V. P. (2012). Effect of nanoparticles on seed germination and seedling growth of Boswellia ovalifoliolata an endemic and endangered medicinal tree taxon. *Nano Vision, 2*(4), 61–68.

Shruti, A., Bage, N., & Kar, P. (2024). Nanomaterials based sensors for analysis of food safety. *Food Chemistry, 433*, 137284.

Siegrist, S. K. (2017). *Assessing the safety of engineered nanoparticles designed for therapeutic use* [PhD thesis, University_of_Basel]. https://edoc.unibas.ch/72177/

Singh, R., & Kumar, S. (2023). Regulatory and safety concerns regarding the use of active nanomaterials in food industry. In I. R. Singh & S. Kumar (Eds.), *Nanotechnology advancement in agro-food industry* (pp. 269–306). Springer Nature. https://doi.org/10.1007/978-981-99-5045-4_8

Srinivasan, C., Saraswathi, R., & Sirisha, D. (2010). Nano-agriculture-carbon nanotubes enhance tomato seed germina-tion and plant growth. *Curent Science, 99*, 273–275.

Sullivan, D. J. (2019). *Natural antimicrobial nanomaterials for food packaging applications*. https://cora.ucc.ie/handle/10468/9471

Suriyaprabha, R., Karunakaran, G., Yuvakkumar, R., Rajendran, V., & Kannan, N. (2012). Silica nanoparticles for increased silica availability in maize (Zea mays. L) seeds under hydroponic conditions. *Current Nanoscience, 8*(3), 902–908.

Tarafdar, J. C., Raliya, R., Mahawar, H., & Rathore, I. (2014). Development of zinc nanofertilizer to enhance crop produc-tion in pearl millet (Pennisetum americanum). *Agricultural Research, 3*, 257–262.

Trusted Science for Safe Food. (2020). *In EFSA strategy 2020* (pp. 4–5). https://efsa.europa.eu/sites/default/files/corporate_publications/files/strategy2020.pdf

Villagarcia, H., Dervishi, E., de Silva, K., Biris, A. S., & Khodakovskaya, M. V. (2012). Surface chemistry of carbon nano-tubes impacts the growth and expression of water channel protein in tomato plants. *Small, 8*(15), 2328–2334.

Wahab, A., Muhammad, M., Ullah, S., Abdi, G., Shah, G. M., Zaman, W., & Ayaz, A. (2024). Agriculture and environmental management through nanotechnology: Eco-friendly nanomaterial synthesis for soil-plant systems, food safety, and sustainability. *Science of the Total Environment, 926*, 171862.

Warheit, D. B., Sayes, C. M., Reed, K. L., & Swain, K. A. (2008). Health effects related to nanoparticle exposures: Environmental, health and safety considerations for assessing hazards and risks. *Pharmacology & Therapeutics, 120*(1), 35–42.

Zhao, Y., Dai, X., Wang, F., Zhang, X., Fan, C., & Liu, X. (2019). Nanofabrication based on DNA nanotechnology. *Nano Today, 26*, 123–148.

Index

V. D. Rajput et al. (eds.), *Contemporary Suitability of Nanobionics in Agriculture*, Nanotechnology in the Life Sciences, https://doi.org/10.1007/978-3-031-82779-2

H

N

P

R

S

The manufacturer's authorised representative in the EU is Springer Nature Customer Service Centre GmbH, Europaplatz 3, 69115 Heidelberg, Germany. If you have any concerns regarding our products, please contact ProductSafety@springernature.com

Printed and bound by CPI Group (UK) Ltd, Croydon, CR0 4YY
30/09/2026
02225598-0009